Surface Geochemistry in Petroleum Exploration

SURFACE GEOCHEMISTRY IN PETROLEUM EXPLORATION

Steven A. Tedesco
Atoka Exploration

Chapman & Hall
I(T)P An International Thomson Publishing Company

New York • Albany • Bonn • Boston • Cincinnati
• Detroit • London • Madrid • Melbourne • Mexico City
• Pacific Grove • Paris • San Francisco • Singapore
• Tokyo • Toronto • Washington

Cover Design: Andrea Meyer, EmDash Inc.

Copyright © 1995
By Chapman & Hall
A division of International Thompson Publishing Inc.
I(T)P The ITP logo is a trademark under license

Printed in the United States of America

For more information, contact:

Chapman & Hall
One Penn Plaza
New York, NY 10119

International Thompson Publishing Europe
Berkshire House 168-173
High Holborn
London WC1V 7AA
England

Thomas Nelson Australia
102 Dodds Street
South Merlbourne, 3205
Victoria, Australia

Nelson Canada
1120 Birchmount Road
Scarborough, Ontario
Canada, M1K 5G4

Chapman & Hall
2-6 Boundary Row
London SE1 8HN

International Thomson Editores
Campos Eliseos 385, Piso 7
Col. Polanco
11560 Mexico D.F. Mexico

International Thomson Publishing Gmbh
Königwinterer Strasse 418
53227 Born
Germany

International Thomson Publishing Asia
221 Henderson Road
#05-10 Henderson Building
Singapore 0315

International Thomson Publishing Japan
Hirakawacho-cho Kyowa Building, 3F
2-2-1 Hirakawacho-cho
Chiyoda-ku, Tokyo 102
Japan

1 2 3 4 5 6 7 8 9 10 XXX 01 00 99 97 96 95

Library of Congress Cataloging-in-Publication Data

Tedesco, Stevem A., 1955-
    Surface geochemistry in petroleum exploration / Steven A. Tedesco.
        P.   cm.
    Includes bibliographical references and index.
    ISBN 0-412-99301-5
    1. Petroleum—Prospecting. 2. Geochemical prospecting. 3. Soil chemistry. 4. Fluids—Migration. I. Title.
        622'.1828—dc20   94-28118
                                                        CIP

Contents

Preface

Initially, I approached this book with a certain trepidation. For several decades, surface geochemistry has been a controversial subject when applied to petroleum exploration. Vertical migration is not a new concept, but the mechanism by which it occurs is still not clear. Quantifying all the different elements from the seal rock to the soil may be an insurmountable task, but we can identify, discuss, and interpret the principal ones. The majority of evidence has not come through laboratory or research investigation but by field observation. Data from laboratory results and from fieldwork are often conflicting. Consequently, there have been steps backward and stumbles forward toward a clearer understanding. Difficulty in re-creating or duplicating conditions of vertical migration in the laboratory will continue to be a problem as we have not really identified all the essential parameters needed to facilitate vertical migration.

Surface geochemistry is a valid tool and is not the ultimate "black box."* Every reasonable tool applied with objectivity will increase success and decrease failure. In petroleum exploration, we hope to achieve an above-average ratio of successes to failures. The view of exploration is like a jigsaw puzzle; every tool is a piece leading to a complete picture and a reward of finishing.

I began my career in the search for base and precious metal deposits; in mining, the concepts and applications of surface geochemistry have always been acceptable. When I was first employed in the petroleum industry, I was uncertain about the viability of surface geochemistry in view of the industry's ingrained concept of the integrity of the "seal" above the reservoir. However, I became convinced of the usefulness of surface geochemistry and questioned the absoluteness of the seal when I was working in an area where production is shallow (400 to 1200 ft). This area made the existing seismic methods untenable and uneconomic. Also, many of the existing data from these shallow wells were unreliable. I was able to utilize soil-gas and radiometrics with good results. But I was still skeptical about the claims

of tremendous successes by various contractors. Unfortunately, many of these successes were related to existing fields. Over time, through association with several companies, I was able to try different surface geochemical techniques with varying results. The overall result was that when a surface geochemical anomaly was not present, there was a dry hole. It was clear, too, that there was very little scientific investigation into these methods.

The lack of published or intensive scientific investigations of significance suggests either an absence of scientific curiosity or the existence of an unwarranted bias by an industry that is riding the downside in the natural resource cycle. Over the years, several of my colleagues have argued that if these techniques are so successful, the major to midsize companies must have thoroughly investigated them. The research groups of these companies are thought always to be looking for new tools. However, this argument does not withstand scrutiny. Working for two major oil companies taught me that few, if any, research results reach the field offices. Seismic technology now pervades the industry and has advanced so that it is quite capable of finding even subtle traps. But the question remains: Is the trap filled with petroleum or another fluid? Some stratigraphic traps are lost in processing, but surface geochemistry can provide a clue that the seismic needs alternative processing to enhance the target's presence. Aside from drilling, surface geochemistry may be the only way to determine whether petroleum is present at depth. In certain cases, seismic can detect gas in the geologic section but, in almost all current onshore exploration environments, it cannot. We also have to look at the present approach to exploration, which tends to specify that if the technique fails the first time, it is not useful. There are still individuals and exploration groups that believe seismic is not a useful tool, in many site-specific or regional instances.

The decision to write this book came about three years ago when I realized that many potential users of surface geochemistry had little understanding of the subject and could not find a book that would give some guidance. Information was disseminated by word of mouth or through publi-

---

*(Black box refers to any technique that is purposely unexplained or kept secret by its developer or owner.)

xi

cations and was typically misleading, if not entirely incorrect. Further, until recently, most short courses on the subject were taught by contractors or academics, both of whom had different objectives than their students. Many contractors are clearly attuned to the strengths of their own techniques and the weaknesses of other methods. Academics are rarely current on the cutting edge and are usually immersed in the esoteric. The students' concern (generally) is how to use a technique logically, objectively, and effectively to find petroleum. This leads to the question of what to consider for a general text that would appeal to the novice but also provide additional information to persons who use surface geochemistry on a regular basis. In addition, the book deals with how to use statistics, adequate grid design, and the effect of soil on collection, analysis, and interpretation, which have become major stumbling blocks for many users of surface geochemical methods. Statistics especially continue to create problems. In many cases, simple statistics suffice if there are a sufficient number of samples; the microseepage either is there or is not there. Current general analytic methods are no longer a problem as their technology has become sufficient to achieve the resolution required.

This book was written not as an absolute text but to bring together many of the divergent methods and ideas. Within this context, there is an attempt to screen out the hype and identify the facts and to suggest areas of research, such as soils, that will achieve the objective of finding petroleum. There are numerous techniques, but several of them seem to "fall out." Either they are too unreliable or there are not enough data to support their use at this time. Economics is also considered because many of the techniques are not as cost-effective as other mainstream geochemical methods, such as soil-gas, radiometrics, iodine, and microbiology.

Surface geochemistry is a complex of scientific techniques, but I have tried to simplify the approaches and discussions. Specifically, statistics present a peculiar problem: They are readily accepted by some, disdained by a few, and never used by many. The statistical methods used in petroleum surface geochemistry were adapted from minerals exploration. To expand on this, minerals (nonpetroleum) exploration typically seeks deposits that may extend from 0.3 m (1 ft) wide to several hundred acres. These deposits, unless they outcrop at the immediate surface, are difficult to detect because of the low mobility of their constituents. Soil-gas methods have been applied to these deposits, but with limited success. On the other hand, petroleum is a highly volatile and mobile group of compounds that generally inundates an area as little as 40 acres and more likely greater than 160 acres. Statistics in mineral exploration were developed to identify very subtle changes in pathfinder elements leading to blind deposits. Therefore, in a real sense, statistics should easily confirm an anomalous data set in petroleum exploration. There is an implication by some that complex statistics need to be applied for the answer. I have approached the matter from another direction; statistics assist in analysis and need to be used only when required. Simple statistics are presented rather than complex formulas. The reader should not misread the chapter on statistics as being completely simplified. Certainly, the use of multivariant and other methods are applicable but should be used *only when needed*. The last three words are important. Just because we have a data set does not mean statistics will make sense of it. There are many factors that affect the data before statistics are brought into play. On the other side, there are those who do not use even simple statistics and thus misinterpret the data. Statistics are a tool and should be used with care. Their misuse can be as detrimental as their proper use can be vital, and sometimes they may not be needed at all.

Attitudes have changed in the 12 years that I have been involved in surface geochemistry for petroleum exploration. When I first advocated the use of surface geochemistry either to management or as a part of investigating a prospect that I was attempting to sell to potential investors, half or more of my presentation and follow-up time was spent discussing, explaining, and defending surface geochemistry. Now, it is rare when I meet with a professional who has had no exposure to the methods. For example: In the spring of 1992, I became involved in a concept play that relied heavily on surface geochemistry finding targets for seismic. The client wanted me to defend the use of surface geochemistry by requiring memos and evidence of its effectiveness. After six months, I learned through an ex-employee of the company that the exploration group were avid believers in surface geochemistry, had been for over five years, and had made some discoveries using it. However, they maintained a low profile in its use to avoid giving away their expertise.

A point that is stressed again and again in this book is that surface geochemistry is a screening tool as, in a sense, all exploration techniques are. The only absolute tool for finding oil is the drill bit. Consequently, when surface geochemistry finds a target that is considered prospective, seismic, or another unrelated method should follow.

Acknowledgments

I want to acknowledge and thank several people who helped in the preparation of this book. In particular I want to thank Dr. Benjamin Collins, who encouraged the project, discussed numerous concepts, helped in the direction of the manuscript, and reviewed its many parts. I wish to thank Tom Swanson, who reviewed and helped balance the text and who added technical material for some of the soil-gas techniques. I thank my partner at Atoka Exploration, Chuck Goudge, who reviewed the manuscript, served as a sounding board for various ideas, and assisted in the chemistry. Through a joint commitment to research and development, Chuck made possible a lot of the investigations presented in this book. But more importantly, he enhanced and increased our ability to find petroleum for many companies and individuals. Special thanks go to Greg Bell, whose graphic talent is well represented in this book, and to Linda Groth, whose editing assistance was incalculable; also to Kenza Dillon and Sally Larson for their help in typing the manuscript.

Most importantly, I wish to thank my loving wife Christine who has supported me in many ways over the years with constant encouragement.

I

Introduction

Many types of geochemical methods have become useful and successful tools for petroleum exploration. Surface geochemical techniques, in particular, have been used with success since the 1930s. Their use has been very controversial, however. This book deals specifically with methods used in surface prospecting rather than in source rock analysis, maturation studies, or basin modeling. Discussions, descriptions, and histories of various techniques attempt to demystify this segment of geochemistry.

The science of geochemistry, in general, has matured considerably since the 1930s and especially in the 1980s. This has been the result of better analytical equipment and techniques, a long history of successful application of exploration (surface) geochemistry in the mining industry, development of sophisticated statistical techniques, and a greater understanding of soil and subsurface conditions and processes. Many recent developments in geologic analysis, such as sequence stratigraphy, seismic stratigraphy and three-dimensional seismic surveys, and maturation studies, are not so much innovations as advances in existing technologies. For the most part, these advances have not delayed, reversed, or stopped the decline in the number of new field discoveries. Recently, geochemistry in petroleum exploration has revolved around maturation studies and basin modeling. In many productive petroleum provinces, there is no question that hydrocarbons have been generated and have migrated. New seismic and sequence stratigraphy techniques can infer and delineate areas where new traps may exist in mature petroleum areas. The question still remains as to whether petroleum is present in these traps. Even though surface geochemistry is thought of as an old technology, it has not permeated the industry as has seismic technology. Many explorationists feel that surface geochemistry is currently going through a "rebirth." The analytical methods finally exist for surface geochemistry to live up to many of the expectations that have been placed on it for so long.

A wide variety of techniques in petroleum exploration have been grouped into a general category called *unconventional methods;* surface geochemistry is classified as one of them. The term *unconventional* is used to describe methods that are not as readily accepted by petroleum explorationists as are subsurface geology and seismic procedures. Conversely, in mining exploration, surface geochemistry is widely accepted whereas seismic techniques are not.

This book is divided into three parts. Each part presents a major category or group of categories concerning surface geochemistry. The first part discusses general hydrocarbon geochemical concepts, aspects of soils and their formation, the history of surface geochemistry, and general theories of vertical migration. The second part defines, discusses, and presents examples of various current surface geochemical methods or of methods used in the past. The third part is a discussion of survey design, statistical evaluation, and data interpretation.

Numerous goals are reflected throughout the book, overlapping one another among the three parts. A general discussion of goals follows.

Surface geochemistry is composed of several techniques that detect chemical changes, particularly in the near-surface soil, atmosphere or in the ocean environment. Discussions of these methods indicate how the changes are expressed and then how they can be described and interpreted. The methods in this book detect hydrocarbon gases, nonhydrocarbon gases, halogens, radioactive elements and their daughter products, trace and major metals, isotopes, radon, helium, magnetic minerals, and Eh-pH. Microbial methods have also been included because some bacteria consume or generate hydrocarbon gases. Each technique will be examined in detail, and examples will be presented, when possible, of actual discoveries made with one or several combined techniques. The significant problems for many users of surface geochemistry are deciding which method will work consistently under a specific set of conditions, determining proper collection procedures and ensuring quality samples, assuring accurate analysis, and obtaining an effective interpretation of the acquired data.

Other techniques that have been classified as surface geochemical but that are not part of this group are: electrical, magnetotellurics, other exotic geophysical methods, satellite imagery, and geomorphics. These will not be included in this text.

There are two categories of surface geochemical methods: direct and indirect. The direct method is the detection and measurement of hydrocarbons (soil gases) that are expelled from the subsurface into the soil substrate. The indirect methods encompass all other forms of geochemical phenomena whose cause and presence are a by-product of soil-gas reactions with the soil substrate and atmosphere. In the literature there are two classical forms of surface expressions that have been identified: the halo and apical anomalies.

Another goal is to discuss the reasons why surface geochemistry should work. One of the first questions is: Will leakage of a reservoir occur if the overlying seal rock is assumed to be impermeable? In recent years, the answer that has come through observation and empirical data indicates that no rock type is completely impermeable and that migration of minute amounts of solids, gases, and liquids through the seal does occur. The once universally impermeable shale has been found to be, under certain conditions, a highly fractured and productive reservoir. Environmental geochemistry has shown that hazardous liquids (such as petroleum) migrate effectively through clays, shales, and evaporates. Therefore, the surface geochemical anomaly is the end result of hydrocarbon leakage from a reservoir at depth. The mechanism for vertical migration is still subject to debate and, at present, no solution is in sight. However, the end result has made the acceptance of surface geochemistry a moot point and no longer a stumbling block to its use. This book seeks to quantify the phenomena at the surface, and determine how to evaluate the data more effectively and approach interpretation with objectivity.

Several years ago, mining explorationists recognized that the anomaly is the end result. They then determined in general terms how the phenomenon works and whether it exists in association with all types of deposits or only specific ones. Or they assumed that the process works and then integrated the methods into existing exploration programs. Conversely, the petroleum industry has sought all the answers to the cause prior to any acceptance of the known result.

A criticism of surface geochemistry implies that these techniques were developed by various countries or companies as a way to pursue exploration *cheaply*. Some critics view inexpensive methods with a disparaging frown as if saving and stretching the exploration dollars are the wrong approach. The Russians, with limited funds, developed surface geochemistry to a high degree in order to prevent spending large sums of money for seismic surveys as the American industry has done. However, no technique is inexpensive if it fails to find petroleum or any other mineral; any technique is inexpensive when it is consistently successful. One of the goals of this book is to present surface geochemistry as a tool and not as a panacea for exploration.

Some have commented that surface geochemical methods have been employed by various small petroleum companies to find marginal to subeconomic fields remaining in mature basin areas. To place this statement in context, we must look at present trends in the mining industry which, like the oil industry, has similar problems of diminishing sizes and grades of new deposits. By the 1940s, most high-grade deposits of base and precious metals that were visible at the surface had been found. The response of many companies at the time was either to mine large, low-grade deposits or to pursue unexplored territories overseas. These strategies worked effectively for a while. However, the result was the same; most of the easily accessible deposits were found and depleted. Subsequently, the mining industry had to develop new methods, such as surface geochemistry, to increase success even though the trend was to find and develop deposits of smaller and smaller size and lower and lower grade. The petroleum industry is undergoing a similar transition; it has found the easy and largest deposits and now has to turn to smaller accumulations. Spending vast sums of money on seismic and large lease plays is no longer effective because the resulting size of the new field discoveries and the ratio of dry holes to successes do not justify this approach. Thus, many companies have turned to newer, less developed and less understood methods, such as surface geochemistry. The petroleum industry is beginning to use surface geochemistry regularly to evaluate various subsurface, Landsat, and other types of leads prior to shooting seismic. This minimizes the money needed for seismic surveys and helps to eliminate costly drilling of prospects that have the least chance of success. A common criticism of surface geochemistry is that, once an anomaly is delineated, there is no way to determine from what depth the leakage is occurring. This is where the explorationist uses other techniques, such as seismic, subsurface geology, and the drill bit to support and complement surface geochemistry.

Choosing the right surface geochemical technique is difficult. This book provides guidelines. Selecting a method that the explorationist will be comfortable with is dependent upon the explorationist and the contractor(s) involved and their collection and laboratory capabilities. Other equally difficult problems are designing a sampling program and interpreting the resulting data that may answer a variety of exploration questions. Often, many surface geochemical programs fail because of the explorationist's lack of understanding of what these methods can and cannot do and how they can be effective.

Additional goals of this book are to convince explorationists to use surface geochemical methods to:

1. Evaluate property and drilling projects prior to leasing or acquisition, and before using expensive geophysical methods such as seismic.

2. Reevaluate a prospect if there is no geochemical anomaly.

3. Select other tools to help confirm or disprove the results of the initial survey.

The geochemical methods will not be as cost-effective if they are used after leasing or only as a final confirmation of seismic, subsurface, or other techniques.

As all tools sometimes fail, so will surface geochemistry. The reasons vary widely: insufficient sampling, inadequate analytical equipment, poor interpretation skills, lack of models, unsuitable soil conditions, other sources of the geochemical anomalies, and complex geologic conditions such as multiple overlapping stratigraphic pays that cause massive leakage and allow little or no anomaly definition. These failures can be addressed and avoided and the problems tempered if the explorationist has the technical and environmental background that this book provides.

Surface geochemical methods may work very well in certain areas, but they may be ineffective in others. Techniques can be best applied:

1. In new basins with no production, where an indication of microseepage would be helpful

2. In areas where the potential reservoirs are stacked

3. In areas where there is either a limited number of potential reservoirs or where the reservoirs are widely scattered

4. In mature basins where many fields are depleted but a few remain to be found

5. To delineate fault and fracture production, which is difficult by seismic or subsurface means alone.

Surface geochemistry is an ineffective tool for:

1. Determining depth to the pay zone

2. Defining the type of petroleum present

3. Calculating productivity of the potential reservoir

4. Delineating targets where overlapping (not stacking) multiple pays exist

5. Determining exactly where to drill

These determinations are best left to seismic procedures and source rock geochemistry.

Surface geochemistry mainly indicates the presence of microseepage, but critics have often demanded of it more precise information such as drill site locations. Many explorationists assume, and will state, that their seismic has found petroleum. This is technically incorrect as seismic has never found petroleum; only the drill bit has. Seismic, like surface geochemistry, high-grades prospects, infers the presence of geologic anomalies similar to existing fields and models, and tries to prevent dry holes. Surface geochemistry is effective in increasing the success rates of drilling by eliminating prospects that indicate no presence of leakage and detecting those that do.

The author has noted with some disappointment that those who have readily accepted surface geochemistry are those with a mining background. The most resistant are those with mainly petroleum-related experience, and they do not generally have a basic background in geochemistry and statistics. Surface geochemistry has often been viewed as seismic has: if the data show an anomaly—fine; if the data do not, then we have to reprocess. However, surface geochemistry is a time-dependent sample of a dynamic chemical process that cannot be subject to similar acquisition and manipulative process techniques developed for seismic. Seismic and surface geochemistry are two dissimilar tools, which attempt to evaluate very different geologic processes that are characteristic of the same specific and aggressively sought-after geologic anomaly: the petroleum accumulation.

Surface geochemistry, like seismic, is a statistical problem requiring the right amount of data, the correct acquisition techniques, and the proper interpretation. However, the solids, liquids, and vapors exhaled from the earth undergo chemical and physical processes that are impacted to varying degrees by time. Therefore, surface geochemistry, unlike seismic, is part of a dynamic ongoing process in which the results from survey to survey may be not exactly the same, only partially repeatable, or not repeatable at all. In this book, the discussions of the various methods confront and mitigate the problems with time and environmental processes by listing advantages and disadvantages of the techniques. The explorationist himself can then determine which method will be most effective for his project.

The acceptance of surface geochemistry as an exploration tool lies not in total proof or understanding of the minute details of the process, but rather in the realization that the earth does not operate under rigid and inflexible concepts or rules. There is strong evidence that the surface geochemical anomaly, as a product of leakage from the petroleum accumulation, does exist. The real and more immediate problem is effective evaluation of the surface expression to determine what its relationship in space is to other geologic and geophysical data or concepts.

1

History of Surface Geochemistry

Introduction

The history of surface geochemical exploration is filled with misconceptions, unsupported claims, misrepresentations, and a poor understanding of the soil and subsurface environments. Surface geochemical methods have not been readily accepted by the United States petroleum industry, but foreign oil industries with limited American influence have pursued this technology aggressively. Recently, economic pressures associated with higher finding costs and lower success rates have caused American companies to reexamine surface geochemical applications in a variety of geologic environments and exploration scenarios.

Surface geochemical methods have found petroleum because of recent technological innovations rectifying several problems that had inhibited their effectiveness in the past. Figure 1-1 is a historical representation indicating the years when various technological methods were applied. Seepage prospecting, the precursor to surface geochemical methods, was employed (along with surface mapping) from the 1880s

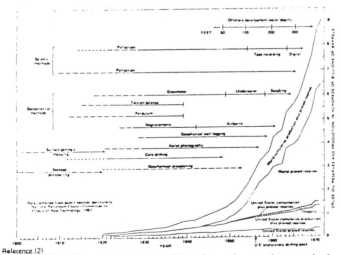

Fig. 1-1 Historical diagram of the time when various methods were implemented in natural resource extraction in relation to crude oil production and proven reserves. (Reprinted with permission of Noyes Data Corp.)

to 1920. Seeps have been recorded as far back as 3000 B.C. In the Baku region of Azerbeijan in the former Soviet Union, burning gas wells predated Christ, and asphalt was used as a building material in 1875 B.C. The Drake well, completed in 1859 in Pennsylvania, and a discovery in Ontario were drilled based on seeps. By the 1920s, most of the known visible oil seeps had been drilled, and the use of this method declined. Geochemical prospecting, the search for microseepage, naturally picked up where macroseepage exploration left off. Developing the technology to be both economical and reliable has taken a considerable amount of time.

A brief history of surface geochemistry is presented in the following sections. Unlike other advances in the petroleum industry, few records exist on the development and implementation of the various techniques. There has been no forum until recently and, thus, the same mistakes and wrong research paths have been taken a number of times.

The Early Years

In December 1929, G. Laubmeyer of Germany submitted to the U.S. Patent Office an application titled "Method and Apparatus for Detecting the Presence of Profitable Deposits in the Earth." The purpose of the patent was to detect hydrocarbons in soil gas as a measurement of petroleum seepage from the subsurface. The patent was issued three years later.

The Soviet Union began conducting surveys around 1930 and, in 1932, Sokolov published the results of this early work in a monograph entitled "Method of Exploration for Natural Gas." The work of both Laubmeyer and Sokolov aroused interest from U.S. explorationists and scientists and, generally, these new methods were accepted. Scientists such as Rosaire, Horvitz, McDermott, and Blau were the most notable supporters of surface geochemistry in the 1930s and 1940s. During the 1930s, several companies were formed to pursue this technology, and some success was reported. Finally, several major oil companies investigated the technique, but the results were mixed.

During the late 1940s, several articles appeared in publications of the American Association of Petroleum Geologists and the Society of Exploration Geophysicists. Discussed were surface geochemical methods, their applications, and their problems encountered in petroleum exploration (Pirson, 1946). At this time, the detection of hydrocarbon gases was used but was limited essentially to the analysis of methane and ethane. Collection of gas via stream sampling was also done but with very mixed results. The use of these methods declined because of the primitive state of laboratory equipment, limited knowledge concerning soil science, and difficulty in repeating the results.

Concurrently, the Russians began to investigate and develop geochemical methods with notable success. The Russians realized the need to address soil problems when using these methods in petroleum and mineral exploration. This has been partly attributed to the agrarian nature of the Russian society vs. the heavily technology-orientated societies in the West. Not only the petroleum industry, but Western industry in general, did not begin to recognize the study of soil as a science until the 1960s.

The Intermediate Years

The 1950s saw resurgence of geochemical prospecting, with numerous articles reporting the successes of this technology. Microbial methods also were being used. An article by Davidson (1982) reported that Horvitz Laboratories, Houston, Texas, had discovered 23 new fields out of 39 prospects drilled. Geochemical Surveys, Dallas, Texas, over the same period, reported that 160 surveys produced 38 new field discoveries. Several authors called for broader acceptance and further investigation of these methods (Johnson, 1963).

Davidson (1982) also reported on Crown Petroleum Company which, in 1957, audited its projects using surface geochemical methods. Crown concluded that geochemistry was six times more effective than any other methods being used. Consequently, the company changed its strategy. Between 1940 and 1960, Crown had been purchasing only 8% of its leases as a result of geochemistry evaluations; by 1960, 65% of its producing leases were the result of surface geochemical exploration. These results were never published, however. The methods once again lost acceptance, mainly because of problems with the technology and declining petroleum prices.

Some major oil companies reinvestigated surface geochemical methods and initiated extensive programs to determine the viability of several different methods, such as radiometrics, microbial, and soil gas. The results were once again mixed, but no data were released on the number of successes vs. failures or how the data were integrated with other methods. Humble Oil's work concluded that the majority of the anomalies were not halo-shaped.

Foreign oil companies also found that results were mixed.

In 1962, Polish, Hungarian, and Czechoslovakian petroleum geochemists stated that they could not find oil using surface methods. In Poland there was correlation between surface geochemical anomalies and faults. A correlation was also identified between the presence of biological activity and a soil-gas anomaly.

The Recent Years

During the 1970s and 1980s, surface geochemistry was rediscovered, and there was an increase in acceptance. Once again, numerous authors published on the necessity of re-evaluating surface geochemical methods, and the use of these methods became more vigorous than before (Garmon, 1981; Gottlieb, 1969, 1981, 1984; Lattu, 1969; Horvitz, 1981; and Davidson, 1982). The domestic petroleum industry flourished in response to heavy demand, high prices, and the Arab oil embargo. New exploration methods were needed. Drilling increased dramatically, but the number of new field discoveries did not increase significantly. As the price of crude oil rose, the need for new technologies grew, and new areas had to be evaluated. When the industry declined in the mid-1980s, surface geochemical technology had an economic advantage and could stretch the exploration dollar. This is reflected in the number of surface geochemical articles written each year (Fig. 1-2). After 1969, the number of articles steadily increased, with a drop in 1986 and 1987, followed by a surge. The data for 1990 are not complete.

As the number of proponents of surface geochemical prospecting increased, so did the number of its detractors. Notably, Hunt (1979) and Price (1986) argued that surface geochemistry is not very successful and that the claims of proponents cannot be verified. Neither of these authors (nor others), however, has ever presented studies or data to indicate where the methods do or do not work. Why have so many large oil companies not embraced these methods? The majority of patents for many of the techniques are held by major oil companies or their subsequent corporate heirs, and their research groups continue to investigate surface

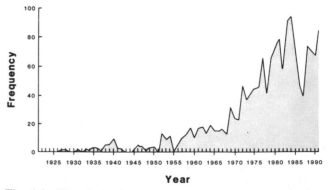

Fig. 1-2 Historical Diagram indicating the number of articles relating to surface geochemical methods in petroleum exploration.

geochemistry procedures. The answer to why large companies do not routinely use these techniques more may often lie not in failures of the methods but in the failure of management decisions to implement them as thoroughly and universally as seismic methods. In terms of drilling, there is a comfort factor in identifying the target on a seismic line. There is difficulty in using a method such as surface geochemistry in which the target depth (no other tools are used for the drilling decision) has to be igneous or metamorphic basement for the source of the anomaly to be located and verified.

Surface geochemistry took a large step forward in the 1980s and 1990s, as it took its place as a practical and applied science exploration tool. What has made this leap possible has been the application of soil science, better collection methods, and more sophisticated analytical techniques. Use of these methods has helped to minimize the percentage of failures and maximize the probability of economic success.

References

Davidson, M., 1982. Toward a general theory of vertical migration, *Oil and Gas Journal,* June 21, pp. 288–300.

Garmon, L., 1981. Born again prospecting, *Science News,* April 25, pp. 267 and 271.

Gottlieb, B. M., 1969. The need for innovation in exploring for oil and gas in the United States, in *Unconventional Methods in Exploration for Petroleum and Natural Gas,* Southern Methodist University, Dallas, TX, pp. 3–6.

Gottlieb, B. M., 1981. Philosophy of symposium II: The need for innovation in exploring for oil and gas in the United States, in *Unconventional Methods in Exploration for Petroleum and Natural Gas II,* Southern Methodist University, Dallas, TX, pp. 5–7.

Gottlieb, B. M., 1984. History and future prospects of the symposia on unconventional methods, in *Unconventional Methods in Exploration for Petroleum and Natural Gas III,* Southern Methodist University, Dallas, TX, p. 3.

Horvitz, L., 1981. Hydrocarbon geochemical prospecting after forty years, in *Unconventional Methods in Exploration for Petroleum and Natural Gas II,* Southern Methodist University, Dallas, TX, pp. 83–95.

Hunt, J. M. (1979). *Petroleum Geochemistry and Geology,* W. H. Freeman and Co., San Francisco.

Johnson, J. F., 1963. Rebellion for reason in evaluating new exploration methods, *Oil and Gas Journal,* March 18, pp. 97–99.

Lattu, O. P., 1969. Keynote address, The long look ahead, in *Unconventional Methods in Exploration for Petroleum and Natural Gas,* Southern Methodist University, Dallas, TX, pp. 7–10.

Pirson, S. J., 1946. Disturbing factors in geochemical prospecting, *Geophysics,* vol. 11, pp. 312–320.

Price, L. C. (1986). A critical overview and proposed working model of surface geochemical exploration, in *Unconventional Methods in Exploration for Petroleum and Natural Gas IV,* Southern Methodist University, Dallas, TX, pp. 245–304.

Sittig, M., 1980. Geophysical and geochemical techniques for exploration of hydrocarbons and minerals, Noyes Data Corp., Newark, NJ.

2

Organic Geochemistry

Introduction

Organic geochemistry as used here is essentially organic chemistry as applied to petroleum exploration. This chapter will present some fundamentals of organic chemistry that are applicable to surface geochemistry: the basic organization and composition of organic compounds and the various reactions and processes these compounds undergo in the earth environment.

Saturated Hydrocarbons

Petroleum is composed of a variety of organic compounds; the most abundant of these are hydrocarbons. Hydrocarbon molecules are made up entirely of carbon and hydrogen atoms. The simplest form of hydrocarbon is the methane molecule, which is a single carbon linked to four hydrogen atoms by tetrahedral bonds. The bonds have a fixed direction and length and are covalent. Complex hydrocarbons always form the four electron-pair bonds per atom; these commonly link to hydrogen as well as carbon, oxygen, and sulfur atoms. The carbon atoms may occur in straight, branched, and combination chains or in rings. These hydrocarbons are also termed *saturated* because all the atoms are linked by one electron-pair bond; a new atom would have to replace a present atom in order to enter the molecule. An increase in chain length causes the hydrocarbon gas to liquify at room temperature and atmospheric pressure. Usually hydrocarbons with 4 carbons (butane) or less per molecule are gases; those with 5 (pentane) to 15 carbons are liquids, and those with more than 15 carbons are solids (paraffin waxes).

Hydrocarbons with chain structures are known as paraffins or alkanes. If the chain is in the unbranched form, the hydrocarbon is called "normal" and is written as "*n*-paraffin" or "*n*-alkane." As chain length increases, the molecular weight and boiling point also increase.

Methane, ethane, and propane are the most common hydrocarbons. Their three-dimensional molecular structures can be depicted on a plane as shown below. Dashes represent electron-pair bonds.

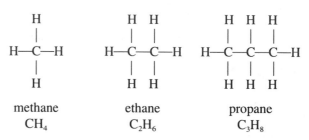

methane
CH_4

ethane
C_2H_6

propane
C_3H_8

A branch may occur at one or more of the carbon atoms in the chain. The hydrocarbon is termed *isoalkane* or *isoparaffin* if the branch is located on a single carbon that is the second carbon in the chain. The possible number of structural forms or isomers increases rapidly with the increasing number of carbon atoms in the compound.

Ring structures are created when the ends of the chain are linked together, thereby eliminating two hydrogen atoms. The rings are known as *napathenes* or *cycloparaffins* and can have one or more side chains and be fused together.

Saturated hydrocarbons are the most common constituent of soil gas, which is measured in the search for leakage from petroleum accumulations. Studies have shown that hydrocarbons of two carbon atoms or greater are usually not generated by biological activity in surface soils or at depth.

Unsaturated Hydrocarbons

The unsaturated hydrocarbons have two or three electron-pair bonds between the carbon atoms, and there are two fewer hydrogen atoms per molecule than in the saturated hydrocarbons. Therefore, new monovalent atoms can be added without displacing any of the present hydrogen atoms.

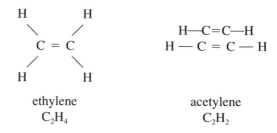

ethylene
C_2H_4

acetylene
C_2H_2

Unsaturated compounds are typically derived from biological activity and thus are not usually analyzed for in surface geochemical exploration. Attempts to relate them to leakage from petroleum reservoirs have had little success.

Aromatics

A special situation arises when double and single bonds existing in a six-member ring are stable and show little tendency to become saturated. This form of hydrocarbon is known as the *aromatic*. The density increase in the chemical bond is greater in a double than in a single bond and, in this case, is spread out evenly through the ring structure. An example is benzene (C_6H_6), which has a relatively low number of hydrogens compared to napathene (C_6H_{12}) and paraffin (C_6H_{14}).

Heteroatoms

A multitude of other elements can bond with a hydrocarbon atom, but the majority of the compounds formed are with oxygen, nitrogen, and sulfur. There are relatively few ways in which these atoms can be structurally arranged in the molecule. Each group is designated as a *functional group* and has special chemical characteristics and consistent, predictable reactions. Many hydrocarbon compounds contain more than one functional group and commonly include more than one heteroatom. Following is a selected summary of some of the more important and common functional groups.

Oxygen-bearing group:

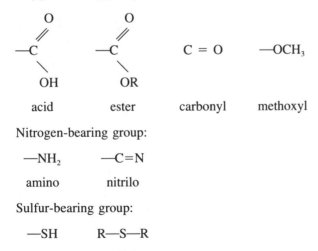

acid	ester	carbonyl	methoxyl

Nitrogen-bearing group:

—NH₂   —C=N

amino   nitrilo

Sulfur-bearing group:

—SH   R—S—R

mercaptan   sulfide

Phenols are very common and are described by the general formula Ar—OH. The hydroxyl group in phenols is attached to the aromatic ring whereas, in alcohols, it is connected to the alkyl group. The OH is more susceptible to substitution in the alcohol group than in phenols. Phenols are more acidic than alcohols and therefore more easily facilitate the weathering process.

Chemical Reactions

Hydrocarbons are constantly undergoing chemical reactions in a variety of pressure and temperature settings. Temperature is the most important factor controlling whether the reaction will occur and the rate of the reaction. Temperatures generally range from those found at the earth's surface to a few hundred degrees, and only in areas of igneous activity are they much higher. Pressure seems to play a minor role. The five most common types of reactions are: (1) oxidation, (2) reduction, (3) elimination, (4) polmerization, and (5) pyrolysis. Only the first three relate to surface geochemical methods.

Oxidation is a common reaction at the earth's surface or in the soil substrate. Organic matter is generally unstable in the presence of oxygen, and hydrocarbons will be ultimately broken down or burn into carbon dioxide and water. However, partial oxidation is the more likely occurrence. Hydrocarbons leaking to the surface will be predominantly subjected to partial oxidation either by chemical or biological processes.

Reduction, the removal of oxygen, is the opposite of oxidation. For example, the conversion of an alcohol to a paraffin results in the generation of a water molecule. A reducing zone typically forms in the strata and soil over a petroleum accumulation.

Elimination, also called decarboxylation, is another form of conversion from an acid group to a paraffin molecule. Esterification is a reaction between a carboxylic acid and an alcohol. This usually results in the elimination of a water molecule. These reactions can occur in either direction, depending on the temperature and concentrations involved.

Alterations of petroleum can occur under a variety of geologic conditions. These types of reactions are (1) water washing and (2) bacterial degradation. Water washing causes the preferential removal of the most soluble compounds, resulting in heavier and heavier petroleum molecules remaining. Tar layers are a product of water washing and, in some cases, these zones are considered critical to the preservation of the remaining crude oil in the trap. Bacterial degradation is a reaction that we will carefully consider in Chapter 11, but it has more application in the near-surface response than to the petroleum trapped in a reservoir.

Characteristics of C_1 through C_6 Hydrocarbons

Discussions in this book will be concerned with only the C_1 through C_6 hydrocarbons because they are the most commonly detected in surface geochemical techniques. The alphanumeric designation refers to the number of carbon atoms in the molecule. The heavier hydrocarbons are usually not analyzed for because many analytical systems are not designed for their detection, because the system's detection limits are not sufficient, or because economics prohibit it.

Methane, CH_4, is denoted as C_1. It is a colorless, odorless gas at room temperature and is the most common hydrocarbon. Methane is insoluble in water, which is a polar solvent, but soluble in nonpolar organic solvents. Methane must be cooled to $-161.7°C$ before it condenses to a liquid at atmospheric pressure and must be cooled to $-182.5°C$ before it forms a solid. It has a specific gravity of 0.424. The molecule is held together by weak van der Waals forces, indicated by the difficulty with which methane is liquefied. Methane can be derived from anaerobic reactions through the complex decay of plants and petroleum deposits.

Ethane, C_2H_6, is typically indicated as C_2. A molecule of ethane is larger than a molecule of methane, and its van der Waals forces are greater. In rare cases, ethane is generated by biological activity but is generally a petroleum-related hydrocarbon.

Propane, C_3H_8, is commonly designated as C_3. Propane, like ethane, can be biologically generated in certain cases, but it is usually derived from petroleum.

Butane, C_4H_{10}, is denoted as C_4. Butane begins the family of hydrocarbons known as the *n*-alkanes. Two different arrangements of atoms are possible. n-Butane is a continuous straight chain molecule in which all the carbons are aligned; isobutane is a branched-chain molecule having three carbons in a row, with a fourth carbon attached to the center one. They have different melting and boiling points; *n*-butane melts at $-135°C$ and boils at $-0.5°C$, but isobutane's melting and boiling temperatures are $-159°C$ and $-12°C$, respectively.

Pentane, C_5H_{12}, is commonly denoted as C_5 and is the next member of the alkane family. Pentane has a total of three isomers. Isomers are compounds that have different chemical and physical properties but the same molecular formula. Isomers are further divided into three groups: (1) structural, which differ in the groups linked to each atom, that is, have different bonding topology; (2) conformational, which have the same bonding topology but are interconvertible; and (3) stereoisomers, which have the same bonding topology but different spatial arrangements. *n*-Pentane, isopentane, and methylpentane have the following configurations and characteristics:

n-Pentane	Isopentane	Neopentane
$CH_3CH_2CH_2CH_2CH_3$	$CH_3CH_2CHCH_3$	$CH_3\!-\!\underset{\underset{CH_3}{\vert}}{\overset{\overset{CH_3}{\vert}}{C}}\!-\!CH_3$
Boiling point: 36.1°	27.9°	9.5°
Melting point: $-129.7°$	$-156.6°$	$-16.6°$

Hexane, C_6H_{14}, is commonly designated as C6. It has a total of five isomers.

Summary

A basic understanding of organic geochemistry is helpful in surface geochemical exploration. Surface geochemistry targets saturated hydrocarbons that are being oxidized and biologically degraded.

References

Barker, C. 1979. *Organic Geochemistry in Petroleum Exploration*, American Association of Petroleum Geologists, Course Note Series No. 10.

Morrison, R. T. and R. N. Boyd, 1974. *Organic Geochemistry*, Allyn and Bacyn, Inc., Rockleigh, N.J.

Waples, D. W., 1985. *Geochemistry in Petroleum Exploration*, International Human Resources Development Corp., Boston.

Wingrove, A. S. and R. L. Caret, 1981. *Organic Chemistry*, Harper & Row Publishers, New York.

3

Soils and Their Formation

Introduction

Understanding soil characteristics and formation is an important but often overlooked aspect of surface geochemical exploration. This chapter describes soil profiles and the influence of climate, parent material, biological activity, time, and topographic relief. Comprehending the relationship between soils and migrating hydrocarbons is critical to the proper collection, analysis, and interpretation of surface geochemical data.

Soil Structure

A layer of unconsolidated material covers most of the earth's surface. The layer is mainly the result of weathering of the underlying bedrock. This aggregate mixed with varying amounts of organic material is called the *regolith*.

Soil, a mixture of mineral and organic material with air and water, constitutes the upper part of the regolith. One of the soil's primary characteristics is its self-evident layering in a vertical section. The layers in each sequence are called horizons and are designated A, B, C, and D in descending order (Fig. 3-1). Further division of each layer into subhorizons is common (for example, A2, C2, etc.). Each horizon differs in composition, texture, and color, and layer boundaries are transitional over 2.5 to 15 cm (1 to 6 in) or are sharp. The term *soil profile* is used when a complete succession is found.

Nahon (1991) has further subdivided soils into lateral and vertical organizations on megascopic and microscopic levels. These pedologic organizations evolve through time as results of both internal and external factors that are present to differing degrees. The first step in weathering is the infiltration of rock by fluids, which begins a process of *self-organization*, or *self-development*, a process that seeks always to achieve equilibrium. The lack of equilibrium drives the weathering process. If geochemical equilibrium were present, weathering would cease. The process of self-organization moves toward equilibrium and is increasingly con-

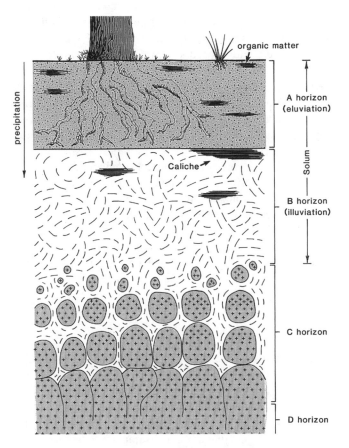

Fig. 3-1 A typical soil profile. The D horizon is the area where bedrock weathering begins by groundwater invasion and subsequent chemical alteration. The C horizon is the layer where large fragments of the bedrock exist in conjunction with coarse-, medium-, and some fine-grain weathered material. In the B horizon illuviation occurs, whereby leached trace and major elements and other compounds are accumulated by descending water. Some organic material is present, plus roots from various plants. The A horizon is the area of eluviation where organic material accumulates and microbial activity is at its greatest. Many elements and compounds are removed by groundwater.

trolled by changes in external factors. Geochemical reactions in the soil direct the weathering process toward equilibrium. For example, a certain combination of rock types and climates leads to the formation of kaolinite and oxyhydroxides. Or small pore size development in the soil causes precipitation of hematite and the dissolution of kaolinite, bauxites, or Ni-rich laterites.

A Horizon

The top horizon, A, usually contains the most humus, the well-decomposed organic matter in the soil. There is a steady decrease in the organic content with depth through the soil profile (Fig. 3-1). The A horizon undergoes extensive leaching, which removes soluble mineral salts and colloidal material to a lower horizon. This top layer is *eluvial*. The process of eluviation occurs as water filters down through the humus and forms weak carbonic and organic acids. Carbonic acid is also formed in the air with CO_2 and invades the soil as rain, which starts decomposition of humic material to form organic acids. The degrading organic matter continually replenishes the acids. These acids move downward, where they react with, and carry in solution, suspension, or colloidal form, a variety of cations and compounds. Subdivision of the A horizon occurs under a variety of conditions, but the explorationist usually does not have the resources for further definition. Sampling of the A horizon must be avoided when certain geochemical techniques are used because the chemical changes and rapid variations in soil composition that occur here render the samples useless and the analytical results nonrepeatable.

B Horizon

The B horizon is illuvial because the leached material from the A horizon accumulates here (Fig. 3-1). This horizon has a prismatic or blocky structure that is caused by high concentrations of iron and/or aluminum oxides in association with organic matter and manganese oxides. Since this horizon is one of accumulation, it is important to determine whether the system is chemically open or closed. A well-drained soil (open system) allows removal of many of the soluble compounds and elements that have been leached from above. Such soils are typically acidic. A chemical system that is closed is characteristic of alkaline soils that are usually found in dry to semiarid areas. There is insufficient water available to remove the soluble material, usually resulting in caliche (iron carbonate) development.

The B horizon is the layer of the profile that is usually sampled in petroleum exploration. It is often subdivided, and each subdivision absorbs elements and compounds differently because of variations in the amount and composition of clay and the presence of iron and manganese oxides. The varying absorption capacity of each subdivision has strong implications for the analysis of soil gas, trace elements, iron minerals, bacteria, and pH/EH. These implications will be explained in later paragraphs and chapters. The A and B horizons together constitute the *solum*, which overlies the C horizon.

C and D Horizons

The C horizon is weathered bedrock (parent rock) at the base of the soil profile. It may have accumulations of calcium carbonate or ferrous iron. The horizon grades downward from loose and partly decayed material to solid rock (Fig. 3-1). The process of bedrock weathering is called *polygenesis*. The C horizon is sampled in terrains where a thin soil cover occurs. The D horizon is the unaltered bedrock.

Climate

Climate exerts the most important influence on soil formation because precipitation and temperature are significant factors in soil development. Soils derived predominantly by climate are said to be zonal. Soils that are affected by factors other than climate, such as biological activity, parent material, and topography, are said to be *intrazonal*.

Soil *erosion* and soil *erodibility* are two distinct terms. Soil erosion in any area is determined by the average rainfall, the vegetation type, and the slope of the terrain rather than by the composition and chemical properties of the soil. Soil erodibility is defined by water infiltration rates, permeability, total water capacity, resistance to rainfall, and resistance to particle abrasion and transport during water runoff. Some soils erode more readily than others, even when similar soil erosion characteristics exist with other soils.

Precipitation controls the amount of soil water present which, in turn, controls the rate and extent of occurrence of oxidation reactions. Climates vary and, therefore, the amount of chemical weathering varies also (Fig. 3-2). Without water there would be little or no chemical weathering. Thus, chemical weathering is minimal in dry regions, both hot and cold, and at a maximum in tropical areas, where water is abundant for leaching and eluviation.

Air and water temperatures are also important. High temperatures allow solution and reactions to proceed more rapidly than low temperatures do. Leaching and eluviation processes decrease significantly in dry, cold areas compared to hot, wet regions. Therefore, temperature and precipitation have an impact on the development of soil profiles, individual horizons, and subhorizons.

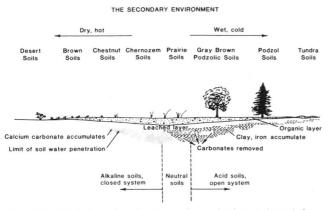

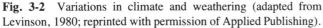

THE SECONDARY ENVIRONMENT

Dry, hot ← | → Wet, cold

Desert Soils | Brown Soils | Chestnut Soils | Chernozem Soils | Prairie Soils | Gray Brown Podzolic Soils | Podzol Soils | Tundra Soils

Calcium carbonate accumulates
Limit of soil water penetration
Leached layer
Organic layer
Clay, iron accumulate
Carbonates removed

Alkaline soils, closed system | Neutral soils | Acid soils, open system

Fig. 3-2 Variations in climate and weathering (adapted from Levinson, 1980; reprinted with permission of Applied Publishing).

Biological Activity

Biological activity aids weathering by oxidizing or reducing iron and sulfur and by developing organic acids through decomposition of plant and animal remains. Therefore, biological activity weakens rock surfaces and makes them susceptible to chemical and physical weathering processes. Mosses and lichens begin the breakdown of the parent rock and add organic matter to the rock particles. Soils evolve as higher- and higher-order plants develop on them. Through a process known as serial progression, decaying plants add more and more organic matter to the soil. The process of organic matter biodegradation is called *humification,* which operates under varying intensities, selective preservation, resynthesis, and transformations. Conversion of organic matter to humus varies between 30% and 50% per year whether in tropical or temperate regions. The organic matter increases the soil's water-holding capacity which, in turn, increases the amount of chemical weathering. Plant development also creates pathways via the root systems through which water migrates down into the soil. Organic acids develop along the roots and cause chemical weathering, which breaks down the soil at greater and greater depths.

Biological activity adds organic material in the form of humus to the soil. The amount of humus has an impact on some types of exploration methods, which will be explained in the chapters dealing with specific methods. The greater the amount of humus, the greater the amounts of migrating hydrocarbons the soil absorbs. The quantity of humus that accumulates is largely a function of climate. Rainfall and permeability control the distribution of organic matter in the soil. Soils that have developed in semiarid grasslands have a uniform distribution of humus through the A and B horizons, but heavily leached soils in the tropics have discrete zones of humic material.

Humus, when undergoing oxidation, degrades into more than a hundred substances. Many of these products are derived from organic matter, based on length of reaction time,

temperature, chemical composition of the water present, pH, EH, and other ongoing reactions. The predominate groups generated are carboxyl groups.

When humus undergoes reduction, several substances are produced. All are phenols and phenol derivatives. The research on humus under reductive processes has been limited compared to work conducted on the effects of oxidation processes.

Most humus is composed of colloidal organic compounds that are either *humic* or *fulvic* acids. These two acid groups vary greatly in their mobility and effect on the soil. The fulvic acids are clearly illuvial and are able to mobilize iron and aluminum readily. In contrast, humic acids tend to inhibit illuviation and stabilize the soil. In soils where humic acids dominate, leaching of metals occurs in the profile, which results in less elemental differentiation. In soils dominated by fulvic acids, leaching of metals is not as prevalent. The amounts of humic and fulvic acids vary considerably among different soil types.

Each soil environment has a unique humic acid composition. Simply labeling a soil environment as a reduction or oxidation environment is too general because of changes in chemical, hydrological, biological, and geochemical reactivity. Therefore, each soil environment can be typed according to its humic components. Variations in humic composition also cause changes in rates of bacterial respiration. Consequently, the type of humic and fulvic acids present may affect the migrating soil gases.

The pH of soils is largely a function of the humus content and is therefore directly related to rainfall. Hydrocarbons migrating through the soil profile will create an area of chemical reduction and subsequently alter the pH. This can cause other chemical changes to occur in the soil profile overlying the petroleum accumulation. Therefore, the soil profile in the area of this reducing zone may not be indicative of the regional vegetation or climate types; there will be a contrast between the local and regional soil profiles. In specific cases, the reducing environment for the regional soil profile may be similar to the soil profile found over a hydrocarbon accumulation, and there is no significant contrast for exploration purposes.

The most common exploration method used to capture and detect migrating hydrocarbons is the collecting of gas in the soil. Biogenic gas, methane (C_1), is a common by-product of biological activity. The amount of methane generated depends on the intensity and type of biological activity. Thus, the amount of organic matter has a direct bearing on the amount of methane present. Currently, there is no direct evidence that heavier hydrocarbons are significantly affected by the bacterial activity whose biodegradation would cause the generation of CH_4.

Biological activity also affects the amount of trace and major elements concentrated in the soil. Organic matter tends to absorb and concentrate certain elements, such as the base metals and halogens. Therefore, element variations occur that are related to the amount of biological activity.

Parent Material

The type of parent material is dependent on geological factors. The permeability porosity, texture and composition of the rock determine its susceptibility to weathering by chemical and physical processes. The more juvenile the soil, the more it will reflect the parent rock. The contribution of the parent material to the soil is certainly critical in exploration. The parent material, if sedimentary, can contribute lithified organic matter to the soil. This altered organic material has undergone lithification through burial, pressure, and temperature changes that cause it to generate various types and amounts of hydrocarbons. The quantity of hydrocarbons being added is highly variable and is dependent on whether the parent material is organically rich or organically lean. Numerous very organically rich and mature source rocks outcrop or immediately underlie the soils in the Nevada Basin and Range Province in the western United States. These rocks have been clearly identified as generating hydrocarbons in significant amounts. Certain geochemical exploration techniques have difficulty distinguishing between anomalies generated by petroleum accumulations and those related to organic-rich shallow soil or parent material.

Topography

Topography contributes to soil development through the control of (1) the rate of surface runoff; (2) the rate of erosion of weathered products, which exposes fresh surfaces; and (3) the rate of removing soluble products through control of the groundwater (Fig. 3-3; see color plate). Topography can create localized variations that dominate profile development over such factors as climate and biological activity. Higher-angle slopes produce soils that are thin and have less well-developed profiles. Mobile elements are easily leached and moved downslope and are reprecipitated in regions of lower relief. Carbonates are often deposited in these lowland regions, and their deposition raises the soil pH. Poorly drained soils in areas of low relief tend to have a higher organic content than soils in well-drained terrain.

Azonal Soil

As azonal, or juvenile, soil has no well-developed horizons and has time as the predominant factor. The time required for a well-developed soil profile to form is largely a function of climate. A soil profile will develop much faster in hot and wet conditions than in a dry and cold climate. Quantitative rates of soil formation have been derived by studying profile development on new geological strata after natural disasters and on moraines of retreating glaciers.

Clay Minerals

Clay minerals are the insoluble products of chemical weathering of silicates that are concentrated in the B horizon. The cation ion exchange capacity of some clay minerals allows them to absorb large amounts of some elements, especially trace metals. Cation exchange capacity varies considerably, and pH affects the quantity of trace metals a specific clay can absorb. For example, montmorillonite absorbs 5 to 50 times more cations than kaolinite. Certain geochemical methods, such as soil-gas detection, rely on the ability of the clays to absorb and retain hydrocarbons that have migrated through the soil. However, variations in pH, clay mineralogy, organic matter, and moisture all interact and determine the amount of absorption.

Soil Analysis or Soil Sample Characteristics

Soil analysis attempts to define several characteristics of a soil sample by measuring these attributes in qualitative terms. The characteristics are typically used to evaluate the soil's potential for farming, engineering, hazardous waste disposal, and so forth. However, these characteristics also affect, to some degree, the hydrocarbon gases that migrate through the soil substrate either on a site-by-site or regional basis. The main characteristics typically analyzed for are:

1. Soil-water content
2. Matrix potential
3. Water-release characteristics
4. Conductivity
5. Particle size
6. Bulk density
7. Liquid and plastic limits

Except for a cursory look (Blanchette, 1989), the majority of surface geochemical papers have not addressed these characteristics by suggesting their impact on collected hydrocarbon data. The author's experience, along with that of others, suggests that the soil's physical and chemical characteristics do influence the data and that they need to be addressed for their effects.

Soil-water content is defined as water that can be evaporated from the soil between 100 and 100°C until there is no further weight loss (Gardner et al., 1991). The choice of temperature range is not a scientific determination. This method removes one or more of the following: free water, water absorbed on the clays, or water in the hydroxyl groups. Water that is structural or in vapor form is not counted in this measurement. Structural water is bonded with the minerals and will not be liberated until temperatures are between 400 and 800°C.

Matrix potential refers to the tenacity with which the water is held by the soil. This is the water that is available to plants in the absence of a high concentration of solutes. The variation in the matrix potential values between different parts of the soil provide the driving force of the unsaturated flow of soil water after allowing for elevation correction (Dirksen, 1991 and Mullins, 1991). Soil water that is in equilibrium with free water is at zero matrix potential. When drying occurs, matrix potential decreases, and large pores empty. As the matrix potential decreases, smaller and smaller pores lose their water, which allows gas migration through the soil.

Water-release characteristics are a measurement that determines the ability of the soil to store water. There is a distinct relationship between matrix potential and water content. The water-release characteristics are unique to a group of similar soils or an individual soil type.

Conductivity of soils determines the ability of clays to absorb or to resist absorption. Therefore, the conductivity of the soil has an impact on the ability of the soil to absorb and retain hydrocarbons as they move through the substrate (Fausnaugh, 1989).

Particle size measurements in a soil and in specific soil layers determine the amount of porosity and permeability present. The particle size is important in that soils that have very fine particle size may have low permeabilities, which allow only small amounts of hydrocarbons to move through or be trapped in the pores. Coarse materials having excellent porosity and permeability allow gases to migrate easily through them. But areas of isolated porosity may not be available to trap significant amounts of hydrocarbons. Large pores are critical for gas exchange in soils, and they have a major affect on chemical, biological, and physical processes. Thus, variations in particle size create porosity and permeability conditions that may maximize or minimize soil-gas migration and capture.

The bulk density of a soil is a measurement of its wet vs. oven-dry mass, which determines the amount of soil moisture present. When a soil is compacted, its bulk density increases. An accurate determination of the bulk density requires knowledge of particle size, which allows calculation of the ratio of porosity to void (Campbell and Henshall, 1991). The application of bulk density is usually used for construction purposes. However, compacted soil interspersed with low- and high-density soils may cause differences in the amounts of soil gas moving through the substrate.

Plasticity is a measurement of the soil that determines its ability to be deformed without cracking in response to stress. A soil has a range of moisture contents within which it is plastic; above this range, it is liquid and, below this, it is considered brittle. There have been no studies to date that have evaluated the relationship between soil-gas migration and plasticity of the soils to determine if there are any effects on surface geochemical data for petroleum exploration. Soil-gas methods have been more effective in dry rather than wet soils, which suggests that brittle soil may give the best results. This seems reasonable because brittle soils have the greatest porosity and permeability, with minimal moisture in the pores.

This introductory discussion of soil and factors affecting its formation and continued development has been brief. However, in any exploration program applying a geochemical technique, the soils present in the area to be surveyed should be investigated.

References

Blanchette, P. L. (1989). Effects of soil characteristics on adsorbed hydrocarbon data, *Association of Petroleum Geochemical Explorationists Bulletin*, vol. 5, No. 1, pp. 116–138.

Bolt, G. H. and M. G. M. Bruggenwert (1976). *Soil Chemistry, A. Basic Elements*, Elsevier Scientific Publishing Co., New York.

Campbell, D. J. and J. K. Henshall (1991). Bulk Density, in *Soil Analysis: Physical Methods*, eds., K. A. Smith and C. E. Mullins, Marcel Dekker Inc., New York, pp. 329–366.

Cawsey, C. D. and P. A. Mellon (1983). A review of experimental weathering of basic igneous rocks, in *Residual Deposits: Surface Related Weathering Processes and Materials*, R. C. L. Wilson, ed., Geological Society (London), Special Publication No. 11, pp. 19–26.

Dirksen, C. (1991). Unsaturated hydraulic conductivity, in *Soil Analysis: Physical Methods*, ed. K. A. Smith and C. E. Mullins, Marcel Dekker Inc., New York, pp. 209–270.

Fausnaugh, J. M. (1989). The effect of high soil conductivity on headspace gas sampling techniques, *Association of Petroleum Geochemical Explorationists Bulletin*, Vol. 5, No. 1, pp. 96–115.

Gardner, C. M. K., J. P. Bell, J. D. Cooper, T. J. Dean, N. Gardner, and M. G. Hodnett (1991). Soil Water Content, in *Soil Analysis: Physical Methods*, eds., K. A. Smith and C. E. Mullins, Marcel Dekker Inc., New York, pp. 1–74.

Levinson, A. A. (1980). *Introduction to Exploration Geochemistry*, Applied Publishing Ltd.

Mullins, C. E. (1991). Matrix Potential, in *Soil Analysis: Physical Methods*, eds. K. A. Smith and C. E. Mullins, Marcel Dekker Inc., New York, pp. 75–110.

Nahon, D. B. (1991). *Introduction to the Petrology of Soils and Chemical Weathering*, John Wiley & Sons, Inc., New York.

4

Concepts of Microseepage

Every triumphant theory passes through three stages: first it is dismissed as untrue; then it is rejected as contrary to religion; then it is accepted as dogma and each scientist claims that he long appreciated its TRUTH.

Karl Ernst von Baer

Introduction

Vertical migration is the mechanism that is used to explain hydrocarbon seepage to the surface. It is the process that causes the surface geochemical manifestation by which we can detect petroleum accumulations using various shallow methods (Fig. 4-1; see color plate).

Macroseeps are the applied and long-accepted precursor of the microseep concept. Macroseepage, which has been documented in various parts of the world (Davidson, 1963; Sittig, 1980; Hunt, 1981), is the visible presence of oil and gas seeping to the surface. Seepages are observed as asphalt and tar residues, such as the La Brea tar pits in Los Angeles, California, the Baku region of Azerbaijan in the former Soviet Union, and the seeps of Venezuela. In the pre-Christian era around Hit, Iran, a thriving bitumen industry existed. In southwestern Trinidad, a large seep called Pitch Lake, which is 606 m across and 45 m deep, has yielded over 25 million tons of asphalt (Davis, 1967). Macroseeps have led to the discovery of numerous oil fields both in known petroleum basins and in new areas. This form of exploration was predominantly used prior to 1900 although oil and gas macroseeps are still major criteria for focusing on exploration target areas such as the Basin and Range Province, Nevada. The seeps of Madagascar provoked international oil company activity, and the California offshore seeps have long pointed to prolific oil production.

A recent source of methane may have coincidentally indicated further evidence for vertical migration. Collins (1992) presented a paper on hydrates in relation to the Messoyakha (western Siberia) and Prudhoe Bay (Alaska) fields. The author's work indicated that the gas hydrate zone found in the shallow surface closely mimicked the productive outline. Further, delta carbon, or del C, evidence indicated that the gas in the hydrate zone was −49.0 for Prudhoe Bay and −48.0 for Messoyakha. Biological methane has a del C −60.0 (see Chap. 10 for explanation). The del C for the gas in Prudhoe Bay is −39.0 and in Messoyakha is −38.5. The author proposed that seepage from the reservoir mixed with biogenic gas at a shallow depth, thus lowering the del C

value, which, he said, suggested methane of predominantly thermogenic origin. That hydrate deposits in both cases are restricted to the outline of the field suggests micro- to macroseepage from the deeper reservoir.

The classification and evaluation scheme of seeps was defined by Hunt (1979). Petroleum seeping to the surface undergoes six possible weathering processes (after Hunt, 1979):

1. Evaporation of volatile hydrocarbons. In the first two weeks, a seep loses all petroleum with a boiling point up to 250°C (C_1 to C_{15}). Over later weeks or months, additional hydrocarbons up to C_{24} are degraded and evaporated.

2. Microbial degradation of petroleum through oxidation.

3. Leaching of water-soluble elements and compounds of sulfur, oxygen, and nitrogen by oxidation.

4. Polymerization of petroleum molecules by eliminating water, carbon dioxide, and hydrogen. This causes the formation of even larger hydrocarbon molecules.

5. Auto-oxidation causing the formation of oxygen-rich polymers. Petroleum components absorb sunlight and oxygen when exposed at the surface for long periods of time.

6. Formation of rigid gel structures (gelation), which may develop for some seeps.

All these processes cause a thickening of the petroleum at the surface. It can be assumed that microseeps, as well as macroseeps, undergo these reactions.

Vertical hydrocarbon movement via the macroseep mechanism for petroleum is:

1. Mass-migrating through open faults and fractures

2. Megaventing along migrational pathways

3. Flowing from overpressurized and breached reservoirs via faults

4. Expelling from currently generating hydrocarbon source rocks (including local intrusive metamorphism of sediments)

5. Arising from bacterially generated shallow gases

6. Occasionally venting from the basement (through basement-penetrating faults) as primordial methane or other gases, such as nitrogen, carbon dioxide, and the nobel gases.

Microseeps are seeping hydrocarbons moving under some of the same processes as macroseeps to reach the surface. These seeps may range in size from the oil-saturated sandstones of Utah's Tar Triangle and the methane-bubbling springs of California's San Joaquin Valley to hydrocarbons moving from pore to pore in a rock formation. Usually microseeps are invisible and so low-level that they cannot be distinguished without modern analytical methods.

Macroseeps and microseeps do not always represent the presence of economic recoverable hydrocarbons at depth, but rather may represent leakage from a temporarily stationary source of petroleum. Therefore, even though the various methods detect hydrocarbons at the surface, the perceived anomaly is not necessarily implying an economic petroleum accumulation. The explorationist must determine if the seepage is just minor venting of petroleum. A definition of the term *petroleum accumulation* as used here implies an accumulation of from one to an infinite number of petroleum molecules trapped permanently or temporarily. It does not suggest either an economic or noneconomic condition of the deposit.

The present theories on microseepage support the premise that a reservoir leaks even though it contains trapped hydrocarbons. The cross-sectional shape of the leaking pattern has conventionally been labeled a "chimney effect." Most chimneys are nearly vertical, although hydrodynamic rationale strongly suggests that groundwater movement and dipping strata should deflect the ascending hydrocarbons. Evidently, the upward movement of gases is so rapid and the buoyancy is so great that the vertical movement vector grossly dominates hydrocarbon transport. Figure 4-1 is the classical representation of this concept. Many authors have noted the lack of offset between the surface geochemical anomalies and the petroleum accumulation. One form of expression of this chimney effect at the surface is reported as a halo pattern directly above the reservoir (Horvitz, 1939, 1954; Davidson, 1963; Duchscherer, 1984; Price, 1986; Klusman, 1989).

Machel and Burton (1991) implied that the term *chimney* is a misnomer and that *plume* is more applicable. However, as applied by Machel and Burton, the plume has no specific direction and is controlled by other hydrologic forces. To date, the evidence indicates that the term *chimney* is more applicable to the surface geochemical manifestation and to its effects between the reservoir and the surface. Groundwater forces seem to have little if any effect on vertical migration.

Many surface phenomena are empirically associated with known deep hydrocarbon reservoirs. Any concept or theory for vertical migration of hydrocarbons must address the following observed conditions (not necessarily in order of importance):

1. Increases in specific type of petroleum-consuming bacteria in association with the geochemical anomaly overlying the accumulation

2. Changes in pH and Eh in the soil substrate

3. Increases in trace and major metals in various strata above the petroleum deposits

4. Increases in halogens, specifically iodine, in the soil substrate and in brines adjacent to the reservoir

5. Increases or decreases in uranium minerals and their daughter products in the soil substrate

6. Increases in magnetic and nonmagnetic iron minerals both in and above a petroleum reservoir

7. Increases or decreases in helium and radon in the soil substrate

8. Increases in hydrocarbon gases from immediately above the reservoir to the surface

9. Cementation of clastic sediments concentrated in zones between the reservoir and the surface

10. Occurrence of high-velocity zones in overlying rock strata

11. Concentrations of iron and manganese clay minerals due to complexing by organic acids in the overlying strata

12. Increases in carbonate minerals in overlying strata

13. Interaction between calcite- and kerogen-derived carbonate causing depletion of carbon 13 in the soil substrate

14. Decreases in gamma-ray response at the surface corresponding to the field

15. Changes in vegetation at the surface

16. Disruption of groundwater sulfate equilibrums by hydrocarbon-consuming and reducing bacteria

Additional Migrating Gases and Fluids

It is often assumed that only hydrocarbons migrate to the surface. If large hydrocarbon molecules can move upward through the system, then other fluids, gases, and possibly some cations in disequilibrium can move vertically as well. The microfracture system does not selectively allow only petroleum to migrate but acts as a conduit for whatever can and does pass through. Documented examples are helium (Pogorski and Quirt, 1981) and uranium (Weart and Heimberg, 1981). These and other migrators should not be ignored during exploration. They may be associated with, or give supporting evidence for, hydrocarbon migration from the subsurface to the surface.

Volume Estimates of Migrating Hydrocarbons

Several authors have tried to estimate the amount of hydrocarbon seepage that occurs on a daily to yearly basis for various reservoirs and the amount of petroleum passing through the system over time. Rosaire (1940) estimated that 10,000 bbl/day reached the surface at the Hasting Field, Texas. This rate suggests macroseepage rather than microseepage and requires a mechanism that allows removal of large volumes of petroleum quickly. Pirson (1941, 1946) and Duchscherer (1981) estimated volumes similar to those of Rosaire. These high volumes would require continual recharging of the petroleum accumulation or the reservoir would be quickly depleted in a short period of time. Measured amounts of hydrocarbons from surface geochemical surveys are in the parts per billion to the parts per million range. Thus it appears that only minute amounts are reaching the surface, and megaventing is the exception. The term *seal rock* is still essentially applicable when used for an unspecific length of time to restrain most of the petroleum accumulated in a trap. However, like any container, it has a few leaks.

Saeed (1991) calculated the rate of migration over time using computer-generated models with a limited number of parameters. The rate was determined to be 10 years from reservoir to surface. However, this conflicts with actual data, which suggest that anomalies can disappear or decrease in intensity within a year after full production is established. Neither author took into account the presence of a microfracture system in their models.

Reservoir Seal Rock

Acknowledging that a reservoir leaks is not inconsistent with the classical concept of the seal rock, which is a major factor in the formation and retention of a petroleum accumulation. A seal rock may have a variety of characteristics that range from retaining the majority of hydrocarbons that migrate into the reservoir to merely delaying their migration on the way to some other point. Trapped hydrocarbons continue migrating in a minor form by seeping through the seal. This leakage is manifested as a geochemical expression at the surface.

A seal rock is usually a shale or evaporite and, in some cases, a limestone or sandstone. The seal quality is determined by the minimum pressure required to displace connate water from pores and fractures by petroleum (Downey, 1984). This is the capillary pressure of the seal, which is a function of the throat radius of the pores and wettability. A decrease in throat radius and wettability increases the hydrocarbon-water interfacial tension and thus increases capillary pressure. The force of the capillary pressure in the seal restricts hydrocarbon entry. Hydrocarbons, or any fluid, must exceed the entry pressure in order to enter the seal.

The effective porosity of the seal rock is generally small. Consequently, hydrocarbons have been thought to leak into the seal rock by diffusion (Downey, 1984). However, an evaluation of different rock types by Aguilera (1980) suggests that all strata are inherently fractured with varying fracture densities. Zieglar (1992) reviewed the hydrocarbon columns for 443 petroleum reservoirs in the Colorado-Montana-Wyoming and California areas. The seals ranged in age from Neogene to Ordovician. It was found that 5% to 10% of the fields had leakage consistent with buoyancy pressures needed to allow migration through the seal rock pores. These fields also tended to have the greatest petroleum column height. In the remaining fields, leakage did not occur along pore throats but rather along fracture, fault, and other types of zones of weakness. Certain rocks, such as sandstone and dolomite, have greater fracture densities than limestones and shales for a specific area. Some shales and carbonate shales have historically been major producers of oil and gas (Landes, 1970). This supports the concept that although a shale may be a seal, it will be fractured to some extent. These fractures can and do act as pathways for the vertical migration of hydrocarbons. The number of fractures in the seal rock over a particular reservoir may be only approximated. Each seal rock's density and pattern of fractures will vary from one accumulation to another. Hydrocarbons migrate more readily into the larger fractures and especially into those that are continuous across the seal rock.

Evaporite sequences are generally considered to be highly ductile and to contain few, if any, fractures. Salt beds typically flow in response to tectonic forces. It has been assumed that tectonic fracturing of evaporites will be healed and not remain open pathways (Gill, 1979). However, the author has not encountered problems in detecting and delineating fields below the evaporites in the Michigan, Paradox, Denver, and Williston basins, all of which have thick salt sections. Evaporite sequences are rarely without shales. The flowage of salt and gypsum could permit these shales to move locally, to fracture, and to act as conduits between the bottom and top of the salt unit.

Petroleum will saturate the seal rock by diffusion over geologic time. How far from the reservoir diffusion will saturate the seal rock depends on pressure, the length of time hydrocarbons are present, and the size of the hydrocarbon molecules diffusing upward. Diffusion will be discussed in greater detail in later paragraphs.

The presence of a fracture or microfracture system is critical to vertical migration theories (Price, 1986). Therefore, a fundamental assumption must be that the rock column contains extensive microfracture systems that are connected between lithologies and that vary in density (Fig. 4-2). The amount, cause, and delineation of a microfracture system are beyond the scope of this discussion.

Migrating hydrocarbons cause geochemical changes in the strata overlying the reservoirs. These changes include increases in amounts of pyrite and magnetite, the presence of

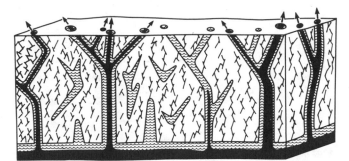

Fig. 4-2 An idealized cross section of fluid migration through the seal rock. Capillary pressure thresholds have been achieved in some of the microfractures but not in others. Oil is colored black and associated brine or water is wavy pattern.

minor amounts of hydrocarbons in overlying nonproductive zones, and the deposition of secondary cements. Perhaps the seal may actually be plugged up by the vertically migrating hydrocarbons directly over the highest concentration of petroleum in the reservoir (Duchscherer, 1984; and Price, 1986). This effective plugging of the migration pathways or the effectiveness of the seal rock, or both, could be the cause(s) of many classical halo patterns. Apical surface patterns suggest to some investigators that migrating hydrocarbons do not effectively plug the migrating pathways above the reservoir.

Groundwater Movement

The groundwater flow in a localized or regional system is also of concern for geochemical explorationists. Critics of vertical migration cite that groundwater flow is predominantly horizontal, reasoning that vertical migration cannot occur because vertically migrating hydrocarbons would be perpendicular to the dominant flow direction. If horizontally moving groundwater was affecting vertical migration, the surface geochemical expressions would be dispersed in the direction of flow and away from the reservoir. Empirical evidence indicates that surface expressions are located directly above both shallow and deep reservoirs (MacElvain, 1963A, 1963B, Siegel, 1974; Horvitz, 1980; and Jones and Drozd, 1983).

Fractals and Fluid Flow

In recent years, geologic systems have been described as fractal in nature (Mandelbrot, 1977; Pyrak-Nolte et al., 1988; Turcotte, 1988A, 1988B; Campbell et al., 1988; and Schroeder, 1991). The concept of fractals is based on a system that exactly repeats itself at different scales. The difficulty in applying fractals to geology or geologic processes is that the geologic section is not a homogeneous system but is rather a heterogenous one. Therefore, geologic

systems or processes may indeed be fractal in nature, but variations and modifications occur at so many levels that the systems are not exact between scales. Pyrak-Nolte et al. (1988) described fluid flow, fracture topology, and fracture density not as random events but as the results of fractal dimensions that are clustering based on predictable equations. If fluid flow and fracture density are indeed fractal, then vertical migration would have to be a partly or totally fractal process. Brownian motion has also been described as fractal (Schroeder, 1991) and may not be the random, unpredictable event it was once thought to be. An underlying problem in applying fractals is that geologic and fluid flow processes are not rigid processes but use concepts and mathematical laws that allow greater flexibility to systems that do not have an absolute end-product or always the same result. Bölviken et al. (1992) have proposed that geochemical regions, provinces, or landscapes are fractal in nature on the basis that topography and ore deposits have been shown to be fractal. Petroleum trends predicting areas of untested but potential accumulations have been identified based on fractal analogs and geometry. A reconnaissance survey represents a regional fractal. Denser or more detailed surveys reveal similar fractals that can be classified as subprovinces. These subprovinces, by further and denser sampling, uncover additional subfractals or levels thereof. Therefore, a predictable hierarchy is created. The overall concept of fractals is applicable to vertical migration or other fluid flow systems in the earth. The flow and mass transport regime can be considered fractal in practice (Fig. 4-3) because similar types of geometries, complexities, and systems are present over a large range of dimensional scales (Hanor, 1987). The flow system is not truly fractal, as presently defined, because of the heterogeneity and inherent complexity of the geologic section.

Groundwater Systems

Fluid, usually but not necessarily water, flows from higher-energy to lower-energy areas as sediments accumulate in a sedimentary basin. Fluid is expelled from less permeable, highly compactible strata to more permeable, less compactible strata during the compaction history of a basin. The flow pattern of any basin is determined by stratigraphy and structures. Fluid flow in regressive sequences is downward from shales and siltstones to the first permeable unit and then landward. In transgressive sequences, fluid flows in all directions but specifically toward the interbedded sandstones and then landward. In alternating units of shale and sandstone, fluid flow is isolated with respect to each specific permeable zone. Basin flow changes when sedimentation ceases, and the main energy for flow movement is the result of elevation from intake areas subject to modification by tectonic events.

Fluid flow is the result of the system's attempt to reach

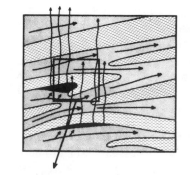

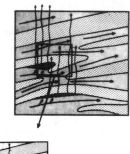

Fig. 4-3 Idealized fractal geometries of processes and stratigraphic sequences are represented and repeated approximately from scale to scale. Horizontal arrows indicate regional fluid flow, and vertical arrows represent vertically migrating hydrocarbons (adapted from Hanor, 1987).

equilibrium (Chapman, 1987). Several factors, chief among them, affect fluid flow: (1) mean grain size, (2) permeability, (3) viscosity, and (4) Darcy's law. Mean grain size is an effective measure of porosity. Permeability is a measure of the size of the connections between pores. Viscosity is a measure of a fluid's internal resistance to flow; the greater the viscosity, the greater the resistance. Darcy's law states that 1 ml of fluid, having a viscosity of 1 cP (centipoise), flows in 1 s under a pressure differential of 1 atm through a porous material having a cross-sectional area of 1 cm^2 and a length of 1 cm.

The regional flow regime can be divided into gravity-, pressure-, and density-driven systems. Gravity-driven flow depends on compaction, buoyancy, and density (Fig. 4-4). Most gravity-driven systems involve fresh water and require differences in elevation to provide energy for the system. Several examples of gravity-driven systems exist in the literature (Toth, 1980; Hitchon, 1984; and Garven and Freeze, 1984).

A buoyancy-driven (or density-driven) system, also called free convection, derives its energy from differences in temperature and salinity (Fig. 4-5). A fluid that is decreasing in density with depth will convect if the buoyancy forces overcome the viscous forces resisting flow (Blanchard and Sharp, 1985; Hanor, 1987).

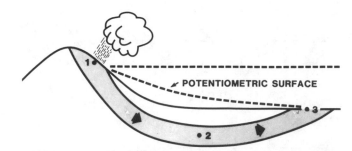

Fig. 4-4 Gravity- or compaction-driven flow. Water extends the system at the outcrop (1) and is driven downward as a result of compaction (2) and discharge (3) (after Hanor 1987; reprinted with permission of the Society of Economic Paleontologists and Mineralogists).

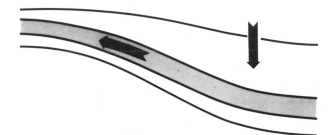

Fig. 4-5 Buoyancy or convection-driven flow. $\Delta T = \Delta \propto$ of hot liquids (after Hanor 1987; reprinted with permission of the Society of Economic Paleontologists and Mineralogists).

Pressure-driven flows are caused by compaction disequilibrium, thermal expansion, and generation of gas (Fig. 4-6). Rates of liquid movement are very slow (Neuzil, 1986). This type of system is supported by data from the Gulf Coast Petroleum Province, where temperature and salinity anomalies are associated with faults in overpressured and productive areas (Hanor and Bailey, 1983; Hanor, 1987).

It has been proposed that all three systems can operate and interact with each other in the same basin. Fluid flow in multiple directions complicates the understanding of the hydrodynamics (Dahlberg, 1982; and Hanor, 1987). Consequently, the question remains: Does fluid flow affect vertical migration? The generally preferred answer is that groundwa-

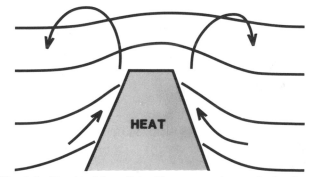

Fig. 4-6 Density-driven flow. Heat causes changes in density of fluids resulting in fluid flow (after Hanor 1987; reprinted with permission of the Society of Economic Paleontologists and Mineralogists).

ter flow has no effect on vertically migrating hydrocarbons because the rate of horizontal fluid movement is relatively slow. Rapid changes seen in the geochemical anomaly at the surface can be caused by removal of petroleum and lowering of reservoir pressure through production. This has been supported by observations of Coleman et al. (1977). Also, repressurization of the reservoir by enhanced secondary recovery gas or liquid injection methods indicates that changes in the amount of hydrocarbons migrating vertically are real-time phenomena. Offset of surface geochemical anomalies relative to the underlying field have been shown to be related to major gas-liquid pathways, such as faults, which funnel migrating hydrocarbons to the surface.

Transport

Three general types of hydrocarbon movement or transport can occur. The first type takes place when the petroleum relocates from the source rocks to the reservoir rocks. The second type occurs when the reservoir is breached, drilled, or flushed and the petroleum moves out of the trap. Vertical migration to the earth's surface occurs all the time. It is the third type or last movement, and it is the one that will be considered here. Many investigators initially suggested that the petroleum was dissolved in an aqueous fluid that was migrating vertically. This concept was a product of early work on the transport of petroleum from the source rock to the trap. Until the 1980s, most explanations of surface geochemical anomalies adhered to this interpretation.

Migration or transport of petroleum from the source rock to the reservoir trap has been extensively discussed in the literature. For example, two concepts dominate the accumulated record and continue to attract support.

1. The transport of hydrocarbons in a dissolved state or in a molecular solution is a result of increased solubility due to high temperatures (Price, 1976, 1979, 1986). Petroleum exsolution occurs when these hot fluids meet cooler fluids at shallower depths. The hot fluids are generally being expelled at the time of basinal compaction, which places time constraints on oil generation and migration. This model requires a fluid to carry the petroleum. The mechanism has a tendency to cause dispersal rather than concentration of hydrocarbons. However, there is considerable evidence that fluids carry large volumes of methane under high temperatures at great depths (deeper than 20,000 ft). Except in rare cases, the temperatures and pressures needed to cause migration do not exist in the shallow basin horizons where a major portion of vertical migration occurs.

2. Petroleum is generated and migrates as a three-dimensional kerogen network moving both vertically and horizontally (McAuliffe, 1979). Hydrocarbon saturation in the fluids ranges from 4% to 20%. A separate condition called *buoyancy* occurs when saturation exceeds 20% along the upper and lower surfaces of the carrier bed. The petroleum is actually moving independently and does not require water flow to carry it.

The ability of water to move decreases dramatically as the rock pores become saturated with hydrocarbons, and the petroleum's ability to migrate increases (Cross, 1979). The water becomes immobile when water droplets are encased in petroleum, and the pores are considered petroleum-permeable. Therefore, petroleum in such a system migrates readily from an area of greater pressure to one of lesser pressure. This type of system does not require that hydrocarbons go into solution.

After hydrocarbons have migrated vertically into the seal of the reservoir, they must move to the surface. The forms of transport that have been suggested are:

1. Diffusion
2. Compactional/Meteoric
3. Colloidal
4. Effusion
5. "Permeation"

Diffusion

Diffusion has been suggested as a possible mechanism of transport for hydrocarbons migrating to the surface (Rosaire, 1940; Kartsev et al., 1959; Stegna, 1961; Baijal, 1962; Siegel, 1974; Donovan and Dalzeil, 1977; Duchscherer, 1980, 1981A, 1981B, 1984; Denison, 1983; and Klimenko, 1976). The mechanism occurs in response to an ion concentration gradient requiring uniform movement in all directions of molecules, atoms, and ions into a vacuum, a fluid, or a porous medium. The theoretical end result is equalized concentrations of the ions in all parts of the system (Fig. 4-7). Diffusion is a spherically dispersive process, with components moving in all directions away from the source. The rate of hydrocarbon diffusion was calculated by Stegna (1961). His results suggest that, in over 310×10^6 million yr of diffusion, a total of 0.008 m^3 of gas will move to the earth's surface from a depth of 800 to 1200 m.

Diffusion rates of hydrocarbons have also been measured by Leythaeuser et al. (1983) using 10-m cores taken from outcrops of organic-rich shales. The concentration of individual hydrocarbons was measured at specific depth intervals to determine diffusive loss. The results indicated that ethane, propane, and butane are more quickly lost whereas the heavier hydrocarbons such as hexane and pentane are lost at a much slower rate. The study showed that the hydrocarbons migrated downward from the overlying siltstone source rock into the sandstone reservoir. Leythaeuser and co-workers calculated that a specific field, the Harlingen in the North Sea Netherlands, would leak all its methane content in 4.5 million yr by diffusion. This would place constraints on

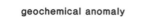

geochemical anomaly

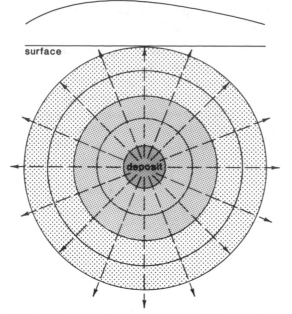

surface

Fig. 4-7 Diffusion or dispersion is a multidirectional process whereby the mobile component moves in all directions in equal amounts in response to a concentration gradient away from the deposit. The amounts of the material being moved become increasingly dispersed or less concentrated with distance from the surface. The ideal geochemical anomaly across such a deposit would be subtle and have an extensive transition zone with respect to background. No abrupt geochemical anomaly would be possible. Colors indicate a decrease in concentration with distance from the source. Dark shade shows the greatest concentrations and light shade the least.

petroleum generation and entrapment not only for this field but for several others because most source rocks would have had to yield their hydrocarbons recently (geologically speaking) or the volatiles at least would be lost. The diffusion coefficient from this study was estimated as $1 \times 10\,e-$ cm²/ s and was considered extremely slow (Klusman, 1989). Hydrocarbons migrating downward are consistent with the diffusion theory.

Kross et al. (1992) recalculated diffusion rates based on Leythaeuser et al. (1983) because, in their opinion, the rates were simplistic. A review of the cap rock/reservoir boundary conditions did not take into account the gas-liquid equilibrium between gas in the reservoir and the aqueous phase of leakage by diffusion. This is still subject to debate, but than the loss is at a more acceptable rate.

Diffusion has been disputed as the primary form of petroleum transport by Hunt (1979), MacElvain (1969), Jones and Drozd (1983), and Price (1986). Hunt indicated that the migrating hydrocarbons transported by diffusion would take millions of years to travel through shales with permeabilities of near zero. Stegna's (1961) and Klusman's (1989) calculations support this very slow rate.

Diffusion cannot explain the sharp outline of the surface

geochemical anomaly. It also cannot facilitate the real-time changes in the surface geochemical anomaly in response to changes in the petroleum reservoir either by pressure decreases due to production or by pressure increases due to enhanced recovery methods. If several million years are needed to create the surface geochemical anomaly because of the slow diffusion rate, the anomaly would not disappear or lose intensity during the time the field is plugged and abandoned. Fossil anomalies would be a common occurrence and, to date, there is no empirical evidence that they exist. Diffusion, as defined, is a spherical process, but pressure constraints force fluids or gases using this mechanism to be dispersed in the vertical or near-vertical direction. The presence of halo anomalies violates this premise of equal dispersion by diffusion. Diffusion suggests that heavier hydrocarbons should be in greater quantities in the soil than have been documented because the large molecular size restricts movement via this mechanism.

Helium, which has very small molecular size, may be the exception to total dismissal of the diffusion theory. The main method of transport of helium is by diffusion (Pogorski and Quirt, 1981). Helium has been clearly documented in association with petroleum accumulations both at the near surface and in conjunction with the reservoir itself. Helium will be discussed in more detail in Chapter 12.

Compactional/Meteoric

It has been suggested that the surface geochemical anomalies are the product of vertical movement due to deep basinal compaction or meteoric-recharge waters. The waters move laterally through the petroleum basin and then escape upward upon reaching fracture zones to the earth's surface. Several authors have proposed complex models (Fig. 4-8) that outline "forced draft" or "deep water discharge" theories (Pirson 1960, 1963, 1969; Donovan and Dalzeil, 1977; Roberts, 1980, 1981; Jones, 1984; and Davidson, 1982, 1984). The theories establish hydrocarbon traps as filters that remove junk molecules or solubilize oil from water and allow the water to pass through the trap on its way to the surface. To date, there is no published evidence for this type of

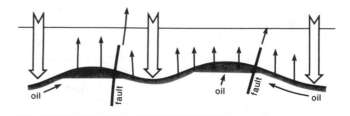

Fig. 4-8 Compactional or outdraft theory in general form. Downward pressure (compaction) causes compression of the reservoir rocks and drives petroleum and associated fluids into lower-pressure areas, or traps. Pressure continues to drive fluids upward from these traps.

phenomenon. The basic premises of the theory violate Darcy's law (Chapman, 1982) by suggesting that large amounts of water can migrate quickly through impermeable shales. This implies that concentrations of hydrocarbons at the surface should be greater than are currently measured.

Machel and Burton (1991) attempted to support basinal or groundwater discharge. Their article was specifically concerned with the formation of magnetic minerals above petroleum reservoirs due to hydrocarbon seepage. They assumed that basinal groundwater flow caused plumes or distortions of the hydrocarbon seepage in the direction of the flow. The article did not discuss several of the known facts concerning vertical migration, such as: (1) The surface geochemical anomaly, except in rare cases, is never offset from the reservoir below. (2) Machel and Burton's model suggests that the anomalies would not be present in recharge areas but would be overabundant in discharge areas. Surface geochemical anomalies exist both in recharge and discharge areas, with no indications that one area is favored over the other.

Jones (1984) supported deep basinal discharge by using an example from Tertiary rocks of the U.S. Gulf Coast (Fig. 4-9). Fluids attempt to achieve equilibrium by migrating from a higher- to a lower-pressure area. Therefore, petroleum migrates laterally under high pressure until it reaches a permeable zone or pathway that allows it to move upward. It will fill this trap and then move laterally until it meets another permeable area to fill or a pathway to the surface. This process may also be termed *effusion*.

A similar type of mechanism is well documented in the establishment of Mississippi Valley–type lead-zinc and fluorspar deposits and specifically in the Fluorspar district of southern Illinois (Hutchinson, 1983). Fluorine-rich brines migrated upward along an extensive fault and fracture system until they were temporarily stopped by a shale or fine-grained gouge in the fracture/fault system (Fig. 4-10). The brines migrated outward into the limestone and replaced calcium with fluorine. The fluids eventually overcame and bypassed the barrier and continued upward. The process

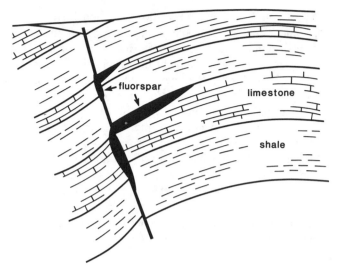

Fig. 4-10 A simplified illustration of a fluorspar deposit located in southern Illinois. Hot fluorine-rich fluids migrate upward along fault and fracture pathways. Upon reaching a temporary barrier, some of the fluid migrates laterally. The flourine replaces calcium (forming CaF_2CO_3). The remaining fluids eventually break the barrier and begin the process again.

was repeated after the fluids reached the next barrier. Petroleum has been found associated with these deposits and sometimes occurs in economic accumulations. However, the Gulf Coast example still does not adequately support basinwide discharge. Rather, it indicates an effusive mechanism along major pathways.

Colloidal

MacElvain (1969) and Price (1986) proposed a different form of migration. They suggested that hydrocarbons move upward as colloidal bubbles (Fig. 4-11). The small size of

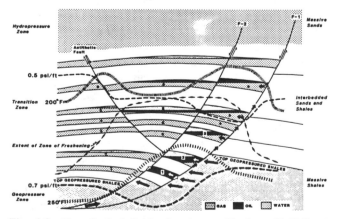

Fig. 4-9 Hypothetical fluid migration profile for a Gulf Coast Tertiary regression depositional system (Jones, 1984).

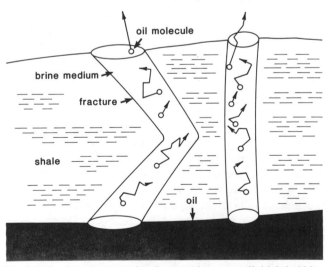

Fig. 4-11 Petroleum molecules moving as colloidal bubbles through a fluid medium in microfractures or micropores. The arrows represent the random nature implied by colloidal movement.

the C_1 through C_4 molecules allows frictional effects to overcome solubility problems by Brownian motion. Brownian motion was observed by Robert Brown in 1827 as ceaseless irregular movement of particles in a colloidal solution (Cross, 1979). Particles are not moving at the same velocity or in the same direction. Einstein in 1905 and Smoluckevskiin in 1906 further quantified the phenomenon as random thermal movement of molecules in a dispersing medium. Brownian movement is caused by the difference in the sum of the impacts received by molecules that hit other particles from various directions. The smaller the particles, the less likely it is that the molecular bombardment will be exactly balanced and, hence, the movement will be intensified and random. Schroeder (1991) indicated that Brownian motion is not random but fractal and therefore can be predicted and quantified mathematically.

MacElvain (1969) and Price (1986) argued that the vertical ascent of colloidal gas bubbles explains the higher and higher concentrations of lower molecular weight hydrocarbons away from the deposit. Also, Price stated that this was the reason the majority of hydrocarbons detected at the surface are in the C_1 to C_4 range (Horvitz, 1939; Rosaire, 1940; Kartsev et al., 1959; Philip and Crisp, 1982; and Price, 1986). They are smaller molecules and are thus more prone to agitation and movement. Price indicated that onshore exploration will find no, or only insignificant amounts of, C_{5+} hydrocarbons present. He assumed that an extensive microfracture system would solve the problems of petroleum migration through impermeable shales.

Price (1986) tested his theory by creating a natural gas atmosphere (hexane) in a sealed container with hexane at 22° t 25°C at 1 atm. After the system had equilibrated, it was tested with gas chromatography. The results indicated that the concentration of hexane was no greater than what would be expected at partial pressure in the atmosphere. However, vertical migration is not occurring at 1-atm pressure or at surface temperature through the rock strata. The experiment is valid only at the soil/air interface. Pressure (in atmospheres) increases with increasing depth into the soil and finally into the rock; temperature gradients vary considerably. Thus, the presence and the volume of heavier and heavier hydrocarbons with depth would be likely.

Since the 1970s, several contract companies have routinely detected C_1 through C_{12} hydrocarbons. Similar relationships are shown when ratios between various hydrocarbons from a surface survey are compared to ratios between the same hydrocarbons extracted from the underlying reservoir. A comparison of ratios would not be a viable exploration method if the higher hydrocarbons were unable to reach the surface in significant quantities. Jones and Drozd (1983) used ratios between the hydrocarbons to predict whether a prospect is similar to models. The models were derived from ratioing of petroleum taken from existing fields. If migrating hydrocarbons were truly colloidal, the amounts of higher and heavier hydrocarbons in solution would decrease steadily away from the reservoir. Ratios of hydrocarbons taken from the soil substrate and from the petroleum produced would not be comparable.

The basic colloidal concept requires petroleum to move in a fluid and not as a dissolved component. This addresses the problem of petroleum's relatively low aqueous solubility of hydrocarbons, which decreases with increasing chain length (Cross, 1979). Estimates of petroleum migration via aqueous dissolution allow only very small concentrations of 12,000 mg/l or less in solution (Dickey, 1975) and, thus, only small amounts of hydrocarbons would migrate. Increasing the temperature raises the solubility of hydrocarbons in water but does not significantly increase the volumes of hydrocarbons that would be needed to fill a reservoir with petroleum or to cause a geochemical anomaly at the surface.

The presence of salt water greatly decreases the solubility of hydrocarbons. The effect known as *salting out* causes the breakage of hydrophobic hydration structures as a result of the formation of stable hydrates by the ionic components of the salt. Therefore, the colloidal method must address the presence and effect of salt water in many of the overlying strata, which inhibit the movement and amounts of hydrocarbons that could migrate.

Effusion

Effusion is defined as the pouring out or ejection of fluid in large quantities, such as an extrusive volcanic flow (The American Geological Institute, 1984). Therefore, this concept (Fig. 4-12), as strictly applied, is in direct conflict with microseepage, which stresses the release or escape of minor amounts of hydrocarbons in relation to the total volume stored in the reservoir. This effusion concept, as determined by Price (1986), is more applicable to macroseepage than to microseepage.

The effusion concept does have application to the vertical migration theory (Rosaire, 1940; Kartsev et al., 1959; Baijal, 1962; Donovan and Dalzeil, 1977; Dennison, 1983; and Duchscherer, 1981A, 1981B, 1984). The massive outflow of volcanic fluids and hydrocarbon macroseeps represents the extreme forms of disruption in the rock column. The majority of petroleum accumulations are not significantly overpressured or overfilled to breach several hundreds or thousands of feet of overlying strata. If fluid and mass transport are fractal in nature, then effusion is present at both a macroscopic and a microscopic scale. Microfractures and sometimes minor faults or megafracture systems may extend vertically through the entire rock column of a basin and allow the escape of some hydrocarbons to the surface. Fractured shales, dolomites, and limestones have proved to be economic reservoirs.

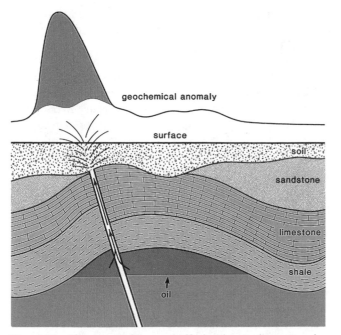

Fig. 4-12 The accepted form of effusion is massive outpouring of fluids (macroseep), in this case petroleum, along an open conduit or fault system where porosity and permeability is essentially infinite. The geochemical anomaly would be visible to the eye.

Permeation

Permeation has been described as a method of vertical migration but does not have a well-defined meaning and is not very specific (Horvitz, 1950; Baijal, 1962). However, it implies spreading through rock, which in itself suggests diffusion with a twist. Permeation, without any clear-cut definition, is not a reasonable method to describe migration.

Derived Assumptions about Vertical Migration

The theory of vertical migration can be divided into five general assumptions.

1. The system must be simple and easily established in numerous geologic basins. A system that requires a set of complex conditions is unlikely to succeed under a wide range of geologic environments. Hutchinson (1983) and Sawkins (1984) described a general model for porphyry copper deposits that was derived from several site-specific models. This eliminated the need to have numerous models. There are certainly differences among deposits, but their origins remain somewhat universal in terms of the overall requirements for emplacement and the general chemical characteristics used for detection. Similarly, a simple system for vertical migration is needed that does not require complex reactions and site-specific conditions and phenomena to

allow it to work. A complex system is likely to collapse easily and become nonfunctional.

2. The hydrocarbons must migrate and disburse in accordance with the first and second laws of thermodynamics. Water does not permeate to the surface because its fluid system is in equilibrium. Conversely, petroleum does leak to the surface because its fluid system is not in equilibrium; the strata over the reservoir are not full of petroleum.

Pressure constraints are generally not addressed in the literature. A gas or liquid migrates from the area of greatest pressure to the area of least pressure. Hydrocarbon gases can transport greater amounts of dissolved heavier petroleum molecules under higher pressures (Price, 1986). This explains why hydrocarbons heavier than C_5 are found in greater abundance at the top of the ocean sediment interface than at the onshore surface; the pressure regime at the top of the ocean sediments is greater than onshore because of the overlying water column. For example, if hydrocarbons reach the top of the ocean sediments in 330 ft of water, the pressure at the interface is 10 atm vs. 1 atm at the onshore surface. Therefore, greater amounts of heavier hydrocarbons should be present on the ocean floor. Unconsolidated sediments, such as in the U.S. Gulf Coast Petroleum Province, have many more fracture pathways and greater permeability than older onshore lithified strata. Thus, the passage of heavier hydrocarbons to the surface may be easily facilitated offshore.

3. The hydrocarbons must have a transport mechanism to the surface. There is a general consensus that petroleum does not move in solution. Colloidal transport has been suggested, but brines will cause salting out of hydrocarbons, making them insoluble. Therefore, it is unlikely that hydrocarbons enter a colloidal state in a brine and are able to move with relative ease. Colloidal movement may be possible in a hydrocarbon wet system. Hydrocarbons dissolved in a water-wet system generally do not move very far. They are easily trapped on the sediment interface in pores and fractures. However, if the system becomes partially hydrocarbon-wet (or less water-wet), petroleum molecules will increase their movement (Cross, 1979). They will be able to migrate more easily and in greater volume as more and more petroleum molecules are forced into the system.

4. There must be pathways or conduits along or through which the hydrocarbons are transported. It has been established that most rocks have a microfracture system, and no rock type is truly impermeable. These microfractures are the pathways for the movement of the hydrocarbons to the surface. The extent of the microfracture system, the type of petroleum molecules migrating, and the connections between the microfractures determine the amount of hydrocarbons reaching the surface. Methane-rich reservoirs, typically gas fields, are more likely to yield high volumes of light hydrocarbons to the surface because of the small size of the

dominant molecules present. This is supported by geochemical surveys over gas fields that yield greater measured volumes of hydrocarbons in the soil substrate than in soils over oil-dominated reservoirs. Thus, heavier and larger-sized molecules will not pass as easily through the fracture system to the surface.

5. All reservoirs do not leak uniformly and do not yield the same amounts of hydrocarbons at the surface. Although it has not been discussed in the literature, the assumption is that leakage volume is a universal constant. This has no basis in fact. Just as there is only a small percentage of extremely large fields in any basin, there is also only a small percentage of large geochemical anomalies and macroseeps in any basin. Therefore, there will be numerous geochemical anomalies of different sizes and shapes.

Conceptual Model

The conceptual model of microseepage takes the best of the existing theories, melds them into a general concept, and quantifies the chimney effect of vertical migration. The migrating hydrocarbons establish an environment that allows transport via buoyancy rather than depending on other factors or fluids as a mechanism. The sharpness of the chimney outline is the result of the establishment of a transition zone that prevents the degradation or bonding of the hydrocarbons by other fluids, which would inhibit further movement. Microfractures and micropores act as conduits for hydrocarbons to travel to the surface.

Characteristics of the conceptual model (Figs. 4-13 to 4-17; see color plates) are as follows:

1. Vertical migration commences when the trap begins filling with petroleum (Fig. 4-13). The trap is essentially a sealed container, but it is composed of heterogenous material with disconformities in the form of microfractures and micropores. The amount of hydrocarbons escaping the typical trap is relatively small. The hydrocarbons begin migrating upward through micropores and microfractures in the seal rock. The first hydrocarbons may be immobilized by the salting-out effect but, as more are pushed upward and into the water-wet area, the system tends to become hydrocarbon-wet. This allows hydrocarbons to increase in mobility and to decrease water mobility. As additional petroleum is added to the system, the salting-out effect is diminished. Therefore, a transition zone is established which is not entirely conducive to hydrocarbon transport but is not entirely water-wet either. No surface expression is detectable at this stage. Buoyancy occurs when hydrocarbon saturation reached 20% of the total fluid volume.

2. As the trap fills, the upward migrating hydrocarbons continue toward the surface along the more conducive microfracture systems as a result of buoyancy rather than in solution or as colloidal fluid (Fig. 4-14). The transition zone is not moving uniformly upward. The trap is still filling, but enough hydrocarbons are present to cause the lower part of the seal to be saturated by diffusion.

The microfractures allow venting, exhaling, or microeffusion of hydrocarbons to the soil substrate by mass migration. Initial transport is probably as colloidal gas bubbles in a water-wet system, but overall movement is due to buoyancy in a partially hydrocarbon-wet system attempting equilibrium. Evidence of this migration is that some oil fields contain shows of petroleum in nonproducing strata overlying reservoir rock. These minor accumulations have been concentrated while migrating upward and represent significant leakage from the reservoir below. This assumes that these overlying reservoirs were not major pathways for basinal hydrocarbon migration as well.

3. The first migrating hydrocarbons reach the surface along the most easily traveled pathways (Fig. 4-15). In some cases, this will be within areas where intense fracturing or open fault systems are present. An initial geochemical response is detected at the surface which, in certain cases, might be visible to the naked eye and be termed a macroseep.

4. More hydrocarbons reach the surface and cause geochemical changes both in the soil substrate and in various rock strata overlying the petroleum accumulation (Fig. 4-16). The hydrocarbons create reduction zones that cause the deposition of various iron-rich minerals (both magnetic and nonmagnetic) and carbonate minerals. The petroleum accumulation is now detectable at the surface over an areal extent similar to the outline of the petroleum accumulation at depth. Changes in the outline and intensity of the surface geochemical anomaly may be due to disruption of the migration front by temporary salting out, pressure changes, closing of certain pathways, opening of new pathways.

The hydrocarbons continue to migrate into the atmosphere upon reaching the soil substrate. The rapidly changing chemical conditions in the soil environment cause the heavier hydrocarbons to drop out and thus cease their migration. Biological activity and variations in moisture content interfere until the lower molecular weight hydrocarbons dominate the measured fraction. This occurs from the top of the water table to the surface.

5. Postdiscovery, after drilling and completion (Fig. 4-17), results in diminishing the surface geochemical anomaly. This may not occur if areas in the field have not been drained efficiently or are not connected to the main reservoir. The pressure forcing the hydrocarbons upward decreases as pressure depletion in the reservoir occurs.

In summary, breached or shallow overpressured reservoirs can be identified as visible effusion (occurring as originally defined) that can be measured. *Macroeffusion* is another word for macroseep or visible oil seep. Vertical migration is a fractal process and not a specific process that changes from level to level. Therefore, effusion occurs at many different scales from microscopic to megascopic levels. Diffusion does occur but is a minor part of the system. Some authors assume that the process of vertical migration is site-specific and ignore certain consistently recognized criteria. Selective and complex processes cannot be valid for a simple, well-documented phenomenon that occurs in several different types of geologic environments and basins.

References

Aguilera, R. (1980). *Naturally Fractured Reservoirs*, Pennwell Books, Tulsa, OK.

The American Geological Institute (1984). *Dictionary of Geological Terms*, R. L. Bates and J. A. Jackson, eds., Anchor Books, New York.

Baijal, S. K. (1962). Geochemical Survey of the Hilbig Oil Field, Bastrop County, Texas, University of Texas, Austin, M.S. thesis.

Blanchard, P. E. and J. M. Sharp Jr. (1985). Possible free convection in thick Gulf Coast sandstone sequences, *Transactions of the Southwest Section* AAPG: pp. 6–12.

Bölviken, B., P. R. Stokke, J. Feden, and T. Jössang (1992). The fractal nature of geochemical landscapes, *Journal of Geochemical Exploration*, vol. 43, pp. 91–109.

Campbell, C. B., S. H. Carpenter, C. A. Carson, E. P. Henderson, K. V. Might, C. M. Richards, R. H. Fluegeman, and R. S. Snow (1988). Fractal analysis in geology: Five example applications, *Geological Society of America Bulletin*, Vol. 20, No. 5, p. 338.

Chapman, R. E. (1982). Effects of oil and gas on water movement, *American Association of Petroleum Geologists Bulletin*, Vol. 66, pp. 368–378.

Chapman R. E. (1978). Fluid flow in sedimentary basins: A geologist's perspective, in *Fluid Flow in Sedimentary Basins and Aquifers*, J. C. Goff and B. P. J. Williams, eds., Geological Society (London) Special Publication No. 34, pp. 3–18.

Coleman, D. D., W. F. Meents, C.-Li. Liu, and R. A. Keough (1977). Isotopic identification of leakage gas from underground storage reservoirs: A progress report, Illinois State Geologic Survey, Illinois Petroleum No. 11.

Collins, T. (1992). Natural gas production from Arctic gas hydrates, presented at the Coal-bed Methane Forum, sponsored by the Gas Research Institute, Oct. 15, 1992.

Cross, S. Yariv H. (1979). *Geochemistry of Colloid Systems*, Springer-Verlag, New York.

Dahlberg, E. C. (1982). *Applied Hydrodynamics in Petroleum Exploration*, Springer-Verlag, New York.

Davis, J. B. (1967). *Petroleum Microbiology*, Elsevier Publishing Company, New York.

Davidson, M. J. (1963). Geochemistry can help find oil if properly used, *World Oil*, July, pp. 94–106.

Davidson, M. J. (1982). Toward a general theory of vertical migration, *Oil and Gas Journal*, June 21, pp. 288–300.

Davidson, M. J. (1984). State of the art and the direction of unconventional oil and gas exploration, in *Unconventional Methods in Exploration for Petroleum and Natural Gas III*, Southern Methodist University, Dallas, TX, pp. 4–8.

Denison, D. (1983). Geochemistry exploration techniques being used to help pin down hydrocarbon prospects in Michigan, *Michigan Oil and Gas News*, Dec. 23, pp. 161–164.

Dickey, P. A. (1975). Possible primary migration of oil from source rock in oil phase, *American Association of Petroleum Geologists Bulletin*, Vol. 59, pp. 337–345.

Donovan, T. J. and M. C. Dalzeil (1977). Late digenetic indicators of buried oil and gas, U.S. Geological Survey Open-File Report 77-817.

Downey, M. W. (1984). Evaluating seals for hydrocarbon accumulations, *American Association of Petroleum Geologists Bulletin*, Vol. 8, No. 12, pp. 1752–1763.

Duchscherer, W. Jr. (1980). Geochemical methods of prospecting for hydrocarbons, *Oil and Gas Journal*, Dec. 1, pp. 194–208.

Duchscherer W. Jr., (1981A). Nongasometric geochemical prospecting for hydrocarbons with case histories, *Oil and Gas Journal*, Oct. 19, pp. 312–327.

Duchscherer W. Jr., (1981B). Geochemical exploration for hydrocarbons: No new trick, but an old dog, *Oil and Gas Journal*, Oct. 19, pp. 312–327.

Duchscherer W. Jr., (1984). *Geochemical Hydrocarbon Prospecting*, Pennwell Books, Tulsa, OK.

Garven, G. and R. A. Freeze (1984). Theoretical analysis of the role of the groundwater flow in the genesis of stratabound ore deposits—I. Mathematical and numerical model, *American Journal of Science*, Vol. 284, pp. 1085–1124.

Gill, D. (1979). Differential entrapment of oil and gas in Niagaran pinnacle-reef belt of northern Michigan, *American Association of Petroleum Geologists Bulletin*, Vol. 63, No. 4, pp. 608–620.

Hanor, J. S. (1987). *Origin and Migration of Subsurface Sedimentary Brines, Lecture Notes for Short Course*, Society of Economic Paleontologists and Mineralogists, Tulsa, OK.

Hanor, J. S. and J. E. Bailey (1983). Use of hydraulic gradient to characterize geopressured sediments and the direction of fluid migration in the Louisiana Gulf Coast, *Gulf Coast Association of Geological Societies Transactions*, Vol. 23, pp. 115–122.

Hitchon, B. (1984). Geothermal gradients, hydrodynamics, and hydrocarbon occurrences, Alberta, Canada, *American Association of Petroleum Geologists Bulletin*, Vol. 68, pp. 713–743.

Horvitz, L. (1939). *On Geochemical Prospecting*, I: *Geophysics*, Vol. 4, pp. 210–228.

Horvitz, L. (1950). Chemical methods, in *Exploration Geophysics*, 2nd ed., J.J. Jakosky, ed., Trija Publishing, Los Angeles, pp. 938–965.

Horvitz, L. (1954). Near-surface hydrocarbons and petroleum accumulation at depth, *Mining Engineering*, Dec., pp. 1205–1209.

Horvitz, L. (1980). Near-surface evidence of hydrocarbon movement from depth, in *Problems of Petroleum Migration*, W. H. Roberts and R. J. Cordell, eds., American Association of Petroleum Geologists Studies in Geology No. 10, pp. 241–263.

Hunt, J. M. (1979). *Petroleum Geochemistry and Geology*, W. H. Freeman and Co., San Francisco.

Hunt, J. M. (1981). Surface geochemical prospecting, pro and con, *American Association of Petroleum Geologists Bulletin,* Vol. 65, p. 939.

Hutchinson, C. S. (1983). *Economic Deposits and Their Tectonic Setting,* John Wiley & Sons, Inc., New York.

Jones, P. (1984). Deep water discharge: A mechanism for the vertical migration of oil and gas, in *Unconventional Methods in Exploration for Petroleum and Natural Gas III,* Southern Methodist University, Dallas, TX, pp. 254–271.

Jones, V. T. and R. J. Drozd (1983). Predictions of oil and gas potential by near-surface geochemistry, *American Association of Petroleum Geologist Bulletin,* Vol. 67, pp. 932–952.

Kartsev, A. A., A. Z. Tabasaranskii, M. I. Subbota, and G. A. Mogilevskii (1959). Geochemical methods of prospecting and exploration for petroleum and natural gas, University of California Press, Los Angeles. (English translation edited by P. A. Witherspoon and W. D. Romey. Original Russian version published in 1954.)

Klimenko, A. A. (1976). Diffusion of gases from hydrocarbon deposits, *International Geologic Review,* Vol. 18, pp. 717–722.

Klusman, R. W. (1989). *Surface and Near-Surface Geochemistry in Petroleum Exploration, Short Course,* Rocky Mountain Association of Geologists, Denver, Co.

Kross, B. M., D. Leythaeuser, and R. G. Schofer (1992). The quantification of diffusion hydrocarbon losses through cap rocks of gas reservoirs, *American Association of Petroleum Geologists Bulletin,* Vol. 76, No. 3, pp. 103–406.

Landes, K. K. (1970). Petroleum Geology of the United States, John Wiley & Sons, Inc., New York.

Leythaeuser, D., R. G. Schaefer, and H. Pooch (1983). Diffusion of light hydrocarbons in subsurface sedimentary rocks, *American Association of Petroleum Geologists Bulletin,* Vol. 67, No. 6, pp. 889–895.

MacElvain, R. (1963A). What do near surface signs really mean in oil finding? *Oil and Gas Journal,* Feb. 18, pp. 132–136.

MacElvain, R. (1963B). What do near surface signs really mean in oil finding? *Oil and Gas Journal,* Feb. 25, pp. 139–146.

MacElvain, R. (1969). Mechanics of gaseous ascension through a sedimentary column, in *Unconventional Methods in Exploration for Petroleum and Natural Gas,* W. B. Heroy, ed., Southern Methodist University Press, Dallas, TX, pp. 15–28.

Machel, H. G. and E. A. Burton (1991). Causes and spatial distribution of anomalous magnetization in hydrocarbon seepage, *American Association of Petroleum Geologists Bulletin,* Vol. 75, pp. 1864–1876.

Mandelbrot, B. B. (1977). *The Fractal Geometry of Nature,* W. H. Freeman and Co., San Francisco.

McAuliffe, C. D. (1979). Oil and gas migration—Chemical and physical constraints, *American Association of Petroleum Geologists Bulletin,* Vol. 63, No. 5, pp. 761–781.

Neuzil, C. E. (1986). Groundwater flow in low-permeability environments, *Water Resources,* Vol. 22, pp. 1163–1195.

Nisle, R. G. (1942). Considerations on vertical migration of gases, *Geophysics,* Vol. 6, No. 4, pp. 449–454.

Pandey, G. N., M. R. Tek, and D. L. Katz (1974). Diffusion of fluids through porous media with implications in petroleum geology, *American Association of Petroleum Geologists Bulletin,* Vol. 58, No. 2, pp. 291–303.

Philip, R. P. and P. T. Crisp (1982). Surface geochemical prospecting methods used for oil and gas prospecting, a review, *Journal of Geochemical Exploration,* Vol. 17, pp. 1–34.

Pirson, S. J. (1941). Measure of gas leakage applied to oil search, *Oil and Gas Journal,* Feb. 20, pp. 21 and 32.

Pirson, S. J. (1946). Disturbing factors in geochemical prospecting, *Geophysics,* Vol. 11, pp. 312–320.

Pirson, S. J. (1960). How to make geochemical exploration succeed. *World Oil,* April, pp. 93–96.

Pirson, S. J. (1963). Projective well log interpretation, *World Oil,* Oct. pp. 166–120.

Pirson, S. J. (1969). Geological, geophysical and chemical modification of sediments in the environment of oil fields, in *Unconventional Methods in Exploration for Petroleum and Natural Gas,* W. B. Heroy, eds., Southern Methodist University Press, Dallas, TX, pp. 159–186.

Pogorski, L. A. and G. S. Quirt (1981). Helium emanometry in exploring for hydrocarbons: Part I, in *Unconventional Methods in Exploration for Petroleum and Natural Gas,* Southern Methodist University, Dallas, TX, pp. 124–135.

Price, L. C. (1976). Aqueous solubility of petroleum as applied to its origin and primary migration, *American Association of Petroleum Geologists Bulletin,* Vol. 60, No. 2, pp. 213–244.

Price, L. C. (1979). Aqueous solubility of methane at elevated pressures and temperatures, *American Association of Petroleum Geologists Bulletin,* pp. 1527–1533.

Price, L. C. (1986). A critical overview and proposed working model of surface geochemical exploration, in *Unconventional Method IV,* Southern Methodist University Press, Dallas, TX, pp. 245–304.

Pyrak-Nolte, L. J., D. D. Nolte, and N. G. W. Cook (1988). Fractal flowpaths through fractures in rock, Geological Society of America, Vol. 20, No. 7, p. A300.

Roberts, W. H. (1980). Design and function of oil and gas traps, in *Problems of Petroleum Migration: American Association of Petroleum Geologists Studies in Geology,* No. 10, pp. 217–240.

Roberts, W. H. (1981). Some uses of temperature data in exploration, in *Unconventional Methods in Exploration in Petroleum and Natural Gas II,* Southern Methodist University, Dallas, TX, pp. 8–49.

Rosaire, E. E. (1940). Geochemical prospecting for petroleum, in *Symposium on Geochemical Exploration, American Association of Petroleum Geologists Bulletin,* Vol. 24, pp. 1400–1433.

Saeed, M. (1991). Light hydrocarbon microseepage mechanism(s): Theoretical considerations, Unpublished Ph.D. dissertation, Colorado School of Mines, Golden, Co.

Sawkins, F. J. (1984). *Metal Deposits in Relation to Plate Tectonics,* Springer-Verlag, New York.

Schroeder, M. (1991). *Fractals, Chaos, Power Laws: Minutes from an Infinite Paradise,* W. H. Freeman and Co., San Francisco.

Siegel, F. R. (1974). Applied geochemistry in *Geochemical Prospecting for Hydrocarbons,* John Wiley and Sons, Inc., New York, Chap. 9, pp. 228–255.

Sittig, M. (1980). *Geophysical and Geochemical Techniques for Exploration of Hydrocarbons and Minerals,* Noyes Data Corp., Newalk, NJ.

Stegna, L. (1961). On the principles of geochemical oil prospecting, *Geophysics,* Vol. 26, No. 4, Aug. 2, pp. 447–451.

Toth, J. (1980). Cross-formational gravity flow of groundwater: A mechanism of transport and accumulation of petroleum (the generalized hydraulic theory of petroleum migration) in W. H. Roberts III and R. J. Cordell, eds., *Problems of Petroleum Migration, American Association of Petroleum Geologists, Studies in Geology,* No. 10, pp. 121–167.

Turcotte, D. L. (1988A). Fractal distributions in geology, scale invariance, and deterministic chaos, *Geological Society of America,* Vol. 15, No. 2, pp. 163–165.

Turcotte, D. L. (1988B). Fractals in fluid mechanics, *Annual Review of Fluid Mechanics,* Vol. 20, pp. 5–16.

Weart, R. C. and G. Heimberg (1981). Exploration radiometrics: Postsurvey drilling results, in *Unconventional Methods in Exploration for Petroleum and Natural Gas,* Southern Methodist Univ., Dallas, Tx., pp. 116–123.

Zieglar, D. L. (1992). Hydrocarbon columns, buoyancy pressures, and seal efficiency: Comparison of oil and gas accumulations in California and the Rocky Mountain area, *American Association of Petroleum Geologists Bulletin,* Vol. 76, No. 4, pp. 501–508.

II

Methods of Microseepage Detection: Direct vs. Indirect

The surface geochemical methods available to the petroleum industry are divided into two groups: direct and indirect. Direct methods are those techniques that measure actual hydrocarbon concentrations that are usually in vapor form in soil, water, atmosphere, or sediments. These direct methods are called soil-gas, hydrocarbon, free-air, or vapor techniques. A similar method is fluorescence, which uses ultraviolet light in the same manner as a mudlogger uses a black-light box to detect shows. Indirect methods are the techniques that infer the presence of petroleum by detecting element(s), compound(s), or chemical settings that are the result of migrating hydrocarbons. Generally, indirect methods are more likely to detect a solid or liquid product than a gas.

All the methods described in Part II of this text can be categorized as pathfinders. The term *pathfinder* comes from the mining industry, where it is used to describe elements or compounds associated with, or part of, the one deposit being sought. There are direct pathfinders (soil-gas) and indirect pathfinders (all others). The direct pathfinders are the hydrocarbons themselves that are assumed to have leaked from the subsurface into the soil, surface water, or air. Indirect pathfinders are those elements or compounds that are closely associated with petroleum. Therefore, we are attempting to define pathfinders that identify areas of petroleum seepage.

There are problems associated with both direct and indirect methods. Because direct technology measures vapors, samples are more difficult to collect, repeatability of results may be uncertain, and data may be hard to interpret. Therefore, the collection and analysis costs for the direct techniques are higher than for the indirect. But, in order to achieve a significant level of confidence, the indirect results may need to be confirmed by direct methods in all or selected circumstances. Indirect methods are generally not as acceptable to the petroleum industry as are the direct methods (excluding fluorescence) because the industry has concluded that there is no absolute certainty that the detected indirect anomalies are related to petroleum accumulations. For some indirect methods, this statement can be made with certainty but, for others techniques, the association with petroleum alone is weighted in their favor.

The majority of explorationists have assumed that soil gas detected in the soil substrate or shallow bedrock is the result of leakage from a petroleum accumulation. However, there are a number of other sources of soil gas; the most obvious is highly organic soils or bedrock near the surface. Organic shales can become part of the soil and release minute amounts of soil gas as a result of weathering or the impact of a probe or auger. This is a major problem in areas such as the Nevada Basin and Range.

Soil-gas changes or anomalies can be due to mineral concentrations as well. In response to a diminishing number of mineral deposits at or near the surface, the mining industry has developed soil-vapor technology to measure changes in the residual soil caused by covered targets. Therefore, hydrocarbon detection is being used to find "concealed," "hidden," or "covered" metal deposits. Hydrocarbons generated by these metal deposits may not be easily differentiated by surface methods alone. Fortunately, many basins that are petroleum provinces seem to lack significant metal deposits in conjunction with hydrocarbons. However, the explorationist should use caution with data from a basin that has both petroleum and metals production.

More localized sources of petroleum may be surface spillage, pollution, and collection contamination. These forms of contaminants are generally avoidable. Spillage and pollution are usually isolated and can be eliminated from the data because they tend to be one-point mega-anomalies. Collection contamination is usually in the container or on the equipment used to collect samples.

As hydrocarbons can be generated by other natural resources, indirect methods will detect petroleum as well as nonpetroleum accumulations. Helium has been associated with uranium, base metals, and geothermal sources. Radiometrics has been related to groundwater changes, topographic variations, uranium and other radioactive elements, and weather and lithology variations. Iodine is associated with migrating petroleum and will accumulate regardless of its source. Radon increases have been identified with earthquake events and uranium and metal deposits. Trace and major element accumulations can be related to base or precious metal deposits, to natural sedimentary processes that concentrate them, and to normal soil weathering. Bacteria that metabolize petroleum will do so regardless of the source. Magnetic minerals are products of numerous reactions and processes and can be directly or indirectly related to metal deposits, concentrations in the soils due to weathering, the presence of associated bacteria, and groundwater. Eh-pH variations are a product of a number of factors such as moisture, soil composition, hydrocarbons in the soil, bacterial action, and climate. Any one of these may control Eh-pH on a regional or site-specific basis.

Despite this list of nonpetroleum influences, surface geochemical techniques used in petroleum exploration have been successful. The explorationist does need to be aware of the other sources and the collection and analytical problems that may affect both direct and indirect methods. The second part of this book outlines each technique, discusses its advantages and problems, and includes case histories (where available). Some techniques may be applicable in a wide variety of geologic and soil environments, whereas others may be more restricted. It is up to the explorationist to apply the methods correctly and evaluate the results.

5

Soil Gas

Introduction

The most common method for detecting and delineating a surface geochemical anomaly associated with a petroleum reservoir is measuring vapor or liquid hydrocarbons that have migrated to the surface or near-surface. The media in which these measurements occur are the soil, atmosphere, ocean water column, ocean bottom sediments, and groundwater or surface fresh water.

There are two methods of collection and analysis: (1) soil-gas or vapor measurements and (2) fluorescence techniques. Soil-gas methods measure the amount of migrating hydrocarbons in the vapor phase. Because some of these methods "liberate" hydrocarbons trapped in the occluded state or temporarily trapped in isolated pores, some of this gas may be derived from the liquid state. There is no indication that the state or phase the hydrocarbons are in is critical to their detection or to the interpretation of the data. However, as will be discussed later, some methods are specifically restricted to measuring either all or a portion of the gas present in a phase, and this can place limitations on the method's use. Fluorescence qualitatively determines the amount of hydrocarbons present in the soils in a similar but more sophisticated fashion than that of a well site geologist using black light to analyze well cuttings.

All methods of collection can generally quantify the amount of hydrocarbons moving through a specific medium over a predetermined time interval during which sampling occurs. These techniques can be divided into onshore, offshore, and airborne methods. All utilize similar forms of analysis, either the gas chromatograph (the most common) or the mass spectrometer. The dominant methods for soil, atmosphere, or water media vary because of the completely different conditions under which collection occurs.

General Information

Several collection methods are available to sample for gases migrating through the soil substrate. Each has advan-

tages and disadvantages. Gas is present in the soil and rock strata in the sorbed, occluded, solute, and free-vapor state. The two most common forms of onshore collection are the free-air and head-space methods. The objective of these methods is to collect gas that is present in the soil at the instant of sampling. Soil disturbances during sampling must be kept to a minimum because a percentage of the soil gas will be released and lost to the atmosphere. Soil gases are highly mobile and volatile. They are affected by rapid chemical changes or reactions which, in turn, impact the ability of sampling methods to collect gases that could identify a geochemical anomaly present in the soil.

Offshore techniques have been more readily accepted because of the ability to acquire undisturbed samples in the water column and visibly identify the seeps. Also, probes are used to penetrate the ocean floor to collect hydrocarbons in sediment cores. These bottom cores, unlike their onshore counterparts, are not as susceptible to rapid geochemical changes that can inhibit or degrade the hydrocarbon gases as they migrate upward. Offshore macroseeps are well documented and are used even today to direct exploration programs. Another consideration in their acceptance is that productive offshore fields tend to be quite large and subsequent leakage is substantial. This is generally not true with most remaining undiscovered onshore fields in the United States. Onshore, many fields that are economic are much smaller and are in areas that either have been heavily explored, in some cases for over a century, or are in areas where petroleum deposits are widely scattered.

Typically, the soil-gas and fluorescence samples are analyzed for methane (C_1) through pentane (C_6). Technological advancements in terms of detection limits continue to develop so that even heavier hydrocarbons can be measured. Some portable methods (such as the free-air method) are restricted to the analysis of ethane (C_2), propane (C_3), and butane (C_4), but many include methane (C_1). It is difficult to analyze for hydrocarbons heavier than C_4 either because they are not in the vapor phase or because the amounts present are smaller than the detection limits of the analytical equipment. If the equipment is sensitive to hydrocarbons in the liquid phase, other methods can detect hexane (C_6) and heavier hydrocarbons, which may be able to identify the source of the petroleum in the reservoir.

Methane is a common gas constituent of the soil and has a variety of sources other than seeping petroleum from the subsurface. Methane is commonly generated by biological activity. Thus, methane alone has not proved a consistently useful exploration tool because of the difficulty in separating the biologic from the petroleum-generated methane. When methane is used in conjunction with heavier hydrocarbons, it can be more useful.

It has often been thought that ethane and heavier hydrocarbons are not generated in significant amounts in the soil by biological activity. However, ethane, propane, and butane are typically present in minor amounts in the soil substrate as a by-product of specialized biological processes that may or may not be present from site to site. Usually, these hydrocarbons constitute less than 10% of the total gases analyzed in a normal soil sample. When these gases, along with other heavier hydrocarbons, are present in greater quantities, it has been observed that methane decreases as a total percentage of gas volume measured. This scenario is typically indicative of the petroleum reservoir. However, soil-gas anomalies have been associated with both productive and nonproductive petroleum accumulations. Further, it has been reported that the presence of C_5 and C_6 in overabundance indicates a nonproductive reservoir at depth or a lateral migration of petroleum. However, this has not been supported by experience or published investigations to date.

There is another group of organically derived hydrocarbons called *olefins*. The two most commonly measured are ethylene and propylene, which are variations of ethane and propane, respectively. Typically, they are not found in significant quantities in a petroleum reservoir but are generated by microbial activity in the soil substrate (Ullom, 1988). Ethylene has been suggested as a potentially useful exploration tool, but its occurrence is related to soil pH and the concentrations of specific trace elements. Cobalt and arsenic promote its generation, whereas iron, manganese, nickle, zinc, and aluminum significantly inhibit its production (Arshad and Frankenberger, 1991). Ullom (1988) and Saenz et al. (1991) advocated the use of ethylene and propylene for petroleum exploration in the form of ratios of ethane/ethylene and propane/propylene as a guide to determining the biogenic or thermogenic origin of the microseepage. Neither ethylene or propylene indicate a relationship between nonproducing or producing areas.

The depth of soil-gas sampling is critical to the successful implementation of a survey. Depths vary from as little as 0.3 m to as deep as 30 m. Sampling is guided by the type of method used and the soil conditions present in a survey area. Many of the published examples are from existing fields. Some of these fields were at or near depletion at the time of sampling. This makes it difficult to compare soil-gas results from existing fields with results from undrilled areas.

An interesting offshoot of soil-gas surveys for petroleum is their applicability in environmental geology and mining exploration. The data collected from soil-gas surveys can be important in these two applications because hydrocarbon detection methods are being aggressively applied. In the Basin and Range Province in Nevada, mining companies are using soil-gas methods to detect the presence of gold deposits that either have organic shales as host or are adjacent to organic shales. In the Viburnum Trend, which is an area of a Mississippi Valley type of deposit in southeastern Missouri, there is at least one lead-zinc body that contains significant amounts of hydrocarbons. Soil-gas and iodine surveys have yielded anomalies that outline the trend and, in some cases, the individual ore bodies. Environmental work uses soil-gas techniques to detect petroleum and chemical leakage or spills that enter into the soil system. Some of these hazardous petroleum accumulations are found in areas that are petroleum-producing basins. However, the hydrocarbon soil-gas measurements from spills are typically in the parts per million to percent range and thus are very distinctive from the analyses related to natural petroleum accumulations. The data acquired across mineral deposits can look very similar to anomalies associated with petroleum accumulations. Therefore, the techniques from any of the applications may be useful in the pursuit of another.

Soil-Gas Composition

Hydrocarbons usually found in the soil are those commonly found in petroleum: the paraffins. They are saturated, have no double bonds, and have a general formula of C_nH_{2n+n}. The typical soil gases in this group that are captured by direct methods are: methane (C_1), ethane (C_2), propane (C_3), isobutane (iC_4), N-butane (nC_4), hexane (C_5), pentane (C_6), and n-pentane (nC_6). Detection of C_8 to C_{12} is becoming more common because of the concern that C_1–C_4 could be generated by biological activity, by the impact of collection tools, or by decaying parent material.

Other classes of hydrocarbons are analyzed for when such methods as the K-V "fingerprint" are used. However, analysis, collection, and interpretation difficulties, plus the higher costs associated with this type of data, have prevented their wide acceptance.

Many hydrocarbon gases in the soil environment are also found in ordinary air. Therefore, the term *soil air* has often been employed in the discussions of soil gas. Davis (1967) reported that soil gases showed an oxygen content of approximately 20% in samples taken from 5 to 6 m deep. Oxygen varies inversely with carbon dioxide, which is typically less than 1% but may be as high as 5% in the soil. Nitrogen has been reported to remain constant at 80% to the depth at which soil-gas exchange occurs with the atmosphere.

Gas exchange between the atmosphere and the soil is an ongoing occurrence. This process can occur to at least several meters in depth and is caused by diurnal changes. During the day, the heat of the sun causes the soil air to expand, and part of it migrates to the atmosphere. During temperature decreases, particularly at night, soil air contracts and draws the oxygen-rich atmosphere into the soil. Rainfall displaces soil air temporarily and suppresses or inhibits its expansion. Barometric pressure fluctuations also influence the exchange rate between atmosphere and soil air.

Changes in the level and composition of soil gas occur constantly because of the factors discussed in the previous paragraphs. Therefore, most techniques collect a gas sample instantaneously from a specific soil layer at a predetermined depth interval. The timing of sampling for various collection methods ranges from instantaneous to extended periods of time. Resampling at the same site a minute, hour, day, month, or year later may not yield the same volume or composition of gas. However, areas that have been identified as background or anomalous will repeat in approximately the same areal extent over a survey area. Time-delay collection methods were, in fact, developed in response to rapidly changing environmental conditions and the inability to repeat specific samples in a survey area. Time-delay methods often fail, however, for reasons discussed later.

Soil Conditions

The geochemical environment of the soil prior to or at the time of sampling has generally been ignored. Therefore, its effects on hydrocarbons migrating through the substrate and thus also on the interpretation of the resulting data have not been accurately assessed. The volume of leakage present determines the effect the geochemical environment has on the migrating hydrocarbons. Large volumes of leakage will generally take control of the soil in the same way a man-made petroleum spill will contaminate it. Minor seepage may have little impact on the immediate environment. Following is a limited list of various geochemical and soil conditions that may affect hydrocarbons in the soil:

1. Soil-gas composition may be affected by indigenous organic matter, whether it is derived from the parent material, biological activity, or plant remains. Organic-rich shales and carbonates that have entered the oil window (in terms of maturation) may be the present-day bedrock surface below the soil layer. The organic matter continues to generate hydrocarbons in various amounts either on a localized or regional basis. Rock fragments separated from the bedrock will release organic matter through weathering. These fragments continue to add additional hydrocarbons during

alteration in the soil. This process is significantly enhanced by the impact of man-made probes and augers. An example of hydrocarbon generation by probe impact occurs in the Chapman Shale (Mississippian) in the Basin and Range, Nevada, and in the Niobrara (Cretaceous) in Ness County, Kansas, where these formations are the surface bedrock. Methods such as time delay were developed to circumvent this problem, but they have not proved effective. Most methods are able to collect gases in the soil under any conditions. The critical factor in sampling is minimizing disturbance of the sample medium during augering or hammering of the probe.

Organic material derived from the decay of plants in the soil generates hydrocarbons, predominantly methane. The greater the volume of organic material present, the greater the volume of soil gas. Peat bogs typically yield large amounts of methane, which has been termed *swamp gas*.

2. Another concern is the indiscriminate sampling of soil or bedrock in a specific area. Bedrock sampling represents an entirely different environment from the soil because the rock is not generally as high in porosity or permeability and is undergoing a multitude of chemical processes. In areas where the bedrock is close to the surface, it is subject to soil- (bedrock-) gas/atmospheric exchange, diurnal effects, moisture, and bacterial attack. Organic matter in the bedrock may release gas. Therefore, it is prudent to avoid bedrock sampling. A concerted effort must be made to collect in the soil unless none is present.

3. Specialized bacteria have been identified that generate ethane and heavier hydrocarbons through biological processes. The impact of these bacteria is still subject to debate.

4. Soil conductivity can be measured. It is caused by the electrical conductivity which, in turn, is controlled by the concentration of heavier hydrocarbons present in the gas phase (Fausnaugh, 1989). The ionic concentration, composition, temperature, and free energy of a solution will affect the migration of hydrocarbons through the soil. Conductivity can either maximize or minimize the effect known as *salting out* (see Chapter 5). Therefore, collected soils that have high conductivity may exhibit lower amounts of heavy hydrocarbons on analysis; those with low conductivity may have higher amounts. The conductivity of soil may vary widely across short distances or remain relatively uniform over large areas. Some methods may have difficulty in detecting hydrocarbons in soils where high-conductivity layers are interspersed with low-conductivity solutions. Therefore, soils with differences in conductivity in the same survey area may need to be identified and corrected so that there is no misinterpretation of the data.

5. Conditions 3 and 4 are interrelated with the moisture content of the soil. Moisture varies locally as well as regionally, daily, monthly, and yearly. Based on the author's experience, soil-gas samples collected at different times

(hours, days, and months apart) are not generally comparable on a quantitative basis, but they remain comparable with respect to background and anomaly repeatability. This condition is acceptable because the quantity of soil gas is not critical as is the repeatability of the presence of the anomaly and background areas. The quantitative incomparability of the data is partly due to the time variations in moisture, which subsequently affect pH, Eh, conductivity, and reduction/oxidation states.

6. The absorption ability of various soils depends on clay composition, which can absorb and release hydrocarbons as a function of cation exchange capabilities (Blanchette, 1989). Analyzing for sorbed hydrocarbons is one of the primary methods for the detection of geochemical anomalies. Chemical composition, matrix potential, permeability, and erodibility are all seldom-used parameters in evaluating the soil for the best collection method. Increasing evidence from environmental geology strongly suggests that these factors have a major impact.

7. Temperature plays a role in controlling rate of gas migration, rates of chemical reactions, absorption, soil conductivity, pH-Eh, and the impact of moisture. Excessively warm, dry soils allow hydrocarbon migration to proceed relatively unimpeded. Soils that are cool and retain moisture in the liquid phase significantly inhibit hydrocarbon migration, so that many other chemical and biological reactions can occur with some portion of the hydrocarbons prior to release into the atmosphere.

The conclusion to be drawn from the preceding list is that we must specifically target the part of the soil that is reducing in nature and that therefore allows the retention and absorption of hydrocarbons. Recall from Chapter 4 that the soil is divided into A, B, C, and D horizons. The A horizon is generally too thin for sampling, and its high chemical variability causes problems in the collection of soil gas. In some cases, this may be the only zone from which soil gas can be collected because of lack of soil development, difficulty in driving the probe to any great depth, or extreme thickness of this horizon. The D zone is bedrock with a partially weathered C/D interface. This zone may or may not retain hydrocarbons, depending on the types of clay, contact with the atmosphere, moisture content, pH, Eh, and temperature. In most cases, this horizon is not reachable without mechanized drilling equipment.

The B and C horizons are the two horizons usually sampled by soil-gas methods. The B horizon is a zone of illuviation and the layer where reducing conditions occur. If it is an open system, soluble compounds and elements are removed downward to be redeposited in the C horizon, where oxidation typically occurs. In a closed system, the boundary between the B and C horizon is usually defined by a hardpan layer.

Recent evidence suggests that soil-gas data from the B and C horizons are quite different for each zone. Apical and halo anomalies will be discussed in later chapters, but one of the causes for these patterns will now be mentioned. Sampling of the C horizon seems to indicate a halo anomaly, but sampling of the B horizon results in an apical anomaly. The reasons for this are not entirely clear. However, it is suggested that, because the C horizon is one of oxidation and is typically saturated with leached elements and compounds from above, the hydrocarbons are typically salted out. They pass through without being sorbed or occluded. The soil gas is concentrated on the edges outlining the accumulation because there is a transition from a very oxidizing to a less oxidizing area. In the B horizon, numerous sites are available for sorption, and incorporation in secondary deposits occurs because the horizon is not salted out.

Therefore a model can be established in which hydrocarbons increase reduction in the illuviation horizon (this is expanded on in Chapter 8). More materials are removed downward and laterally and are deposited in the C horizon. Near the edges of the microseepage, conditions begin to diminish for both the B and C horizons. This allows absorption and occluding of hydrocarbons, to a lesser extent in the B horizon and to a greater extent in the C horizon. The optimum sampling horizon is the B zone, which should indicate a more accurate outline of the petroleum accumulation at depth. The C horizon can be used for sampling but, in either case, the same horizon must always be sampled to minimize influence on the data.

Other Sources of Soil Gas (Hydrocarbons)

Soil-gas surveys are being utilized in the exploration for gold, uranium, and base metal deposits (Fisher, 1986; Jaacks, 1991). The migrating hydrocarbons and subsequent soil gas relating to these types of deposits exhibit no relationship to an existing petroleum accumulation. Organic matter has been thermally altered by the proximity of the hydrothermal solutions that formed the mineral deposits, and it continues to engender petroleum (Hulen et al., 1990). Generation of hydrocarbons by bacteria in these types of deposits can also occur. The petroleum migrates with hydrothermal fluids or groundwater into the rock containing the metal deposits. The hydrocarbons continue to leak to the surface based on the concept of microseepage.

There are several areas of the continental United States where oil seeps have been found associated with petroleum trapped with base metal, precious metal, or nonmetallic mineral deposits, such areas as the Fluorspar District, Southern Illinois; White Pine Mine, Upper Peninsula of Michigan; Powder River Basin, Wyoming; Viburnum Trend, Missouri; Basin and Range Province, Nevada; Galena District, Illinois and Iowa; and Kankakee Arch, Indiana. This source of hydrocarbons is not usually a major concern in a petroleum basin, but explorationists should be aware of this association when working in basins with many types of natural resource targets.

Gas Chromatographs

Gas chromatography is the most common method for analyzing soil gas. Four types of gas chromatographs are used as detectors (D): the thermal-conductivity (TCD), the flame-ionization (FID), the electron-capture (ECD), and the photonization (PID). The TCD detects changes in thermal conductivity of the carrier gas flowing over it. Typically, the carrier gas is helium, which is thermally conductive. There are two filaments. The carrier gas passes over one, and the sample and carrier gas pass over the other. As the gas passes through the detector, variations in the thermal conductivity of the gas are recorded, indicating changes in the amounts of hydrocarbons in the sample. The sensitivity of the method is dependent on the differences in the thermal conductivity between the carrier gas and the sample.

The FID measures electrical current flow, usually in the 100- to 300-V range, across a hydrogen flame. There are two electrodes; the hydrogen flame burns at one, and the other is the collector electrode. The hydrogen at high temperatures is ionized, which allows an electrical current to flow. When hydrocarbons enter the flame, they burn readily and increase the ionization. This is amplified and recorded. The FID technique is the most common form of gas chromatography because of its sensitivity in measuring small amounts of volatile organic compounds.

The ECD method uses solute molecules, such as halogen compounds, that capture free electrons. The electrons are produced by B-radiation from a radioactive source, which is usually nickel-63. The carrier gas is typically nitrogen. The ionization of the gas causes a decrease in electrical flow. This type of detector is highly sensitive and selective.

The PID method is becoming increasingly popular because it does not require a hydrogen flame and it has a lower oven temperature. It is relatively inexpensive and is portable. The disadvantage of the PID is its lack of sensitivity compared to other detectors.

The important element of the gas chromatograph is the column. When the gas sample is passed through the column, the constituents of the sample being sought are detected. The type of hydrocarbons determines the composition of the column and the speed at which the target volatiles will be detected. For example, detecting man-made volatiles and petroleum hydrocarbons requires different detection times. Therefore, a column is specifically designed to detect naturally occurring petroleum hydrocarbons or specific man-made volatile organics.

Quality-control methods are always initiated, whether the gas chromatograph is in the field or at a permanent laboratory. Typically, blanks and standards are run prior to beginning analysis of any field samples. Depending on the number of samples to be run, blanks and standards are run periodically to determine if the equipment needs to be recalibrated or a new column installed.

Soil-Gas Methods

Free-air, head-space, deabsorption, acid extraction, time-delay, groundwater, and fluorescence methods all directly sample the soil or sediment substrate for hydrocarbons. Airborne and offshore methods sample different media. Soil-gas methods attempt to capture hydrocarbons in a free, sorbed, occluded, or solute state or a combination of these states. The form of soil-gas capture (or sampling) defines under which conditions—of soil, weather, terrain, and environment—a particular method will be effective. Misapplying a method occurs when the hydrocarbons are predominantly present in a specific phase in the soil across part or all of the survey area but the method chosen cannot capture that phase.

The following section is divided into onshore, airborne, and offshore methods of sampling. All methods are described, as well as some of the factors affecting their use and the advantages and disadvantages of each.

Onshore: Free-Air or Soil-Probe Method

The first documented method used to measure hydrocarbons migrating through the soil column was the free-air, or soil-probe, technique (Figs. 5-1 and 5-2).

The method extracts gases that are moving through the well-connected interstitial (especially large) pores rather

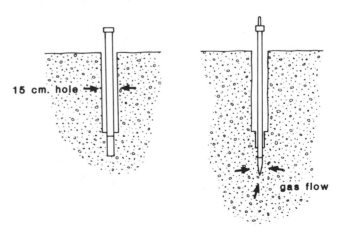

BARREL SAMPLER   **GAS SAMPLER**

15 cm. hole

gas flow

Fig. 5-1 Free-air method: A barrel or probe is driven into the ground. The gas is drawn into the barrel or probe near the tip, which is located in the soil. The gas is drawn upward by plunging action. A sample of gas is taken after the interior of the probe is vacated of atmospheric gases.

GAS SAMPLER

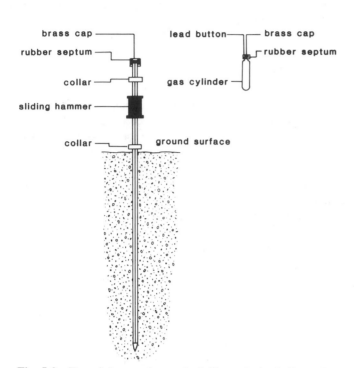

Fig. 5-2 Free-air hammering method: The probe is similar to the one described in Fig. 5-1 except that the hammer is part of the probe and not a separate mechanism.

than seeking gases that are temporarily trapped in secondary mineral deposits or absorbed onto clays and other compounds. The gas is collected in the vapor phase and not in the sorbed, solute, or occluded states. The impact of the auger liberates some of the gas from these other sources and, as such, is collected and becomes part of the sample.

Sampling Technique. The technique requires driving a hollow steel probe or barrel 1 to 3 m into the soil. The hollow portion of the probe has a measured volume. When the probe reaches a predetermined depth, a syringe is inserted into the septum, and the probe's hollow portion is purged of atmospheric gases. Theoretically, the gases then entering the evacuated chamber are from the soil itself. A second sample is taken by syringe and is ejected into a sample container, which is generally in a vacuum container, shown in Fig. 5-2. The containers used for collecting the gas at the surface vary from simple syringes to double-sealed containers. New technology in gas chromatography allows analysis of the gas on-site. Otherwise, the samples are taken to a stationary or portable laboratory having a gas chromatograph.

Factors Affecting Use. Several environmental and procedural factors must be taken into consideration when the free-air method is used:

1. Location of the water table is critical for this technique. Samples taken below the water table yield little or no gas and may not necessarily be representative of the gas migrating through the soil profile. If the substrate is saturated from the surface downward, the probe must be driven down until it reaches relatively "dry" soil. In some regions and at certain times of the year, this may be impossible.

2. The soil's organic content should be determined (in a relative sense). Hammering the probe through soils that are high in organics (either from biological activity or from weathering of the parent material) may generate hydrocarbons that can cause false or overly optimistic results.

3. Migrating gases are retained in different amounts in the various soil layers depending on the types of clays present. Many investigators have ignored this important fact and have assumed that any predetermined depth will suffice for sampling. This typically results in data that are difficult to interpret because of a failure to adequately define the soil layer or layers that were repeatedly and randomly sampled.

4. Moisture content is generally not uniform across a survey area. If one part of the area is wet and the other part is dry, data variations may be the result of changes in the soil moisture content. Slight moisture variations should not cause concern, but different soil types will have different moisture-retention abilities.

5. Changes in barometric pressure cause purging or suppression of gas in the soil. If the barometric pressure drops during the survey, gas can migrate to the surface more easily. Increases in barometric pressure cause reduction in the volume of soil gas that can migrate, and the atmospheric/soil-gas interface is lowered. The sampling depth would then have to be increased. Because pressure and moisture vary through time, data obtained from samples collected at the beginning of the survey may be quite different from data obtained at the end of the survey. To correct for this, a base station(s) should be used where samples are collected before, during, and at the end of each survey day to determine diurnal variations in soil-gas/atmospheric exchange.

6. The depth at which atmospheric exchange occurs is critical. In arid regions, this can be several tens of meters deep but, in humid climates, it may be only a few millimeters. Perhaps the probe cannot be driven deep enough to reach below the soil/atmospheric gas exchange boundary.

7. The sample chamber into which the soil gas is drawn must be properly purged to remove all atmospheric gases. An adequate seal must be created around the probe to prevent atmospheric contamination from being drawn down into the sample chamber.

8. The sample container must be effectively sealed and then taken promptly to the laboratory for analysis.

9. Possible contamination by petroleum residues from equipment manufacturing, from previous samples, or from

poor maintenance in the probe interior must be determined prior to sample collection.

The preceding factors indicate that a clear understanding of the soil environment and of the limitations of the free-air method is needed to implement an effective soil-gas sampling program and subsequent data interpretation. Typically, many of the problems are overcome in quality-control methods that are standard in the environmental industry. If these control methods are not implemented, there must be cause for concern. In some cases, the quantities of leaking hydrocarbons from the petroleum reservoir will be sufficient to render insignificant the influence of local conditions on the data. On the other hand, if leakage is minor or suppressed, local conditions will probably have a major impact on the soil-gas program. False anomalies also become a problem if other shallow sources of gas exist in the soil or bedrock.

Advantages/Disadvantages. The free-air method has several advantages.

1. The sampling procedure is very mobile. One person can walk and carry enough sample containers and associated equipment to survey a relatively large area.

2. The method is excellent in rugged terrains where the soils are shallow, coarse-grained to cobble size, and the parent material is a major component of the surficial material. In this type of area, a method cannot be used that requires absorption of a gas on a clay fraction or that depends on trapped carbonate minerals.

3. Portable gas chromatographs can be used directly in the field or close to the survey area. This limits time lost in transporting the samples to the laboratory. Fast analysis allows rapid resampling and detailing of an anomalous area.

There are several disadvantages to the free-air method.

1. The gases that are pulled into the hollow portion of the probe are assumed to be coming from the soil itself. This may not be true, and probably the ratio of soil gas to atmosphere will never be known. In clayey/silty soils that are moist to wet, a partial to complete seal around the probe is likely to prevent the atmospheric gases from being drawn downward into the soil and then up into the sample chamber. In dry, loose, sandy soils, the collected gas could be partly to wholly atmospheric because atmospheric mixing is a function of the large pores and permeability of the soil.

2. It is difficult to identify which layer in the soil profile the probe has entered. It is also difficult to determine if sampling of the same soil layer has occurred from point to point across the survey area.

3. Typically, the method is useful only for ethane (C_2) to butane (C_4). Some percentages of the heavier hydrocarbons are usually in the gas phase, and they may move as quickly as the lighter gases because of the vacuum created

by the purging of the probe. Heavy hydrocarbons in the gas phase are typically in minute amounts and may not be detected, depending on the sensitivity of the analytical equipment. More sophisticated analytical equipment will detect methane and C_5 or heavier hydrocarbons, which increases the viability of this method.

4. Samples are useful for only a one-time analysis. Obtaining the same results at the same site at a later time can be difficult because the conditions under which the original sampling occurred are not likely to be present.

5. This method is greatly affected by weather conditions, changes in barometric pressure, soil temperature, amount of water saturation in the soils, and location of the water table.

6. The results of this method can vary depending on the competence of the person(s) collecting the samples and on the integrity of the sample containers.

In summary, the free-air method, despite several disadvantages or factors that affect its use, has proved to be a very viable and useful tool in many areas. Excellent data can be obtained when a well-documented, quality-control procedure is instituted and there is a clear understanding of soil environment. In other areas, it is not as effective as the methods described below.

Onshore: Head-Space Method

The head-space technique measures hydrocarbons absorbed onto, or loosely bound by, the clay and organic materials in the soil. The method attempts to overcome the difficulty in capturing free soil gases that migrate rapidly through the substrate. Gases are trapped temporarily in isolated pores of the clays or organics. The soil sample is placed in a sealed container of measured volume. The soil gas migrates to the top of the container either through time, by the application of outside heating, or by agitation (Fig. 5-3). The collected soil gas is assumed to be from a sorbed state and not from the action of the auger or hammer. However, the more mechanized and destructive the augering system, the more likely it is that some of the hydrocarbons captured and detected are a product of this action on the soil. This technique collects soil gas that is less susceptible to the rapid geochemical and weathering variations that can occur across a survey area during free-air sampling. However, it does not eliminate or even minimize some of the problems that affect the free-air method.

Sampling Techniques. A hole is drilled, either with a hand or power auger, to a specified depth. This depth is usually a set distance of 0.5 (2 ft) to 30 m (100 ft) from the surface. Normally, the target depth of the drill is not related to a specific soil horizon or layer, even though it should be. The

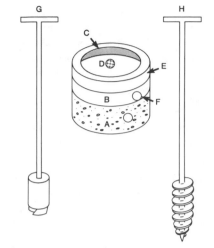

Fig. 5-3 Head-space container: A is the soil collected; B is an exact quantity of distilled water; C is the lid, sealed with nonpetroleum epoxy; D is a drop of the sealant placed on the lid to act as a septum; E is the dead or head space of the container where the gas is collected; F is steel or glass marbles. Augers: G is one type of auger with a set of teeth and an open container to allow soil to enter; H is another type of auger that can be either hand- or power-driven; the spiral riser tapers from top to bottom to allow easier extraction of the soil sample.

drill bit is brought out of the hole, and the soil or bedrock material that has clung to the end is removed as the sample. Some drill bits have core barrels at the end that can collect an intact (but not undisturbed) soil sample. The amount of soil removed from the bit is of predetermined volume and should remain consistent across the survey. Transfer of the soil to the container is done as rapidly as possible to prevent hydrocarbons from escaping. The amount of soil will vary from a few grams to several hundred grams, depending on the size of the container. The interior volume of the container must be known. A measured amount of distilled water and a bacteriostatic agent are added to the container. The bacteriostatic agent prevents microbiological oxidation of hydrocarbon gases prior to analysis. However, some users suggest that there is evidence that the bacteriostatic agent may also affect the soil gas and should not be used. Experience indicates that this is not true. One or two glass marbles or stainless steel bearings (optional) may be added before the container rim is sealed. The marbles or bearings enhance agitation of the sample in the laboratory, which results in a quicker and almost total release of the hydrocarbon gases from the soil. After the container is closed, the space between the lid and the container is sealed by using a nonorganic sealant to prevent leakage. The sample is then taken as quickly as possible to the laboratory for analysis.

The sample is placed in a water bath and heated at a set temperature. The amount of time in the warm bath varies and is more a matter of experience than scientific determination. Over time, agitation of the soil liberates the gas, which migrates into the empty volume of the container. This gas

accumulation area is known as the head-space. The sample is removed from the bath, and a syringe is inserted into the head-space where the gas has collected. A set amount of gas is extracted and then injected into a gas chromatograph. An alternative form of agitation is a mechanical shaker, similar to a device used to shake paint. This method is more destructive, and the released gas may also be coming from the occluded state. The results would then be a total soil-gas analysis. In addition, agitating organic-rich soils, especially if the parent material is thermally mature, releases additional hydrocarbons into the head-space.

Factors Affecting Use. The critical factor with the head-space method is sample integrity. This can be summarized in three areas.

1. The gas can be lost as a result of improper sealing from the time of collection up to heating in the warm bath or shaking. The amount of gas that escapes will affect the interpretation. The escaping gases may be replaced by atmospheric gases migrating into the sample container. The sealant, if it contains any hydrocarbons, will contaminate the sample and render it useless. Contaminants are usually propane, butane, pentane, and hexane.

2. Inconsistent measuring of soil and/or fluid placed in the sample container affects volumetric calculations of the gas that is analyzed. Samples that vary in volume may not be comparable unless a correction can be applied.

3. Storage of the containers for more than a couple of weeks is not recommended. Leakage and atmosphere exchange will eventually occur, regardless of the effectiveness of the container seal.

Advantages/Disadvantages. The clear advantage of the head-space method is the acquisition of an actual sample of the soil containing the gas. At the collection site or in the laboratory, the soil sample can be described and analyzed for its grain size, clay and silt content, matric potential, Eh-pH, carbonate and organic amounts, and also any unusual characteristics that may affect the hydrocarbon gases in the soil. If sampling integrity is maintained, this method minimizes contamination by atmospheric gases.

The head-space method has three main disadvantages.

1. An undisturbed sample is not obtained. Gases liberated in varying amounts are dependent on the soils and the impact of various sampling methods. These methods vary from hand to truck-mounted augers or hammers that can penetrate from 1 to more than 30 m in depth. All these methods disrupt the soil to different degrees. The greater the impact of the auger on the soil, the more likely it is that gases will be liberated before the sample is sealed in the container.

2. Flexibility is limited because of the need to have containers, distilled water, an auger, a bacteriostatic agent, nonorganic sealants, and marbles or bearings in the field.

The amount of equipment requires a vehicle, which usually limits access in agricultural areas at certain times of the year (mainly fall and spring) and in rugged or mountainous terrain.

3. In farming regions during fall and spring, fertilizers and insecticides are sprayed or injected into the ground using petroleum-based carrier fluids. This places various refined hydrocarbons into the soil that will be added to the gases already migrating through the soil substrate. These additives usually affect samples that are taken from a depth of less than 2 m.

Onshore: Absorbed Hydrocarbons Methods

As gases migrate through the soil, some of the hydrocarbons are trapped during the formation of carbonate cements or are absorbed onto clays. The formation of carbonate cements and the ability of the clays to absorb hydrocarbons are functions of the soil environment. The cements can remain stable over long periods of time. As long as chemical conditions remain relatively constant, the carbonates will not be dissolved and, thus, trapped gas will not be released. Therefore, collecting a sample does not require special precautions other than proper labeling of the sample bags. The acid-extraction and the absorption methods were designed to alleviate the problem of gases escaping during collection, transport, and storage. The methods determine either (1) the amount of hydrocarbons absorbed onto the clays or (2) the amount of hydrocarbons that have been occluded into carbonate cements deposited on the clays.

Sampling Technique. Deabsorption and acid extraction samples are acquired by drilling a hole in the soil as in head-space sampling (Horvitz 1969, 1972). After the auger is brought to the surface, the sample is placed in a cloth bag. There is no concern about gases escaping prior to analysis because they are trapped on clays or in cements in the soil. The soil sample is dried at room temperature and sieved. The coarse fraction is discarded, and the fine fraction is set aside for analysis.

For the acid extraction method, a measured portion of the fine soil fraction is placed in a sealed tube with acid that will convert the carbonates to CO_2 and water, and any trapped hydrocarbons will be liberated (Fig. 5-4A). The hydrocarbons rise to the top of the sample chamber, where they are extracted by syringe. The sample of gas is then injected into a chromatograph for analysis. Results are reported in microliters of gas per gram of clay. A variation on the collection of the gas is to absorb the water and the gases and cool the mixture to $-196°$ C. Methane and CO_2 are extracted and burned. The amount or ratio of methane to CO_2 is compared. The remaining hydrocarbons are ana-

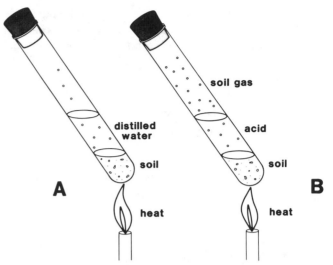

Fig. 5-4 Acid extraction and deabsorption methods: A. Test tube filled with a predetermined volume of fine fraction soil and hydrochloric acid and then sealed. The HCL will dissolve the carbonates that have trapped the hydrocarbons. B. Distilled water replaces the HCL. The sample container is heated, which drives the gas out of the soil and into the head space. The top of the container has a septum for extracting the gas.

lyzed by a gas chromatograph. This method was common prior to technological advances in the gas chromatograph.

The deabsorption method of releasing gas is to place a measured portion of the fine clay fraction in a chamber with two ceramic balls. Shaking the container desegregates the gas, which moves toward the top of the chamber (Fig. 5-4B). The sample is then placed in a warm bath for a set period of time. It is later removed, and the gas is extracted for analysis.

Factors Affecting Use. The deabsorption method attempts to minimize loss of gas during collection. The technique does allow repeat analysis as long as sufficient sample remains. However, several factors are critical to the success of this method.

1. The soil composition and the particular horizon sampled should both be known. One horizon or soil layer should be the only zone sampled throughout the survey. The sorption of hydrocarbons is dependent on the composition and absorption capability of the clays present. Soils rich in carbonates yield hydrocarbons that were part of the original parent material in addition to those trapped by secondary carbonates. Therefore, soils derived from a carbonate terrain are not conducive to this method (Weissenburger, 1991). The amount of carbonate material and the types and percentages of clays present in each soil layer vary, and thus it is critical to sample the same interval at each collection site.

2. The acid extraction method requires the presence of carbonates, but in some soils there may be little or no carbonate material present. Noncarbonate or carbonate-poor soils

are the product of climate, the parent material, and/or the Eh-pH of the soil. If soil is acidic, no carbonate cements will be deposited, and therefore no hydrocarbons will be occluded. The amount of hydrocarbons absorbed is a function of the absorption capacity of the clays and the conductivity of the environment.

3. There seem to be only small amounts of gas absorbed by clays, or perhaps this method is not effective in liberating sufficient quantities of gas.

Advantages/Disadvantages. The deabsorption and acid extraction methods have an important group of advantages over other soil-gas techniques. The methods are not affected by (1) sudden barometric pressure changes, (2) moisture variations, (3) sampling inconsistencies, or (4) loss of sample due to breached containers. (5) A repeat analysis is possible. These five factors are critical for a technique that measures vapors. (6) Samples can be collected from almost any type of drill hole that does not use a mud or water system for drilling.

Disadvantages with these methods are that they require extensive soil profiling to determine types of clays present, organic matter content, water capacity and absorption characteristics, and the variations in the amounts of carbonates present. In acidic soils, the acid extraction technique has proved ineffective because secondary calcite being deposited is quickly dissolved, which releases the hydrocarbons. This technique was found to be unusable in all or portions of the Forest City, Illinois and Michigan basins, where the soils have little or no secondary carbonates in the fine fraction. Parent material that is rich in carbonates and organics will release hydrocarbon gases on reaction with the laboratory acids or with simple disaggregation. The data can thus suggest anomalous conditions that do not exist. The technique is useful in coarse or cobble-dominated soils where the unweathered parent material is a large portion of the soil profile. The method requires either the presence of optimum and constant carbonate material or a clay composition with a measurable and relatively uniform absorption capability.

Absorbed Hydrocarbons Surveys: Discussion

A number of studies using the deabsorption and acid extraction methods have been published in the literature. Locations are in the Texas Gulf Coast and Florida; Cooper Basin, Australia; Olds-Carolina area, Alberta, Canada; and the eastern part of the Michigan Basin in Ontario, Canada. The Texas and Florida surveys correlated well with existing production and produced, at the time of surveying, several large undrilled anomalies. The survey in the Olds-Carolina area had different results, indicating no correlation with existing production. The conclusion was that the gas was syngenetic (McCrossan et al., 1972). The soil-gas anomalies had a distinct relationship to the carbonate fragments of

glacial origin that had been transported from the foothills west of the study area. This led to the conclusion that the absorbed hydrocarbons, as well as others, were not applicable to the area of Canada covered by glacial till. This is probably erroneous because no other soil-gas methods had been tried for this study.

A Cooper Basin survey also concluded that the acid extraction method was not usable (Devine and Sears, 1975, 1977). The authors used extensive statistical analysis to arrive at their conclusions. A more recent survey in North Africa (Weissenburger, 1991) can be summarized as follows: all the surface geochemical anomalies were related to a specific type of carbonate that was derived from the bedrock terrain, and the anomalies did not have any association with seeping hydrocarbons. But there was no production in the region to correlate with the survey, nor was any drilling done within the geochemical anomalies.

Another series of surveys was carried out over six oil fields, three in Ontario and three in Alberta, with similar mixed results (Debham, 1969). There was no strong correlation between the acid extraction surveys and the fields. The anomalies changed in strength or presence with variations in acid strength, and there was no confirmation from other surface geochemical methods. Price (1976) argued that the acid extraction method had failed because of the lack of a consistent and uniform anomalous shape associated with the six producing fields, and actually some of the fields had essentially no anomalies. The fact that both apical and halo anomalies existed should not be a concern. The six fields studied are in different areas, and each field represents a different type of trap. In Ontario, Colchester Field, discovered in 1959, is a dolomitized Trenton chimney; Morpet Field, discovered in 1954, produces from the Silurian A-2 carbonate; and the Gobles Field, discovered in 1959, produces from a stratigraphic trap of Cambrian age. The three fields in Alberta were also dissimilar in the ages of productive strata and in the types of traps. All the fields were "old" at the time of the geochemical survey. The lack of anomalies at some of the fields is probably due to depletion of the reservoir and reduction in vertically migrating hydrocarbons. The presence of any anomaly should be considered an excellent indication that some reservoir pressure still exists.

In summary, these techniques can be of limited use in some areas. Problems stem from their reliance on the development of carbonates or on the absorption by clays. They are dependent on the suitability for development of the Eh/pH conditions of the soil, soil clay composition, moisture, and temperature. They are not as effective as free-air and head-space techniques and should be used with caution.

Onshore: Time Collection Methods

This form of soil-gas collection was designed to eliminate many of the effects that plague the previous three methods in terms of soil-gas fluctuations, soil profiling, and container integrity. Time collection methods attempt to recreate the soil environment in which the gas migrates after emplacement of the sample equipment. The detection or collection, or both, of gases occurs over an extended period of time or once at a later date. Proponents contended that this provides more accurate representative volume and composition of the gases migrating through the soil at a sample location. In practice, however, numerous other factors come into play that affect and, in many cases, negate the usefulness of these time collection methods.

Time Sampling Techniques. Two methods, "K-V fingerprint" and time-delay, have been developed to implement the time collection technique. The two methods approach the sampling and analysis from two separate and distinct directions. The K-V fingerprint method is a comparison technique. Areas are selected that have hydrocarbon leakage similar to that of the prospect area. Data from the known production and dry holes are called a *training set*. A wide spectrum of hydrocarbon types is collected and compared to data from the prospect area. The time-delay method is essentially an extended free-air method that allows for restabilization of the soil environment over time and for extraction of gas at a later date.

Time-Delay Method. A hole is augered to a specific depth, typically above the water table as in the free-air, headspace, and acid extraction methods. The hole is thoroughly cleaned of loose soil. Gravel or sand is placed in the lower 0.3 m (1 ft) of the hole (Fig. 5-5), along with a small-diameter copper tube that extends to the surface (Fontana, 1988). The coarse material creates an artificial reservoir and consists of material that does not have any clastics that could generate hydrocarbons or interfere with gas flow. Bentonite or a low-permeability material is placed on top of the gravel to prevent or minimize atmospheric contamination. The material is packed down to reduce any porosity and permeability that developed in the bentonite. The end of the copper tube at the surface has a rubber septum attached to allow drawing of a gas sample. The system is then purged with a syringe until an amount of gas equivalent to the volume present in the copper tube is extracted. The sample site is left for a period of time. The duration may depend on access to the sample site, time of year, or a predetermined time interval to allow normal seepage conditions to reestablish. After the time interval has elapsed, a sample is withdrawn from the copper tube using a syringe through the septum. The sample is then taken to the laboratory for analysis by chromatograph. A portable gas chromatograph can also be used.

K-V Fingerprint Method. This method relies on attempts to "fingerprint" and use pattern-recognition methods to identify and compare data from known petroleum fields with data from presently nonproductive areas. This method has been termed *integrated* because it is supposed to detect all the hydrocarbons that are migrating through or have saturated the soil at the sample site. A glass culture tube is buried in a hole that has been augered from 0.2 to 0.4 m deep (Fig. 5-6). The tube contains activated charcoal or

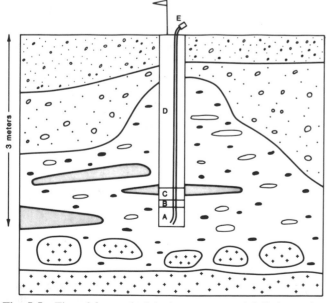

Fig. 5-5 Time-delay method 1. A, coarse material; B, bentonite or similar material; C, additional sealing material; D, backfill; E, septum; F, survey flag. The copper tube extends to the surface. A gas sample is drawn from it after the soil environment has stabilized over time.

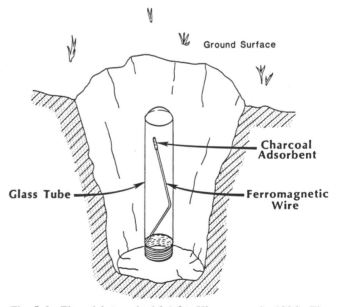

Fig. 5-6 Time-delay method 2 (after Klusman et al., 1986). The container is an inverted glass tube placed in an excavated hole. The base of the container is open. Inside the container is a ferromagnetic wire with a charcoal absorbent. The charcoal absorbs hydrocarbons that migrate into the container. The absorbent is removed after retrieval of the container and is analyzed by a mass spectrometer (reprinted with permission of the Institute for the Study of Earth and Man, Southern Methodist University).

some other type of absorbent on a ferromagnetic wire. After burial, the container is flagged and left for a period of one week to several months before it is retrieved (Heemstra et al., 1979; Klusman et al., 1986).

In practice, the survey starts by identifying petroleum flux in relation to a dry hole and a producer, which constitutes a "training set" of data. The number of samples in the training set has never been specified. The samples from a survey across a prospective area are compared to this training set to determine if a viable petroleum target may be present. The K-V fingerprint method identifies many of the different types of hydrocarbons that have seeped into the soil substrate and have collected on the sampling medium. The technique utilizes ratios of specific hydrocarbons in the produced petroleum and then tries to identify those ratios in the prospect area.

Analysis is done by thermal mass spectrometry, which can detect hydrocarbons C_2 through C_{17}, alkanes and cycloalkanes, alkenes, and cycloalkenes, nepthalenes, aromatics, and substituted aromatic compounds up to mass 240 in the operating mode of the detector. The data can be interpreted using cluster and discriminate analysis. The mass spectrometer will generate numerous lines of complex ionic and molecular fragments. Pattern-recognition methods seem to work best and are most effective for interpreting the data (Klusman et al., 1986). These methods aid in comparing samples from a known producing area to those from a survey area. Similarities in the patterns from the two areas indicate that the petroleum leaking into the soil is derived from the same source and imply a possible accumulation in the survey area. The sample can also be analyzed using a gas chromatograph. Fig. 5-7 is an example of anomalous and nonanomalous samples for this type of data.

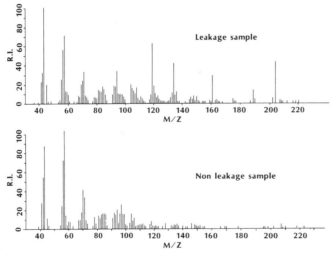

Fig. 5-7 Examples of nonanomalous and anomalous K-V fingerprint methods. The upper mass spectra are from a leakage sample. The lower mass spectra are from a nonleakage area (reprinted with permission of the Institute for the Study of Earth and Man, Southern Methodist University).

Factors Affecting Use. The first concerns in using the delay techniques are the time required for sampling and the high cost of collection per sample. These tools are not useful when only a short period of time is available to evaluate a project, prospect, or reconnaissance area. These techniques are more expensive than any of the other onshore surface methods. Costs are higher because a longer period of time is spent at the sample site, more raw materials are needed, mass spectrometry is expensive, and the sample site must be revisited one or more times at a later date.

The soil environment is critical to the success of these techniques (Heemstra et al., 1979). The initial augering disrupts the soil. If organic matter derived from weathered rock fragments is present, it may generate or increase the expulsion of hydrocarbons with the impact of the drilling tool. These rock fragments will continue releasing unspecified amounts of hydrocarbons for an uncertain time interval, and they may still be releasing when the sample is collected. Water saturation of the soil from precipitation or a rising water table and variations in moisture through time can suppress migrating hydrocarbons and interfere with collection. Moisture will also be absorbed onto the collector and again interfere with the gas collection. The potential weather conditions for the time of year and location of the groundwater table must be considered when the depth of the absorbent collector and the base of the copper tube is chosen. Heemstra et al. (1979) also recommended that radon and free-air gas measurements be taken in conjunction with this type of survey to confirm any results.

Disadvantages/Advantages. There are two assumed advantages of time collection techniques.

1. They will obtain an undisturbed sample by allowing the soil to return to equilibrium over a period of time.

2. They will allow the acquisition of a continuous or integrated sample from the same site over a long period of time. This eliminates fluctuations in soil gas and results in more accurate representation of the hydrocarbons migrating through the soil.

Time collection methods have several disadvantages.

1. The costs of collection procedure are extremely high. There is no evidence that either of these methods are more or less effective than the free-air or head-space techniques in detecting leakage.

2. More time is required to complete the survey.

3. The methods do not overcome the problem of false data generated by organic-rich soils that have been impacted by the auger or drill. Some soils that are derived from a thermally mature and organically rich substrate (regardless of the length of time) and that are impacted seem to generate hydrocarbons continually. If the hydrocarbons are generated from sampling procedures, natural seepage from the bed-

rock, or bacteria, it would be difficult to segregate the types and the quantities of hydrocarbons that are derived from petroleum seepage.

4. Animal activity (for example, cattle, elk, and rodents digging up the copper tubing and other material related to the sample site) and people tend inadvertently to destroy sample sites even in remote areas. A high loss of samples should be expected.

5. Changes in groundwater levels may cause flooding of the gravel pack. Local surface flooding may cause water to migrate downward.

6. Published examples of the ferromagnetic or K-V fingerprint method clearly indicate problems in defining existing fields and possible targets. Data interpretation is very difficult, requires experience, needs numerous training sets, and relies on sophisticated statistics to be effective. In one case, this method defined sand dunes in southeastern Colorado rather than identifying existing production or potentially new targets.

Onshore: Groundwater and Stream Methods. Sampling of groundwater is a possible way to evaluate large areas quickly for petroleum potential. Vertically migrating gases and fluids pass through numerous aquifers on the way to the surface. Some of the gases will be trapped or temporarily delayed in these zones. Gas in water samples from wells can yield reliable hydrocarbon data because most well water comes from greater depths than soil sampling and is not generally subject to fluctuations in soil environment. Sampling streams is another collection method. Hydrocarbons seeping into the soil may be dissolved in the migrating groundwater or trapped in the sediment. Eventually, the gases enter the surface drainage system. The Soviets and a small number of Western sources have suggested this method for regional evaluation.

Water Sampling. A container is filled with a specific volume of water from a well, spring, or stream. The gas that is dissolved will eventually migrate into the head-space area. It can be extracted in a fashion similar to the head-space method and then analyzed by a chromatograph.

Another form of sampling is to obtain a core of the bottom sediments of the stream. As with other onshore methods, the sediment has to be classified in terms of clastic, organic, and grain size of the material. Sediments composed largely of clay are considered the best because coarse material will not retain gas as well.

Factors Affecting Use. The main requirement is to have a sufficient number of wells, springs, or flowing streams to sample in order to determine adequately where a potential prospect or project area may be located. It is helpful and perhaps essential to know how each water well was drilled and completed, but this information may be difficult to ob-

tain. The pipe in the well may have a lubricant at the intersection of each pipe joint that will add hydrocarbons to the well water. These wells must be eliminated from any survey.

Advantages/Disadvantages. The time collection method is advantageous in a basin having limited geologic data and no previous production or drilling but having an extensive drainage, water-well, or spring system. Sampling a regional water-well system might be useful in populated areas where other methods may not work because of extensive altering of the soil by man. Fig. 5-8 is an example of a groundwater survey carried out in the Lena-Vilyuy petroleum region, Vilyuy Basin, Siberian Craton, in the former Soviet Union (Filatov et al., 1982). The strongest anomalies are shown along the Tyung River, where 42% of the gas present in the samples was methane. Carbon isotope analysis indicated that 10% to 30% of the methane present was of thermogenic origin. Additional anomalies are indicated in the other two rivers, Linde and Lina, where gases were composed of 45% and 36% to 50% methane, respectively. Heavier hydrocarbons were reported as well. Other surveys in the Soviet Union were conducted in the Cis-Verkhoyansk, Lena-Anabar, Nepa-Botuobinsk, and Baykit-Katanga petroleum areas. There is no indication of success in finding new production based partly or totally on these surveys. Filatov et al.

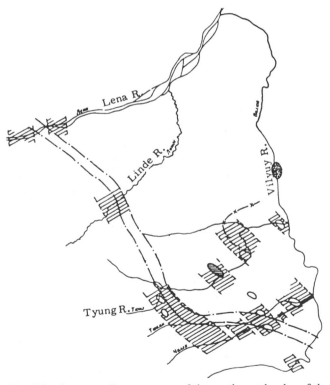

Fig. 5-8 Stream sediment survey of the northwest border of the Vilyuy Basin, in the former Soviet Union (after Filatov et al., 1982). The large diagonal cross-hatching represents areas of gas seepage identified in the streams. The smaller ovals represent gas condensate fields.

seem to use methane as the lead tool in conjunction with carbon isotope analysis in this form of exploration.

Groundwater sampling has several disadvantages.

1. It depends on an extensive water-well or drainage system. The former is generally not available in unexplored basins, and man-made pollution can be a problem even when a satisfactory number of sample sites exist.

2. Exposed bedrock material could be leaking hydrocarbons into the streams through normal weathering processes.

3. The volatile nature of hydrocarbons suggests that the lighter hydrocarbons will either be removed because they are in the gas phase or be biologically decomposed in the groundwater. The heavier molecules will undergo rapid biological removal due to their nonmobility. There has been no documented evidence of plumes of hydrocarbons in the soil's downdip from surface geochemical anomalies. These types of plumes or dispersion anomalies are typically associated with metal and nonmetal deposits. This suggests that, unlike high metal concentrations in the soils, high hydrocarbon concentrations are quickly removed. This also makes it difficult for sampling to pinpoint potential targets.

Onshore: Fluorescence Methods

Fluorescence as an exploration tool was first documented by the Soviets (Kartsev et al., 1959). This method is based on the concept that, as petroleum leaks into the soil, the lighter hydrocarbons continue upward into the atmosphere but the heavier hydrocarbons are deposited in the soil substrate. A spectrophotofluorometer is used to analyze these soil samples for hydrocarbons in a manner similar to a black light used to analyze well cuttings for traces of hydrocarbons. The results are essentially the same, but the spectrophotofluorometer is able to measure more precise amounts of each hydrocarbon present for a specified level of detection. Researchers have found that the petroleum occurring in the soil substrate typically exhibits a set of wavelengths similar to those of the hydrocarbons from nearby producing fields (Fig. 5-9). The method is generally unable to detect reliably the quantities of lighter hydrocarbons present. Fluoroescence is usually restricted to detecting heavier hydrocarbons that remain in the soil because they are in the liquid rather than the gaseous phase.

Sampling Technique. The soil sample is collected, dried, sieved, and separated into two factions. One of the factions is discarded (normal sampling procedure), and the other is analyzed for specific or several hydrocarbons (Hebert, 1984). The spectrophotofluorometer is set to detect in the ultraviolet region of 265 to 500 nm, but a narrow range is generally chosen in order to simplify analysis and interpretation. Light hydrocarbons are typically found in the 300- to 350-nm range and heavier hydrocarbons in the 350- to 400-

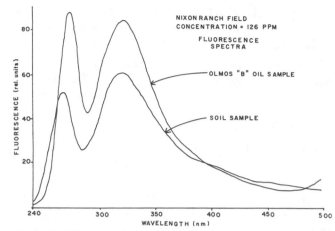

Fig. 5-9 Fluorescence profile example (from Hebert, 1988; reprinted with permission of the Association of Petroleum Geochemical Explorationists).

nm range. The wavelengths greater than 400 nm are attributable to the presence of soil waxes. The method of analysis is similar to gas chromatography in that it uses peaks to recognize a specific pattern that is characteristic of the petroleum(s) produced in the survey area. The procedure is similar, in some respects, to the K-V pattern method. Fig. 5-10 illustrates a pattern of Olmos B oil superimposed on a pattern of petroleum extracted from anomalous surface soil samples (Hebert, 1984).

A set of base samples is established that has specific amounts (in ppm) of hydrocarbon(s) present. These standards are used to calculate the amount of hydrocarbons in each survey sample after correction for sample weight, dilution, and instrument settings. With this method, the resulting amounts of hydrocarbons are not as accurate as

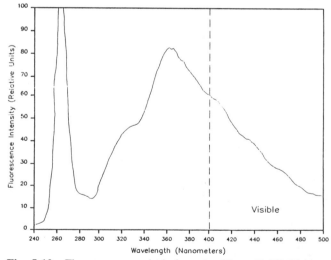

Fig. 5-10 Fluorescence analysis from the Olmos B Oil (Hebert, 1984). Note the similarity in spectra patterns between the petroleum in the soil and that taken from a producing well (reprinted with permission of the Institute for the Study of Earth and Man, Southern Methodist University).

with most other soil-gas methods, but the data have more definitive qualitative values. The data are analyzed by means, modes, standard deviations, and threshold values and are plotted and contoured on a map.

A recent example using fluorescence methods was presented by Calhoun and Burrows (1992) to predict the results of 14 wildcats in the Permian Basin. The authors specifically utilized naphthalene, phenanthrene, and anthracene to compare the produced petroleum to that found in soil. Naphthalene is a two-ring benzene hydrocarbon, and anthracene and phenanthrene are three-ring benzene hydrocarbons. These are aromatic hydrocarbons and are less likely to be degraded by microbes in the soil compared to alkanes such as ethane and propane. The aromatics tend to accumulate temporarily until they are removed by weathering processes. Seventy sample stations were located over a 1500-ac area around the prospect. The emission wavelength was scanned over a 270- to 530-nn range. The three hydrocarbons chosen are readily identifiable on the resulting graph. The authors used a laser and photon counter to produce the results. Since the resulting data are relative and not quantitative, a model was established by utilizing a dry hole in the area of each of the 14 prospects. The three hydrocarbons were averaged around each wildcat and dry hole location to determine whether a prospect was a discovery or a dry hole. The wildcat utilized the average of seven samples and the dry hole utilized the average of three samples. Of the 14 prospects, 36% (5) were predicted to be productive. The actual results were that 7% (1) remains undrilled, 57% (8) were dry, 28% (4) were marginally productive, and 14% (2) were considered good productive discoveries. Of the five predicted producers, two were discoveries, one was marginal, and two were dry. Of the dry holes, three were marginal, one is undrilled, and the rest were dry. The conclusion is that, in these cases, fluorescence was more accurate in predicting dry holes. The concept that the petroleum in the soil is exactly reflective of the potential reservoir at depth is partly true. The problem is that no quantities were reported, just a qualitative presence. Therefore, minor accumulations, weathering, and degradation of petroleum cannot be screened out utilizing the fluorescence methods.

Advantages/Disadvantages. The advantages of the fluorescence method are:

1. Sampling of the soil does not require minimizing substrate disturbance to prevent hydrocarbons from escaping.

2. The technique directly analyzes hydrocarbons that have been retained in the soil.

3. There is a strong correlation between data from the petroleum being produced and the interstitial soil-gas data in many of the examples presented by Hebert (1984).

The disadvantages of this method are:

1. Quick analysis of the sample is required to prevent the loss of hydrocarbons.

2. Typically, only one hydrocarbon type is analyzed for, even though the resulting data may indicate that several petroleum species are present.

3. Dry gas areas may not be amenable to this method since the lighter hydrocarbons leave the soil sample relatively quickly.

4. Sample processing requires a sophisticated procedure that can introduce errors into the analysis.

5. In areas where two or more petroleum types are being produced, there may be difficulty in determining which types are present. Several petroleum types may be leaking to the surface, and separating them and interpreting the data may be impossible.

Airborne Methods

Airborne methods comprise all techniques that sample the atmosphere for hydrocarbon gases escaping from the earth's surface. Sampling is conducted either near the surface with a truck-mounted "sniffer" or at low altitude with a remote gas sensor on an airplane or helicopter (Fig. 5-11). An analyzer continuously monitors the air's hydrocarbon content while the aircraft is in flight. Air samples can also be collected and taken to the laboratory for more detailed analysis. Recently, lasers have been employed on airborne platforms to analyze for hydrocarbons (Fig. 5-12). In some cases, lasers are used to detect hydrocarbons in the same way fluorescence methods work onshore by exciting the hydrocarbons (Fig. 5-13).

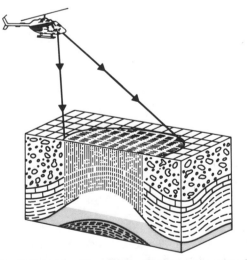

Fig. 5-11 Helicopter with a sniffer. The red is the area of petroleum seepage from the reservoir to the surface (after Barringer, 1981; reprinted with permission of the Institute for the Study of Earth and Man, Southern Methodist University).

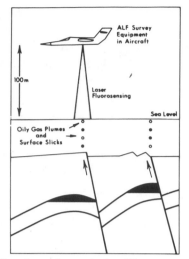

Fig. 5-12 Flowchart of a typical sniffer survey (Schumacher, 1992; reprinted with permission of the Rocky Mountain Association of Geologists).

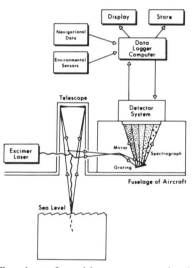

Fig. 5-13 Flowchart of an airborne survey using laser methods for hydrocarbon gas detection in offshore areas (Schumacher, 1992; reprinted with permission of the Rocky Mountain Association of Geologists).

Sampling Techniques

A truck-mounted sniffer has a boom 1 to 4 m above the vehicle to avoid contamination by the engine exhaust. The term *sniffer* refers to a mechanism that simply samples the air for hydrocarbons through an intake vent and associated sensor. Samples can be taken in a stationary or continuously moving mode.

A helicopter or airplane has an extended boom in front, in back, or between the skids of the aircraft. The sensing system sends out radio frequency pulses. The receiver or detector measures the characteristic signature frequency from the various low molecular weight hydrocarbon gases. The locations and relative intensities of these gases are

shown on a CRT video display screen, and anomalous areas are recorded on maps.

The speed of the aircraft must be slow, and its height above the earth's surface must be minimal to increase the chance of detection. Thus, a helicopter has an advantage over an airplane, especially since it can hover over possible anomalous areas. A disadvantage is that the helicopter produces winds that can disperse the gases before their detection.

Factors Affecting Use

The success of this method is very dependent on wind and ground conditions, which may suppress hydrocarbon leakage into the atmosphere. Generally, this method is of limited use except for detecting large accumulations both offshore and onshore, in new areas that have little or no production, and in areas with few drilling tests or little direct evidence of the presence of petroleum. Airborne sampling onshore in the United States is not as extensively used as it is internationally because few of these large reconnaissance areas or large fields remain in the United States. Small petroleum accumulations will generally not indicate leakage anomalies a short distance above the surface because their hydrocarbon concentrations become quickly diluted. Anomalous areas indicated by airborne sampling must be validated by soil sample analysis.

Advantages/Disadvantages

Airborne methods have an excellent advantage in regional reconnaissance surveying. An aircraft with a remote sensing system can survey thousands of acres a day if the land is flat or rolling and relatively clear of thick vegetation. The results allow rapid preliminary evaluations of frontier areas.

The disadvantages are that it is dependent on large volumes of leakage which, at present, seem to be associated only with very large fields in new areas. Even a slight wind can cause dilution and dispersion of an anomaly over a large area so that it may not be detectable or definable. High humidity can prevent gases from rising to flight altitude. Certain vegetation, such as crops and forests, can interfere with the transmission and identification of signals. Sometimes the hydrocarbon signatures cannot be recognized and separated from natural and cultural background levels in new areas.

Offshore Methods

Detecting hydrocarbons offshore has been a standard part of exploration programs for some time. Several major offshore oil accumulations such as the Santa Barbara Channel area, California, have visible petroleum seeps rising from the ocean floor. These seeps have formed asphalt mounds on the sea floor and are very numerous around Point Conception. The mounds form an irregular east-west trend that coincides with a series of faulted anticlines. The Ventura Basin, California Coal Oil Point seep area, is reported to release 50 to 70 bbl of oil a day. Other areas of seepage include the U.S. Gulf Coast, the Red Sea, the Arctic Circle of Offshore Alaska, the Gaspé Peninsula of Quebec, and the South China Sea. Many of the offshore fields now being discovered do not necessarily have visible seeping petroleum, but the volumes leaking offshore are significantly greater than those detected onshore. Dunlap and Hutchinson (1961) made a statistical study of known seeps and petroleum accumulations in southern Louisiana. Their study concluded that using seeps was more successful in finding an accumulation than geologic and geophysical methods. The seeps were considered relatively direct to the surface because the faulting angle is 60% down to 3030 m (10,000 ft). Therefore, the venting of petroleum has only minor displacement. Dunlap and Hutchinson also concluded that the closer to the seep drilling occurred, the larger were the petroleum reservoirs that were found. The target size, in terms of petroleum in place, is also significantly greater for offshore than onshore fields. Generally, the variety of problems created by the onshore soil environment do not seem to exist offshore. Therefore, collection and interpretation of data are easier and quicker.

Sampling Techniques

Sampling is relatively simple. It is assumed that the hydrocarbons leaking from the ocean floor form a plume that is either directly (relatively) above the accumulation because of the lack of currents or is shifted down current. In some cases, a hydrocarbon anomaly is not detected unless sampling is done at the sediment/sea interface or in the sediment itself.

In one type of offshore method (developed by ARCO in the 1950s), a ship collects water samples along a grid either continuously or at specific locations during a sniffer survey (Fig. 5-14). This ship tows a "fish" with a water pump in

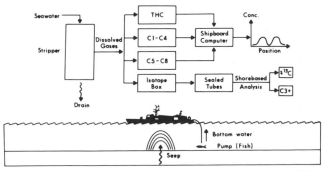

Fig. 5-14 Flowchart for a typical offshore gas survey (Schumacher, 1992; reprinted with permission of the Rocky Mountain Association of Geologists).

depths below the wave base. The pump supplies a steady stream of seawater for samples. The gas phase is separated from the seawater and is analyzed by chromatography onboard the ship. This allows rapid resampling of all or selected parts of the survey if necessary. An airplane or helicopter can be used in a manner similar to a "fish." Figure 5-15 is an example of a sniffer survey across the East Cameroon Block 330, Gulf of Mexico, in the United States.

In another method, piston cores of the ocean floor sediments are collected. Cores are quickly retrieved to the surface, sealed, and frozen for later analysis. Gas is extracted from the core and analyzed. This method is similar to that used for head-space gas analysis. Samples can be analyzed using a gas chromatograph (Sackett, 1977), infrared analyzer (Dunlap et al., 1960), or laser. Bottom-water samples are also useful in detecting seeping hydrocarbons (Sweeney, 1988).

Factors Affecting Use

A steady hand at the ship's wheel and a good navigational system are necessary. As smaller and smaller petroleum accumulations are sought in areas like the U.S. Gulf Coast, offshore methods (especially water sampling) may prove ineffective without more sophisticated equipment and tighter grid spacing.

The bottom sediments are typically zoned in descending order: an aerobic, a sulfate, and a methane-generating horizon. In the methane zone, bacterial activity is high, causing the generation of biogenically derived hydrocarbons (predominantly methane). In the overlying sulfate zone, hydrocarbons are decomposed by the reaction:

$$2CH_2O + SO_4 = \rightarrow 2CO_2 + H_2S$$

Minor seepage of hydrocarbons will be destroyed, and any sampling in this zone will not detect anomalous amounts. In the aerobic zone, hydrocarbons are further degraded by the reaction:

$$CH_2O + O_2 \rightarrow CO_2 + H_2O$$

It is essential for bottom cores to penetrate to the methane zone unless there is high leakage of hydrocarbons.

It was originally thought that analyzing for methane was sufficient to determine the presence of an oil accumulation (Dunlap et al., 1960). However, Sackett (1977) found that 13 of 14 delineated seeps were of biogenic origin. Sackett then concluded that methane analysis should be discarded in favor of C_3 through C_6. A similar survey done in the North Sea by Faber and Stahl (1984) concluded that the methane in samples was not of biogenic origin but was derived from petroleum sources at depth. Therefore, the contradiction suggests that methane is usable, but sophisticated analysis is needed to determine its origin. If C_2 through C_6 are collected during the survey, then the origin of the methane present will be obvious.

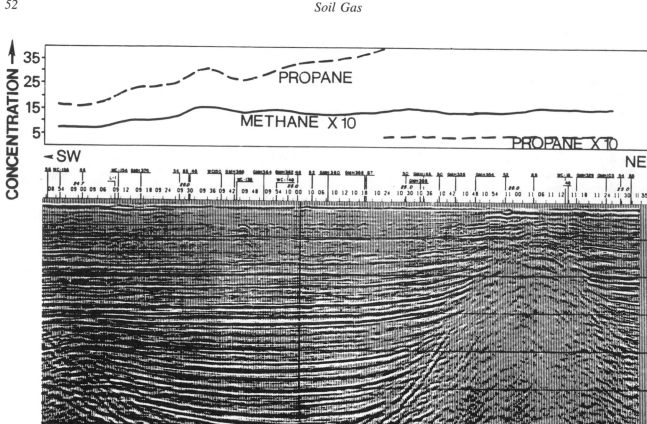

Fig. 5-15 A sniffer survey and associated seismic profile, offshore Gulf of Mexico, E. Cameron Block 330, United States (Weismann, 1980).

Advantages/Disadvantages

Offshore methods, like airborne methods, are best used for rapid reconnaissance surveying of large areas rather than for evaluating individual prospects. Data from many worldwide locations indicate that the marine sniffer is very reliable for distinguishing between nonproductive and productive trends. Of course, the words *nonproductive* and *productive* are relative terms.

The petroleum industry generally prefers bottom-sediment samples to water samples because the former give a more accurate measure of the hydrocarbon seep flux. An advantage of the sniffer as it moves through the water is that it gives continuous readings. Seep gases are detectable in the water for a few miles from the emission point because the gases spread into a plume down current, and continuous sampling should intersect it. Bottom-sediment sampling can miss a seep if a sample point is too far from the emission location on the sea floor. Detection of hydrocarbon plumes is often done in conjunction with seismic acquisition. It is relatively very easy to correlate seepage to a geophysical

structure or anomaly. Significant amounts of gas in the section also cause an effect on the seismic data that is easily seen. The costs of a sniffer survey are considerably less than bottom-sediment sampling.

A disadvantage is that cultural and natural contaminants will give false positive indications of a seep.

Coastal waters vary in normal hydrocarbon content, and resolution of what concentrations are anomalous and what are not may be difficult. Discharge by rivers will cause fluctuation in hydrocarbon data in the near offshore. Hydrocarbons seem to concentrate in the vicinity of the thermocline, whose location varies with the season. Methane concentrations vary with salinity, seawater composition, density, oxygen content, and biological activity. The depth limit of the sniffer is unknown and, therefore, it may be unusable in some marine conditions.

Application to Environmental Science

In recent years, the use of soil-gas methods to detect volatile organic compounds that are causing contamination in the shallow soil or bedrocks has become increasingly popular. The methods, as applied here, detect a wide range of volatile petroleum-related compounds. There is a series of significant differences between applications for petroleum exploration and for environmental science:

1. In the detection of contamination by man-made causes, the amount of gas present is typically in the parts per million to as high as percent of total gas present. Petroleum exploration typically detects parts per billion, sometimes parts per million, and the literature has not indicated any surveys in the parts per thousand onshore. When these methods are applied in environmental science, the volumes collected are significantly higher than anomalous values detected in petroleum exploration.

2. In environmental science applications the types of hydrocarbons detected expand considerably since many of the contaminants are man-made by-products. If the type of contaminant is known, the detection equipment can be calibrated to identify it specifically in the soil gas. The environmental industry seems to prefer the detection of halocarbons over petroleum hydrocarbons because of the former's rapid volatilization out of groundwater. Environmental literature indicates degradation by oxygen of petroleum occurs quickly, whereas degradation of halocarbons are not as rapid. Halocarbons have been used to detect contaminants as deep as 100 m. Rapid variations in the hydrocarbon gases trapped in the soil profile have been noted, whereas halocarbons are seen consistently through the soil strata. This inconsistency in sampling could be corrected, however, and hydrocarbons can be useful if more sensitive equipment is utilized.

Methods and applications in environmental science display several similarities to those of petroleum exploration:

1. In environmental applications, the methods are used primarily to determine if contamination exists and the areal extent of contamination. The depth and true extent of contamination is left to remedial action. Therefore, plume detection, like identification of microseepage for petroleum exploration, is a precursor to more sophisticated methods.

2. Forms of detection are similar except that terminology is different. Passive methods are those requiring burying an absorbent, as with the K-V method described earlier. Active methods are the free-air type, with a portable gas chromatograph to obtain real-time analysis, direct ongoing investigations, and minimize unnecessary sampling and laboratory analysis. Passive methods are not generally used because they require considerable increased costs, time, and effort. They do not provide the quantity of each contaminant, and thus cannot distinguish between a minor or major pollution problem. Active methods are the most commonly used methods because they can delineate the plume quickly in the field.

3. Evaluation methods identify the plume by the use of simple statistics, ratios between various halocarbons, and petroleum hydrocarbons.

4. Soil conditions can affect detection. Several problems exist with soil-gas detection for plumes or man-made contamination. Areas of water recharge tend to dilute the contaminants, causing loss of volatiles and resulting in poor delineation of the entire plume. Water vapor or saturation of the soils tend to render these methods ineffective until the soils "dry out." Therefore, like petroleum exploration, the soil environment can control the ability to detect the plume as with petroleum methods. Little work has been done to attempt soil profiling to determine optimum layers for sampling.

It is important to determine a suitable target compound for plume detection. The determination of the target's suitably is based on the compound's volatility, stability, and aqueous solubility. The optimum compounds are those that have a boiling point less than 150°C, low aqueous solubility, and relatively good resistance to degradation. Compounds with boiling points greater than 150°C are typically detected only when significant amounts are present, leaving a residue in the soils. Compounds that are miscible with water are also not useful. Petroleum hydrocarbons are assumed to degrade rapidly when exposed to oxygen. Thus, depths to sample are considered optimum when soil gas can be sampled 5 ft from the water table. In many cases, this is considered uneconomic and, therefore, halocarbons are used instead. Halocarbons, such as dichloroethene and its isomers, degrade slowly in the time needed to effect sampling from soil gas. Acid extraction and head-space methods are not as commonly used but could overcome some of the problems encountered with free-air methods that dominate environment science applications. An alternative to petroleum or halocarbon detection is the use of CO_2 to detect the plume. Because of degradation, elevated levels of CO_2 have been detected above the plume, but this method is in its infancy. Indirect methods have not been used except when they are known to be part of the contamination problem although they might be useful since they are by-products of volatile organic material migrating through the soil.

Soil-Gas Discussion

Soil-gas or soil-vapor methods (excluding fluorescence) are usually considered to be the best direct surface techniques for measuring the presence of migrating hydrocarbons. The methods are based on the assumption that the collected and analyzed hydrocarbons have migrated from a reservoir at an unspecified depth. Consequently, the explorationist's determination that an economic target is present at a particular site is influenced by the amounts of hydrocarbons present, the soil environment, and the relationship between the sample density of the survey and models of existing fields. The term *economic target* is typically used interchangeably with *drillable target*. The justification for drilling the target cannot be resolved on the basis of surface geochemistry alone. It must be arrived at by comparing the undrilled (or prospect) anomaly's data to date from existing models and by integrating the geochemical data with geologic and geophysical methods.

Explorationists typically do a step-by-step evaluation of data from the point of collection to the final interpretation. This step-by-step appraisal is not necessarily written down, consciously invoked, or the same from prospect to prospect or explorationist to explorationist. There are numerous questions that have to be asked after collection and before interpretation. For example: Was the soil consistent throughout the survey area? Was the same soil layer sampled or were multiple layers sampled? Was the sampling crew experienced or inexperienced? What period of time elapsed between collection and analysis? When was the last time the equipment was calibrated or checked for contamination? Many of these initial questions have nothing to do with: Did we see anomalous values somewhere across our prospect area? Or more likely: Did the geologic/geophysical concept hold up or was it condemned? However, the initial questions are vital to answering the subsequent set of questions. Unless some of the questions are answered positively, doubt will be cast on the answers to the remaining set of questions. Therefore, we have to consider not only what affects the interpretation of data for the determination of drillable prospects but also the integrity and the limitations of the mechanical procedures used in obtaining the data. These same concerns also apply to indirect methods.

The first step in any soil-gas survey is determining which technique to use. In preceding sections, this has been discussed to a limited extent. The advantages/disadvantages of one technique vs. another are based on terrain, soil, weather conditions, and personal preference. The purpose of any technique is to measure the presence of hydrocarbons at a particular site in the soil. Gas is present either in the vapor phase, moving along the permeable pathways, trapped temporarily in interstitial pore spaces, absorbed onto clays, dissolved in water, or occluded in secondary carbonate cements or oxide coatings. All these forms are fragile and are quickly removed in a normally evolving soil geochemical environment. Therefore, the first step in sampling hydrocarbons in any physical state is to collect an *undisturbed* sample.

Obtaining an undisturbed sample of soil is possible using sophisticated coring techniques that are prevalent in engineering geology. These coring methods may result in higher-quality data, but they are expensive and time-consuming. Therefore, in oil exploration, there may have to be a trade-off between inexpensive acquisition with lower-quality data and costly acquisition with higher-quality data. Airborne and water methods do collect an "undisturbed" sample, but they have numerous disadvantages that soil methods attempt to correct.

The depth at which a sample is taken is a major source of controversy. There is significant evidence that the deeper the sampling, the more likely it is that higher volumes of hydrocarbons will be found (Saenz et al., 1991; Richers and Maxwell, 1991). We have to look at several different factors that affect microseepage from the seal rock to the soil. Several authors have assumed that seepage is specifically related to faults and fracturing and does not necessarily imply the presence of a petroleum accumulation at depth. A second assumption is that soil conditions have minimal or no impact on the migrating soil gases, and a third is that soil-gas volumes will be similar to those of the model. All these assumptions usually refer to conditions in a few specific cases and cannot be applied everywhere. For example, some small stratigraphic traps like the D Sand in the Denver Basin, Silurian reefs in Michigan and Illinois, or the Minnelusa of the Powder River, have no known associated faulting and fracturing. Therefore, relying on the first assumption would dictate that soil-gas methods would not detect these types of accumulations; the microseepage would be minimal, and identifying the subtle anomalies would be difficult. The third assumption is also false, as there is no evidence to date that the total volume of microseepage is a true indicator of productivity or potential. Several breached or barren structures in North America have large soil-gas anomalies that look similar, if not better than, those of the models. Increasing volume with depth is a consistent pattern seen typically over large fields that are leaking high volumes of hydrocarbons. Small fields have fewer hydrocarbons trapped and seem to leak fewer hydrocarbons. If no fracturing system exists, a field of significant size could yield a subdued or subtle anomaly at the surface, or a small field that lies in an area with extensive fracturing could indicate extensive leakage. In an area where high-volume leakage occurs, the concern about the effects caused by soil may be moot. However, in an area where leakage is small, which is true of most U.S. onshore areas today, the effects of soil on migrating gases give cause for concern. In the case of two proprietary studies in which the author was involved, specific soils had to be mapped, identified, and targeted to obtain consistent results using soil gas.

Of the direct methods, measuring the actual hydrocarbon gas concentrations in the soil is the most viable one and has become the most common technique used for surface geochemical exploration. Problems that must be dealt with are the impact of the soil environment, other sources of soil gas, and sampling the vapor phase. These difficulties can be solved or tempered by having a thorough knowledge of the soil and the geology of the survey area and by maintaining the integrity of the sampling equipment and technique. The soil-gas results can then be enhanced and confirmed by other methods.

Hydrocarbon Ratios and Predicting
Petroleum Generation

The use of ratios in hydrocarbon data evaluation was developed as a way to predict the type of petroleum present at depth (wet gas, dry gas, or oil) and as a filter to eliminate false indications of anomalous conditions. Jones and Drozd (1983) compiled a large soil-gas data set taken from dry gas, wet gas, and mixed petroleum areas. They observed differences in the ratios of heavier hydrocarbons to methane and among the heavier hydrocarbons themselves. When plotted on a pixler plot (Fig. 5-16), the data indicate the petroleum characteristics of each sample and also demonstrate a close correlation with the petroleum being produced.

An excellent use of ratios is in a data set where there are large variations in volumes of leaking hydrocarbons. Ratioing data eliminates samples that exhibit large volumes of leakage but are not actually anomalous. Thus, samples in a survey area that have anomalous amounts of hydrocarbons show similar characteristics in the ratios as well. A data set that exhibits anomalous values based on total volumes but does not show anomalous ratios when compared to background data is probably a false indication.

The formula for ratios is expressed by the following equation:

$$(A) \; C_1/C_2 \; \text{or} \; C_n/C_{n+1} = \text{Ratio}$$
$$\text{or}$$
$$(B) \; C_2/C_1 \; \text{or} \; C_{n+1}/C_n = \text{Ratio}$$

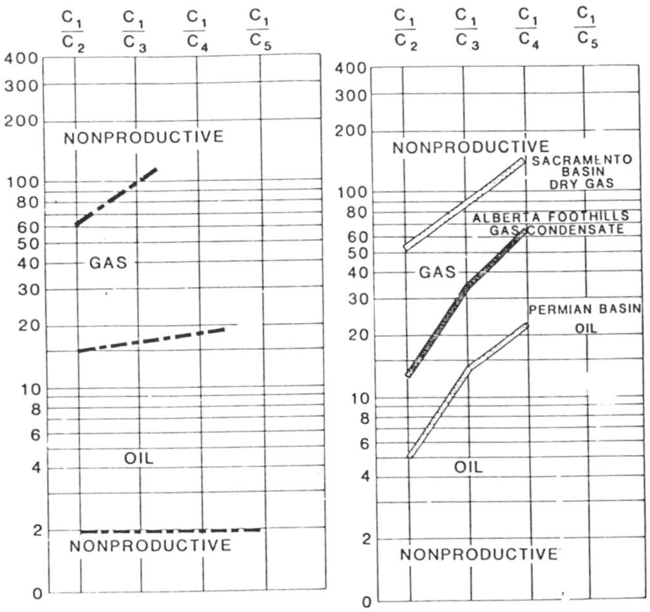

The lower the ratio (reverse in case B), the greater the likelihood that the sample is anomalous. An example of ratios is given in the case histories below.

Ratios have also been used to determine the potential commercial or noncommercial nature of a petroleum reservoir (Yasenev, 1986). It has been found that large volumes of the heavy hydrocarbons and smaller ratios of the lighter hydrocarbons generally imply economic viability. However, this is not always the case, and ratios between hydrocarbons can vary considerably from basin to basin. This use of ratios has been applied in the United States with limited results. The one consistent conclusion is that the greater the volumes of heavy hydrocarbons (butane, hexane, and pentane), the more likely it is that a significant accumulation is present. Chakhmakhev and Vinogradov (1980) used this method to identify the type of gas condensate pool that was present. Their work was later expanded, and they found a similar relationship to oil fields. The use of ratios in helping to evaluate a soil-gas survey is well documented. However, predicting productivity is using the data for a purpose it was not designed for and, to date, cannot seem to fulfill.

The application of ratios has been taken a step further. By comparing hydrocarbons from models of different petroleum areas, predictions can be made in terms of the type and possible target reservoir in a prospect area. Surveys across fields that no longer display anomalous volumes of hydrocarbons will continue to display anomalous ratios of hydrocarbons during the later stages of depletion.

In recent years, soil-gas surveys have been used in an attempt to evaluate the source rock potential of an area, region, or basin. These surveys have utilized both the interpretation of total volumes of gases and ratios of gases to identify areas where petroleum is generated. Landrum et al. (1989) presented a study of two valleys in Montana using the head-space method and which used a paint shaker to disaggregate the gas from the soil sample (Fig. 5-17). The study assumed that the data are comparable between the two areas. No data from producing areas were presented as a model. The study concluded that one valley had excellent petroleum potential and the other generally did not (Figs. 5-18 and 5-19). Figure 5-20 illustrates a mean score method that is a form of data filtering represented by the formula:

$$Xi = [(X - \text{mean})/\text{standard deviation}] \times 100.$$

There was no indication that the study evaluated and compared the soils or bedrock of the two areas. The two data sets were separated by 220 km. The head-space (paint shaker) method used is especially destructive, and probably some of the hydrocarbons analyzed were generated from the absorbed and occluded states, organic matter, and rock fragments in the soils. Landrum et al. used ratios to determine areas of oil, dry gas, or wet gas. The problems with the study were that it ignored several elements that affect soil-gas sampling and data interpretation and that it attempted to compare two distinctly different areas without

Fig. 5-16 Pixler plot of hydrocarbons. The left side is gas ratios determined from well logs. The right side is soil-gas ratios (Jones and Drozd, 1983; reprinted with permission of the American Association of Petroleum Geologists).

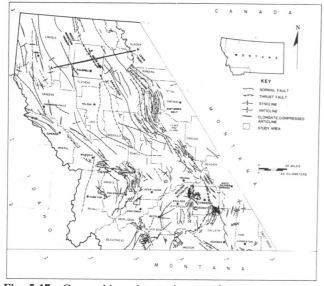

Fig. 5-17 Geographic and tectonic map of western and central Montana, showing the locations of the Flathead and Townsend Valley study areas (Landrum et al., 1989; reprinted with permission of Elsevier Science Publications).

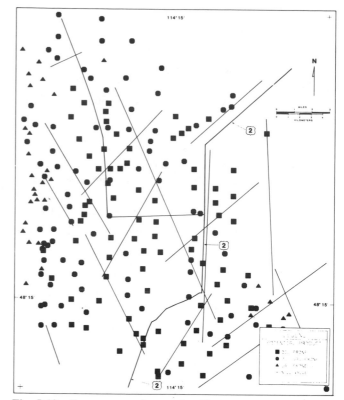

Fig. 5-19 Propane to methane ratios (C_3/C_1) in the Flathead Valley for each sample site location (Landrum et al., 1989; reprinted with permission of Elsevier Science Publications).

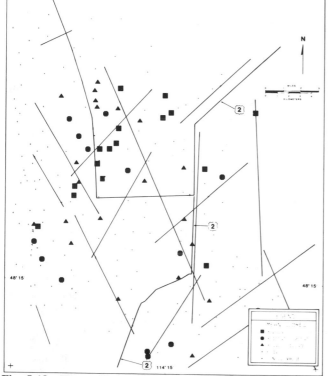

Fig. 5-18 Ethane (C_{2+}) and heavier hydrocarbon concentrations for the Flathead Valley with posted mean score values (Landrum et al., 1989; reprinted with permission of Elsevier Science Publications).

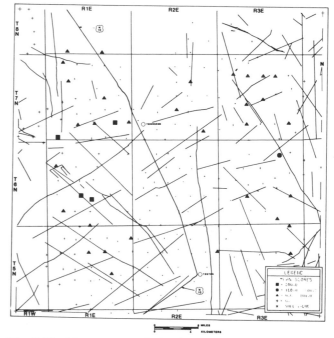

Fig. 5-20 Mean score analysis of the Flathead and Townsend valleys, Montana, with sample locations (Landrum et al., 1989; reprinted with permission of Elsevier Science Publications).

models on which to base conclusions. Samples from the southern area yielded higher hydrocarbon volumes than those from the northern valley. This does not mean that the northern valley has any less potential. Many hydrocarbon anomalies are subtle either year-round or at different times of the year, depending on the moisture content of the soil or a lack of fracturing from the seal rock to the surface. One could conclude that larger potential fields might exist in the southern area, but the data could simply represent massive leakage or breached traps, or both.

Dickinson and Matthews (1991) conducted a similar study to evaluate the potential or Cretaceous source beds in the Overthrust Belt, Utah and Wyoming. The free-air method was used, and samples were collected over a period of two years. No soil evaluation was presented. There were existing fields in the study area, and sample spacing was over 1 km in density. The study had no base stations and did not attempt to normalize the data to allow samples taken during different seasons (let alone different days) and under different soil conditions to be compared with any degree of confidence.

A bottom-sediment survey in the North Sea was carried out by Faber and Stahl (1984). From the data, they determined areas that were poor, good, or excellent in terms of source rock generation. The survey results showed a good correlation between the productive areas and areas of good to excellent source rock potential (Fig. 5-21). However, there was no discussion comparing hydrocarbon leakage in the poor and good source potential areas and whether it actually indicated the existence of petroleum accumulations.

These types of studies are well intended, but they attempt to evaluate source rock potential, zones of petroleum migration, and source rock maturation using methods that were not designed for these purposes. It is assumed from the conclusions of the preceding studies that the rocks in the basin have generated petroleum and caused it to migrate. The question could be: Did the seeping hydrocarbons migrate in from another area and then leak out? The Salina Basin, Kansas, is an example of an immature basin, as currently defined by source rock studies. The numerous oil fields found at its southern edge represent oil that has migrated over 300 km from Oklahoma to the south. However, using the line of thinking described by Landrum et al. (1989) and Dickinson and Matthews (1991), we could make the assumption that the basin source rocks were mature. These studies generated a lot of data, but they cannot reach their described conclusions.

Case Histories

Hastings Field, Brazoria County, Texas

The Hastings Field is an excellent example of reservoir depletion affecting the soil gas anomaly at the surface (Horvitz, 1985). The first survey in 1946 (Fig. 5-22) indicates anomalous values of ethane plus heavier hydrocarbons in a

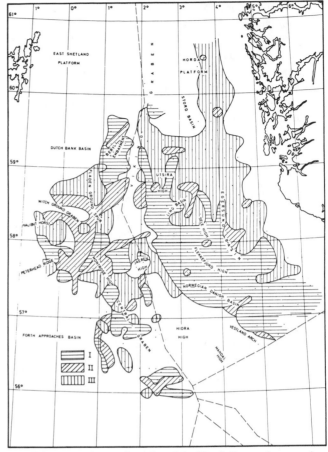

Fig. 5-21 Location and results of the North Sea study assessing the source rock potential with an overlay of present production. Area type I: poor source rocks; Area type II and III: medium to good source rock areas (Faber and Stahl, 1984; reprinted with permission of the American Association of Petroleum Geologists).

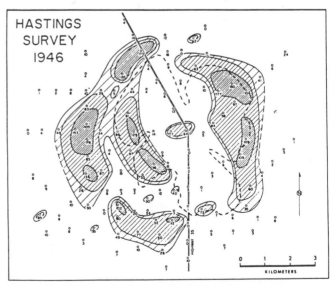

Fig. 5-22 Hastings Field, Brazoria County, Texas, first hydrocarbon survey, 1946. The hydrocarbon is ethane in parts per million (from Horvitz, 1969; reprinted with permission of the Institute for the Study of Earth and Man, Southern Methodist University).

halo pattern around the field. The second survey was done in 1968, and ethane and heavier hydrocarbons data indicate that only isolated anomalies remain except in the northeast corner of the field (Fig. 5-23). We can conclude that vertical migration has declined significantly in response to the loss of reservoir pressure. The Hastings Field strongly suggests that explorationists avoid using fields with a long history of production as models on which to base exploration.

Offshore Louisiana

Horvitz (1981) presented a sniffer survey carried out offshore on the Louisiana Gulf Coast from 1973 to 1974. The area is located about 100 miles southeast of New Orleans. The 256 samples were collected in water depths ranging from 318 ft at the northwest corner to 1860 ft in the southwest corner of the survey. Sample spacing was approximately one-half mile along north-south profiles. All data are in parts per billion. Horvitz interpreted the data to imply a halolike feature, based on ethane and heavier hydrocarbons, in the southeast corner of the map (Fig. 5-24). Methane and pentane surveys basically confirmed this interpretation (Figs. 5-25 and 5-26). Subsequent drilling in 1975 just off the northeast corner of the map led to the discovery of the Cognac Field. At the time of Horvitz's article, the field was not yet producing.

Jace Field, Kiowa County, Colorado

The Jace Field, located in section 1, Township 18 South, Range 42 West, Kiowa County, Colorado, was discovered in 1989 and produces from the Stockholm Member of the Morrow Formation (Fig. 5-27). The discovery well is lo-

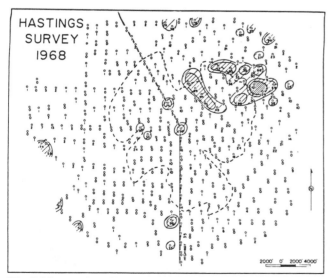

Fig. 5-23 Hastings Field, Brazoria County, Texas, second hydrocarbon survey, 1968. The hydrocarbon is ethane in parts per million. Note that the anomaly has all but disappeared, probably because of depletion of the reservoir, which decreased the amount of migrating hydrocarbons to the surface (from Horvitz, 1969; reprinted with permission of the Institute for the Study of Earth and Man, Southern Methodist University).

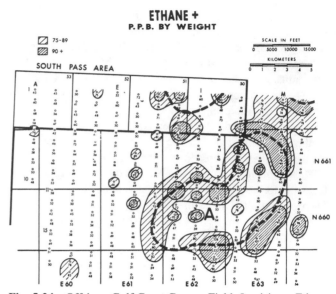

Fig. 5-24 Offshore Gulf Coast Cognac Field, Louisiana. Ethane (C₂) plus heavier hydrocarbons. The anomalous ethane values form a halo pattern (Horvitz, 1981; reprinted with permission of the Institute for the Study of Earth and Man, Southern Methodist University).

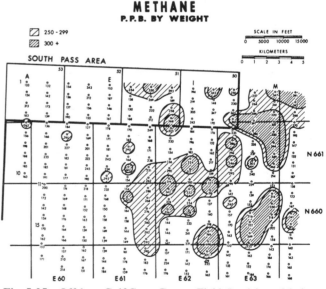

Fig. 5-25 Offshore Gulf Coast Cognac Field, Louisiana. Methane (C₁) results (Horvitz, 1981; reprinted with permission of the Institute for the Study of Earth and Man, Southern Methodist University).

cated in the NE NE of section 1 and had an initial potential of 150 bbls of oil per day. The second well drilled in the SE SE section 1 had an initial potential of 700 bbl a day. The well's production declined very rapidly, and it appears to have drained a limited part of the reservoir. An offset is currently producing insignificant amounts of oil.

Soil-gas (acid extraction) and iodine surveys were performed at the same time the second well (SE SE of Section 1) was being drilled. Pentane and hexane data, summarized

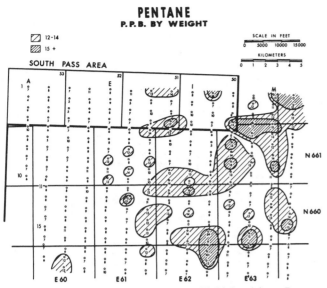

Fig. 5-26 Offshore Gulf Coast Cognac Field, Louisiana. Pentane (C_6) results form a halo anomaly (Horvitz, 1981; reprinted with permission of the Institute for the Study of Earth and Man, Southern Methodist University).

in Figs. 5-28 and 5-29, indicate anomalous patterns similar to the iodine survey results (Fig. 5-30). It was observed that the anomalous soil-gas samples contain increased amounts of carbonates. One of the noted effects of microseepage is an increase in carbonates overlying the reservoir.

Figure 5-31 shows the projected productive channel trend interpreted by Adams (1990) based on seismic and subsurface well data. The geochemical surveys did not support this interpretation. There was a considerable amount of drilling along the projected channel trend, but all the holes were dry. Either there was no sandstone or it was wet. Wells drilled after the soil-gas/iodine surveys are indicated in Fig. 5-32 by larger well symbols. Subsequently, several wells have been drilled in and around the anomaly and have extended production. The drilling has modified the geologic and geophysical model, which now generally matches the original geochemical interpretation.

Patrick Draw Field, Wyoming

The Patrick Draw Field is located in Township 18 North, Range 99 and 100 West, Sweetwater County, Wyoming. In

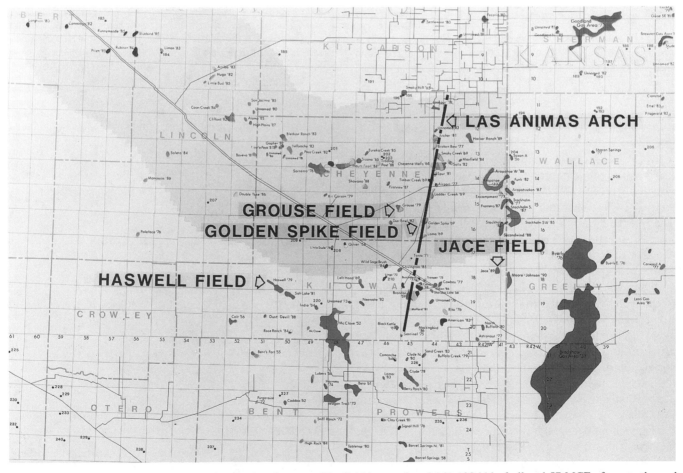

Fig. 5-27 Jace Field, Kiowa County, Colorado, location map. The field has produced 161,139 bbl of oil and 57 MCF of gas to the end of December 1991 (reprinted with permission of Promap Corporation).

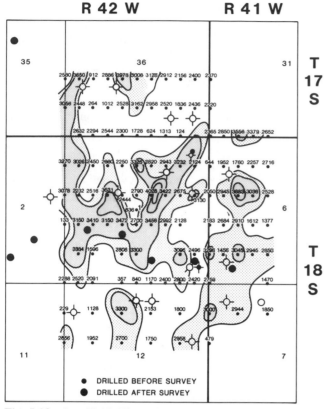

Fig. 5-28 Jace Field, Kiowa County, Colorado. Pentane Survey results (in parts per million) as analyzed by the acid extraction method. Contour interval is 2500, 3000, and 3500 ppm (printed by permission from Atoka Exploration).

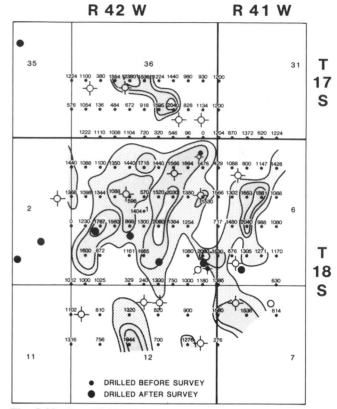

Fig. 5-29 Jace Field, Kiowa County, Colorado. Hexane survey results as analyzed by the acid extraction method, in parts per million. Contour interval is 1200, 1400, and 1800 ppm (printed by permission from Atoka Exploration).

recent times, it has been the subject of several geochemical, geobotanical, and Landsat studies (Fig. 5-33). Inadvertently, it has probably been one of the most controversial geochemical surveys presented in the literature. The field was discovered in 1959 and has produced over 53 MMbbls and 135 BCF of gas from the Almond and Lewis pay zones. The reservoir trap is stratigraphic with no known faulting in the subsurface. A good correlation of Landsat, surface geochemistry, and geobotany data identified surface leakage in the area, but leakage was not always clearly identifiable with the field. Several papers concluded that all the surface phenomena were related to the field (Richers et al., 1982; Richers and Weatherby, 1985; Richers et al., 1986). However, the results of the initial paper have been questioned by Scott et al. (1990) because of the lack of any clear correlation of the geobotanical and geochemical data with the existing field. Also, many of the data sets presented were statistically small or insignificant, allowing skepticism over obtaining a reliable interpretation.

The soil-gas sampling was done using three different methods: free-air, desegregated extraction, and acid extraction. Results of these surveys are presented in Fig. 5-34. The data do not clearly define the field. The field was surveyed 30 years after discovery, and the present anomaly is defining

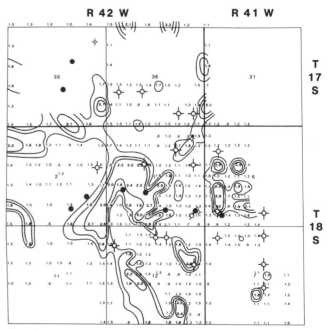

Fig. 5-30 Jace Field, Kiowa County, Colorado. Iodine survey with postdrilling results. Values are in parts per million of iodine. Contour intervals are: 1.6, 1.8, and 2.0 ppm (printed by permission from Atoka Exploration).

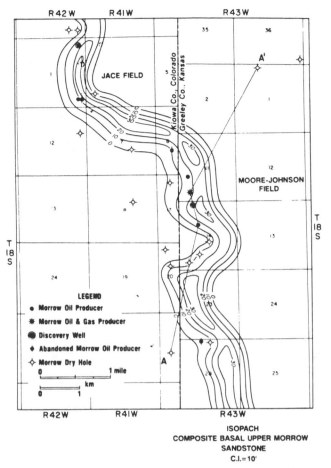

Fig. 5-31 Jace Field, Kiowa County, Colorado. Isopach of composite basal upper Morrow sandstone (from Adams, 1990; reprinted with permission of the Rocky Mountain Association of Geologists).

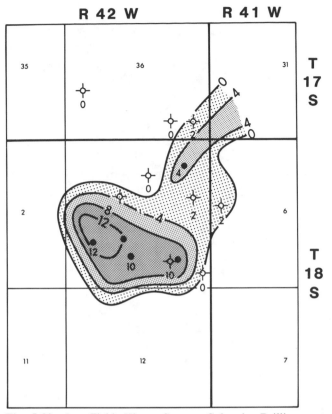

Fig. 5-32 Jace Field, Kiowa County, Colorado. Drilling post Adams (1990) and subsurface geologic interpretation.

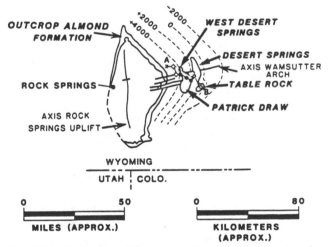

Fig. 5-33 Patrick Draw, Sweetwater County, Wyoming, location map (from Richers et al., 1985; reprinted with permission of the American Association of Petroleum Geologists).

only areas that are still productive and leaking, which are on the updip part of the field (Fig. 5-35). The disaggregated and acid extraction data do not seem to define the field at all. A sagebrush and a Landsat anomaly were also found in association with the field. The sage anomalies have been identified near the eastern edge and center of the field. No sage anomalies have been identified in areas associated with other fields. The Landsat anomaly covered most of the field. We can conclude at this point that the diminished surface phenomena are caused by reservoir pressure depletion as at the Hastings Field in Texas. A new and less depleted field in the area should have been surveyed for comparison purposes.

Arp (1992) attempted to resolve the problems concerning the surface geochemistry and specifically the geobotany. He contended that the reservoir was originally underpressured. Based on his historical analysis, the sagebrush (geobotany) anomaly or kill area was not due to climatic factors but was a response to overpressure caused by the waterflood. The pressure of the waterflood caused increased vertical migration in the area of the gap cap. This increased the reducing

conditions, which subsequently stressed and then killed the sagebrush. Although faulting has been suggested, all the various sources state that there are no identified faults associated with the production. Arp did not present any new data, and he restates that sagebrush kills like this are not restricted or inherently found only in association with oil fields. Underpressured reservoirs can cause surface geochemical anoma-

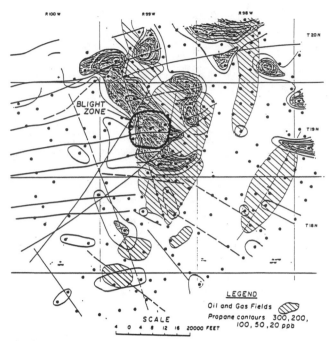

Fig. 5-34 Patrick Draw, Sweetwater County, Wyoming. Propane data from four samples and the blighted area (Richers et al., 1986; reprinted with permission of the American Association of Petroleum Geologists).

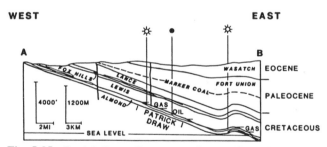

Fig. 5-35 Patrick Draw, Sweetwater County, Wyoming. Patrick Draw geologic cross section (Richers et al., 1985; reprinted with permission of the American Association of Petroleum Geologists).

lies far more intense than those found in an overpressured area or over a waterflood field. Therefore, the identification of underpressured or overpressured reservoirs does not indicate a direct correlation with geochemical results. Rather, intensity of a geochemical anomaly at the surface is related to the volume of leakage and the density of pathways present from the reservoir to the surface. Assumptions about the original pressure and possible intensity of the original surface geochemical anomaly would be mere speculation. No data have been presented that detail the soil composition, matrix potential, clay minerals present, organic matter percentage, and so forth. There is the possibility that a uranium roll-front deposit is the cause. Stratigraphic deposits such as this are notorious for having a surface geochemical anomaly that clearly outlines the updip edge of the field. The sage anomaly is restricted to a small part of the top and center of the field and actually extends for some distance

out of the field. This brings into question the real effect of the waterflood on the surface geochemical anomaly.

Questioning the various authors' lines of reasoning brings up the more important query as to why so much effort has been spent on a depleted field that could cause explorationists to make erroneous decisions. The Patrick Draw survey example is presented as a warning. Fields that have undergone significant depletion through production and waterflooding, regardless of their size or other attributes, do not yield reliable geochemical results. Why use geobotanical and Landsat anomalies if they have numerous other causes and rarely are related to petroleum seeps? The methods that were used here should have been employed at several fields rather than on an isolated case.

Dolphin Field, Williston Basin, North Dakota

The Dolphin Field is located in Township 160 and 161 North, Range 95 West, Divide County, Williston Basin, North Dakota (Fig. 5-36). The field was discovered in 1986 and has an estimated reserve of 6 MMbbls of oil and 7.5 BCF of gas. The field was not discovered using geochemical methods but is an example of the effectiveness of soil-gas and fluorescence methods in sandy soils (Horvitz and Ma, 1988). The sample spacing is more at the reconnaissance level than at the stage that could define the areal extent of the field.

Maps of the ethane plus heavier hydrocarbons for both sand and clay and for fluorescence have outlined the field in relative terms based on the limited number of samples (Figs. 5-37–39). The florescence spectrum used for detection was at 365 nm for both sand and clay samples.

Ballina Plantation Prospect, Louisiana

The Ballina Plantation Prospect is located in Township 8 North, Range 8 East, Concordia Parish, Louisiana (Fig. 5-40). Figure 5-41 is a structure map on top of the Tew Lake Sand, and Fig. 5-42 is the fluorescence survey (Herbert,

Fig. 5-36 Dolphin Field, Williams County, North Dakota, location map (Horvitz and Ma, 1988; reprinted with permission of the Association of Petroleum Geochemical Explorationists).

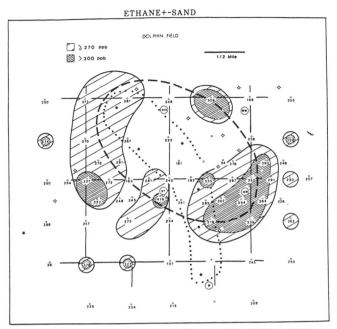

Fig. 5-37 Dolphin Field, Williams County, North Dakota. Ethane (C₂) plus heavier hydrocarbons (in ppb) for sand (Horvitz and Ma, 1988; reprinted with permission of the Association of Petroleum Geochemical Explorationists).

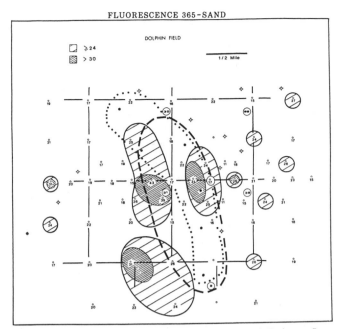

Fig. 5-39 Dolphin Field, Williams County, North Dakota, fluorescence results of the 365-nm wavelength for sand fraction of the soil (Horvitz and Ma, 1988; reprinted with permission of the Association of Petroleum Geochemical Explorationists).

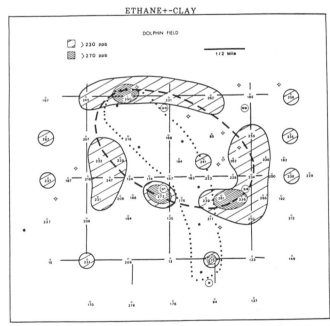

Fig. 5-38 Dolphin Field, Williams County, North Dakota. Ethane (C₂) plus heavier hydrocarbons (in ppb) for clay (Horvitz and Ma, 1988; reprinted with permission of the Association of Petroleum Geochemical Explorationists).

Fig. 5-40 Ballina Plantation, Louisiana, location map (Hebert, 1988; reprinted with permission of the Association of Petroleum Geochemical Explorationists).

1988). Herbert recommended drilling a test well based on the fluorescence survey. The well had shows but was plugged and abandoned. The survey was noteworthy for the following reasons: (1) An exploration decision was made based on a survey that did not determine the limits of the anomaly that was to be drilled. (2) An area of dry holes was as anomalous as the prospect. If this dry hole anomaly

was indicative of a halo for the existing field, it was not supported by infill work that was later presented.

Northern Denver Basin

A reconnaissance survey using the K-V fingerprint method was carried out in portions of Townships 5 and 6

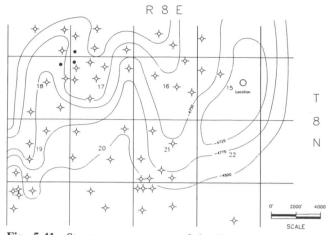

Fig. 5-41 Structure map on top of the Tew Lake Sandstone; contours are, in fact, below sea level (Hebert, 1988). The closure in section 15 is the structure targeted for the fluorescence survey and drilling (reprinted with permission of the Association of Petroleum Geochemical Explorationists).

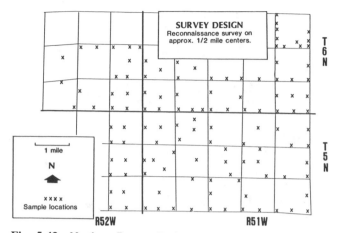

Fig. 5-43 Northern Denver Basin. Location of samples for K-V fingerprint survey (from Klusman et al., 1986; reprinted with permission of the Institute for the Study of Earth and Man, Southern Methodist University).

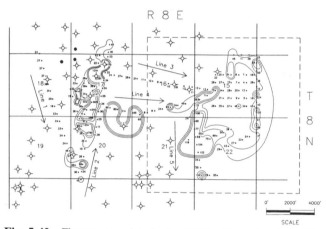

Fig. 5-42 Fluorescence data for the Ballina Plantation (from Hebert, 1988). The Ballina Field is located in the upper right-hand corner and is the only field on the map. Concentrations are in parts per million. Contour intervals were based on statistical analysis (reprinted with permission of the Association of Petroleum Geochemical Explorationists).

North, Ranges 51 and 52 West, Washington and Logan counties, Colorado, in the northern part of the Denver Basin (Figure 5-57; see color plate). Sample locations are shown in Fig. 5-43 and are approximately one-half mile apart. The analog or model was indicated to be 10 miles north of the survey. Production in this portion of the basin is from the Cretaceous D and J sands in association with structural closures and noses. The D sand was specifically targeted in this survey. An analog field was used (Fig. 5-44). Figure 5-45 indicates the anomalous area in the reconnaissance survey that is similar to the analog field (Klusman et al., 1986). Figure 5-46 is the hydrocarbon flux alkanes (C_2–C_{10}) data indicating areas of anomalies similar to the analog area. Maps 7-46 and 7-47 were combined to identify coinciding

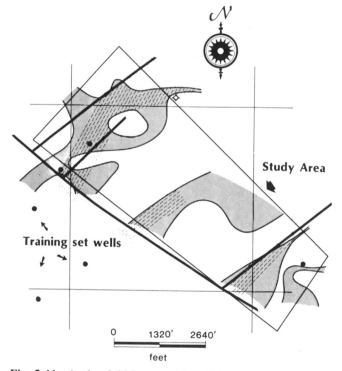

Fig. 5-44 Analog field for use with K-V fingerprint (from Klusman et al., 1986; reprinted with permission of the Institute for the Study of Earth and Man, Southern Methodist University).

anomalous areas. The resulting map is shown in Fig. 5-47. A detailed survey was later carried out to high-grade some of the anomalous areas.

A discovery well was indicated based on the survey results. Figure 5-48 is the interpreted structure on the D sand prior to the discovery; Fig. 5-49 indicates the D sand structure after discovery. However, the well was finally designated as a dry hole. Additional drilling has indicated no production in the area.

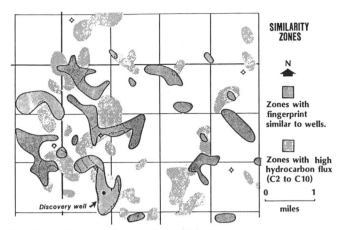

Fig. 5-45 Northern Denver Basin, K-V fingerprint method, similarity map (from Klusman et al., 1986; reprinted with permission of the Institute for the Study of Earth and Man, Southern Methodist University).

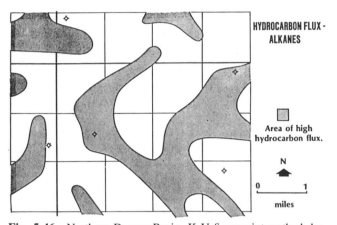

Fig. 5-46 Northern Denver Basin, K-V fingerprint method, hydrocarbon flux alkane map (from Klusman et al., 1986; reprinted with permission of the Institute for the Study of Earth and Man, Southern Methodist University).

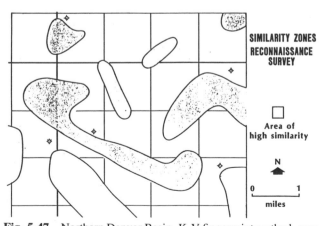

Fig. 5-47 Northern Denver Basin, K-V fingerprint method, combined results from the survey and the discovery well (from Klusman et al., 1988; reprinted with permission of the Institute for the Study of Earth and Man, Southern Methodist University).

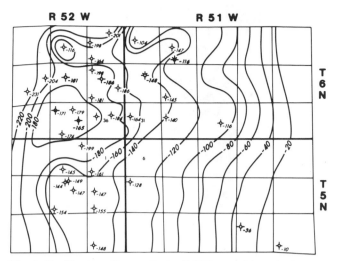

Fig. 5-48 Northern Denver Basin, K-V fingerprint method, structure on top of the D sand prior to drilling.

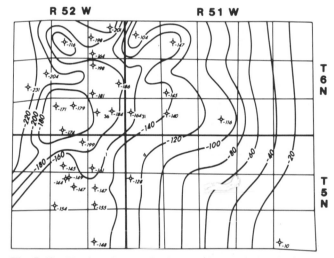

Fig. 5-49 Northern Denver Basin, K-V fingerprint method, structure on top of the D sand post drilling and up to 1991.

In this example, it is difficult to assess the comparative results between the survey and the model because no hard data were presented. Therefore, the pattern of the model could not be compared to that of the survey to determine any critical differences and similarities. The shapes of the anomalies that were identified at the analog field do not seem to fit a regular or identifiable pattern that corresponds in a general sense to the geometries of producing reservoirs in the area. The subsequent drilling is a disappointment, considering all the anomalies present (Fig. 5-47).

Alabama Ferry Field, Texas

The Alabama Ferry Field is located in Leon and Houston counties, Texas, and was discovered in 1983. The field produces from Glen Rose carbonates at 2637 m (8700 ft) to 3182 m (10,500 ft) and has an estimated recovery of 55 MMbbls and 200 BCF (Pollard, 1989). This example was part of a series of single profiles presented by Saunders et al. (1991) to show a correlation between soil-gas and soil magnetic susceptibility measurements. Initially, an airborne hydrocarbon survey was carried out (Fig. 5-50) over all open fields and clearings. No data were presented, just a general outline. A second survey was conducted using a free-air method along a 24-km (15-mile) profile line with a sample spacing of 0.32 km (0.2 of a mile). The soil-gas and magnetic susceptibility (see Chapter 8 for description) results (Fig. 5-51) suggest anomalous conditions in the center of the field. Portions of the survey indicate no soil-gas anomalies over the producing region. The recentness of the discovery, the relatively large size of the field, and Texas state laws imposing allowables on production would preclude depletion of any anomalies in the producing areas. The difficulty with this type of survey is that it represents a small geochemical profile of a very large productive area. Background data are essentially lacking, but this was compensated for by using mathematical methods described by Malmquist (1978). However, using statistics does not compensate for collecting insufficient data to identify the range of background values. It is difficult to ascertain true background values, the actual outline of the anomaly, the resulting intensity of the soil-gas leakage, and the actual association of magnetic susceptibility to the soil-gas method (or any other method) without an extensive data base or a frame of reference.

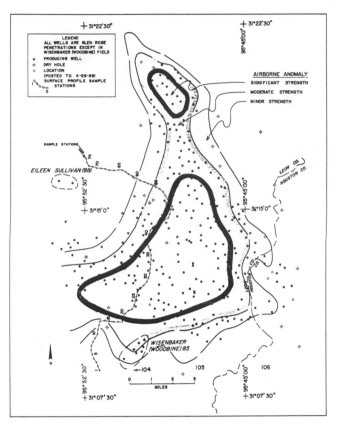

Fig. 5-50 Alabama Ferry Field, Leon and Houston counties, Texas, airborne anomaly and profile location (from Saunders et al., 1991; reprinted with permission of the American Association of Petroleum Geologists.

Lishu Gas Field and an Unnamed Oil Field, Huanghua Depression, People's Republic of China

Two interesting examples of soil-gas and fluorescence surveys were presented for the Lishu Gas Field and an unnamed oil field located in the People's Republic of China (Xuejing, 1992). Fig. 5-52 indicates the location of both fields. The Lishu Gas Field was surveyed using soil-gas and fluorescence methods as well as Delta C, mercury, and iodine. Methane, ethane, and the fluorescence data (Fig. 5-53) indicated anomalies with different areal extents. The methane anomaly was interpreted as a partial halo, but it did not clearly define the field. The ethane anomaly was more apical, covered a larger area than the present production, and is typical of soil-gas anomalies elsewhere. The fluorescence indicated two anomalies, one located on part of the field and the other on trend but in the lower corner of the map. Interestingly, the mercury correlated well with the fluorescence, the iodine with the ethane, and the Delta C with the methane (Fig. 5-54).

The unnamed oil field is located in the Huanghua Depression (Fig. 5-52), and both fluorescence (Fig. 5-55) and Delta C data (Fig. 5-56) were presented. The fluorescence map indicated a large anomaly, with the oil field lying in the

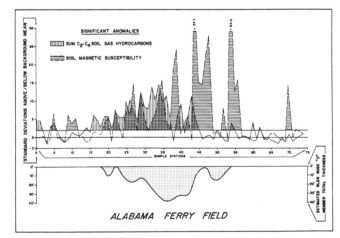

Fig. 5-51 Alabama Ferry Field, Leon and Houston counties, Texas, profile of standardized anomalies of hydrocarbon and magnetic susceptibility data along the profile and the estimated total thickness of the Glen Rose "D" members (Saunders et al., 1991; reprinted with permission of the American Association of Petroleum Geologists).

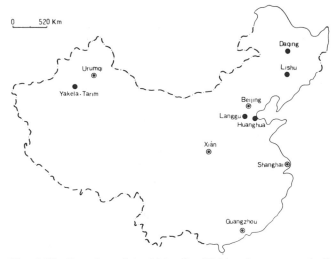

Fig. 5-52 Location of the Lishu Gas Field and an unnamed oil field in the Huanghua Depression, People's Republic of China (after Xuejing, 1992; reprinted with permission of Elsevier Scientific Publication).

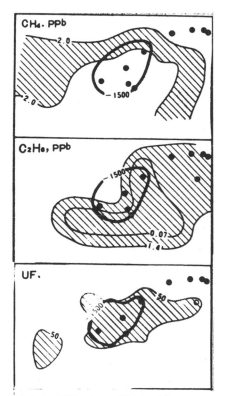

Fig. 5-53 Methane (C_1), ethane (C_2), and fluorescence geochemical maps over the Lishu Gas Field, People's Republic of China (after Xuejing, 1992; reprinted with permission of Elsevier Scientific Publication).

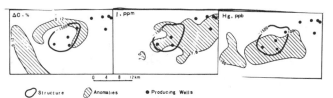

Fig. 5-54 Mercury, iodine, and Delta C maps over the Lishu Gas Field, People's Republic of China (after Xuejing, 1992; reprinted with permission of Elsevier Scientific Publication).

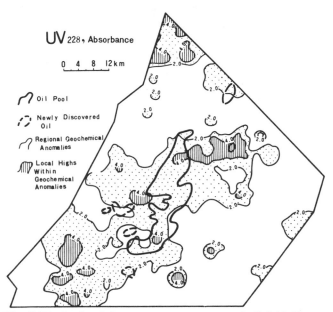

Fig. 5-55 Regional fluorescence map, unnamed oil field, Huanghua Depression, People's Republic of China (after Xuejing, 1992; reprinted with permission of Elsevier Scientific Publication).

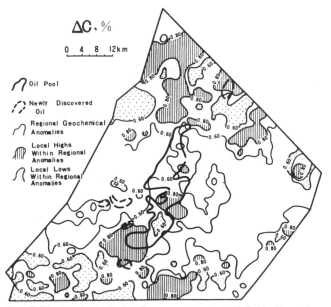

Fig. 5-56 Regional Delta C map, unnamed oil field, Huanghau Depression, People's Republic of China (after Xuejing, 1992; reprinted with permission of Elsevier Scientific Publication).

67

center. Several stronger anomalies are located at the top and lower corner of the map. The anomalies lying over the field and lower corner of the map correspond well. It is unfortunate that the location of sample points and associated values for both examples were not presented.

Silo Field, Laramie County, Denver Basin, Wyoming

The Silo Field is located in Townships 15 and 16 North, Ranges 63, 64, and 65 West, Laramie County, Denver Basin, Wyoming (Fig. 5-57; see color plate). The field was discovered in 1981 and produces from fractures in the Niobrara Shales. Permian salts were removed, and the overlying strata collapsed, which caused the fractures to develop (Fig. 5-58). This was followed by localized generation of petro-

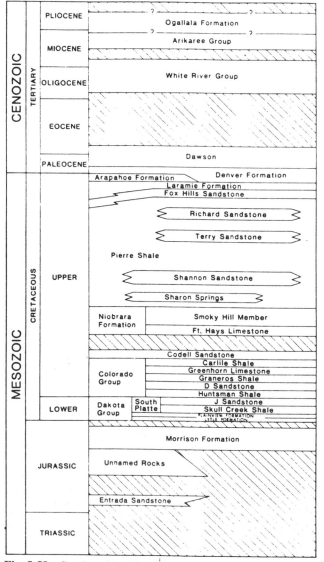

Fig. 5-58 Stratigraphic column for the Cretaceous, Denver Basin, Nebraska (Tainter, 1984; reprinted with permission of the Rocky Mountain Association of Geologists).

leum that filled the fractures (possibly because of an isolated thermal event). This filling caused an increase in radioactivity (Fig. 5-59; see color plate). Production is somewhat erratic (Fig. 5-60; see color plate). A small percentage of wells have produced over 100,000 bbl of oil, but the majority have produced less than 30,000 bbl. In the late 1980s to 1991, there was increased interest and success with horizontal drilling, and the D-J Basin in Wyoming saw several of these types of wells drilled.

A reconnaissance free-air soil-gas survey was carried out in 1990 to determine the viability of this surveying technique in identifying the field and possibly other fields as well. Ethane (Fig. 5-61; see color plate) and propane (Fig. 5-62; see color plate) indicate the results of the reconnaissance survey. The survey was successful in that the technique was able to identify the field. Typically, fracture plays are thought of in terms of identifying one closely spaced group of fractures or a single fracture. However, based on the wide sample spacing, this survey suggests that numerous fractures exist across a large area.

Pipeline Leakage Evaluation, Western United States

As previously discussed, there are other uses for surface geochemical methods used in petroleum exploration. A survey collected 106 soil-gas samples over several pipelines in a town in the western United States. The free-air method was used, and the samples were analyzed by an FID gas chromatograph. Sample locations were controlled to some degree by man-made facilities. Figure 5-63 (see color plate) indicates the results of the survey. Leakage and contamination of the soil section generally indicate linear anomalies associated with the pipelines. The extent of the leakage is minor, but this type of survey can be used as an efficient means of identifying deteriorating sections of a pipeline and allowing remedial action to occur.

XYZ Gas Station Seepage Evaluation, Western United States

The XYZ gas station located in the western United States (the name and location are not identified at the owner's request) was evaluated based on the assumption that the underground tanks were seeping hydrocarbons into the surrounding soil and groundwater. Fifty-one samples were collected using the free-air method. Figs. 5-64 through 5-66 (see color plates) indicate the results of the survey. *n*-Butane (Fig. 5-64) indicates an apical pattern centering around the gas station. The hexane and ratio of the *n*-butane to benzene-toulene (Figs. 5-65 and 5-66; see color plates) indicate similar apical anomalies centered over the gas station. Based on the mapping of the various hydrocarbons, it is evident that leakage is occurring out of the underground storage tanks into the soil and possibly the groundwater. The owner of the station corrected the problem as a result.

Iron Springs Laccolith, Southwestern Utah

The Iron Springs laccolith located in southwestern Utah is related to a thrust fault during Sevier deformation. Van Kooten (1988) presented a study that encompassed seismic, soil-gas (two surveys), and subsequent drilling of an anticline below the laccolith. The rocks in the anticline were found to be overmature, and no hydrocarbon accumulation was found. The study concluded that hydrocarbons were diverted away from the trap and that the mapped seepage is the end product of vented petroleum. Figs. 5-67A and 5-67B represent the first survey of 119 soil-gas samples taken at 5 ft for ethane and butane. Figs. 5-67C and 5-67D represent a later survey of 506 samples for ethane and butane. Both surveys indicate similar strong anomalies in the same area. Seismic mapping indicates that the test well was drilled on the flank and not on top of the structure. Further, if the anomaly was considered a halo, then it was essentially tested. If the anomalies are considered apical, they are untested. There are productive fields in southwestern Utah that are hydrodynamically controlled and lie on the flanks of

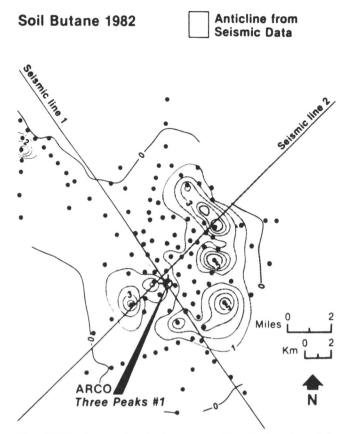

Fig. 5-67B Butane (ppm), first survey, Iron Springs laccolith, southwestern Utah (reprinted with permission of the Geological Society of America).

structures. Therefore, testing the top of the structure may not be the most advantageous option.

There is also a clear relationship between ethane and butane in both surveys. Even though the data are not presented, there seem to be a 5:1 (first survey) and a 7:1 (second survey) ratio of ethane to butane across the entire area. Because there is no distinctive change in the ratios, the concept of increased leakage at the top of the structure is possibly related to increased fracturing. This suggests enhanced normal leakage with no potential significance for indicating a petroleum accumulation. Therefore, with numerous pathways present, the leakage may represent an increase in background rather than a seepage anomaly related to an accumulation or thrusting.

Summary

The most common techniques in surface geochemical exploration have been the direct methods that analyze the actual hydrocarbons that have migrated into the soil. However, hydrocarbons in the vapor phase are difficult to collect and analyze. Many of the problems associated with collecting, analyzing, and interpreting the data can be resolved or

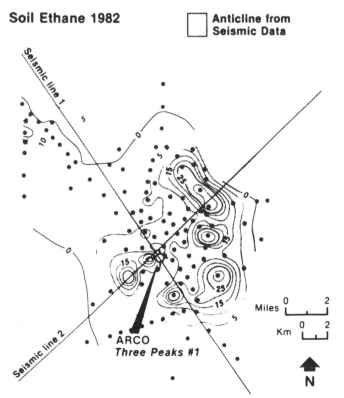

Fig. 5-67A Ethane (ppm), first survey, Iron Springs laccolith, southwestern Utah (Van Kooten, 1988; reprinted with permission of the Geological Society of America).

Soil Ethane 1983   ☐ **Anticline from Seismic Data**

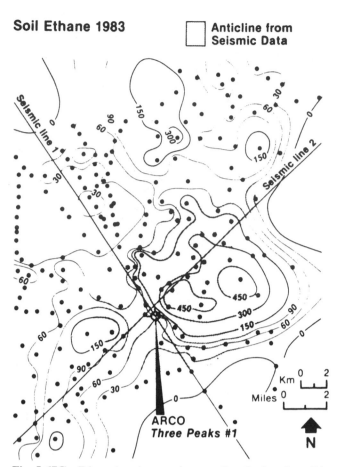

Fig. 5-67C Ethane (ppm), second survey, Iron Springs laccolith, southwestern Utah (reprinted with permission of the Geological Society of America).

Soil Butane 1983   ☐ **Anticline from Seismic Data**

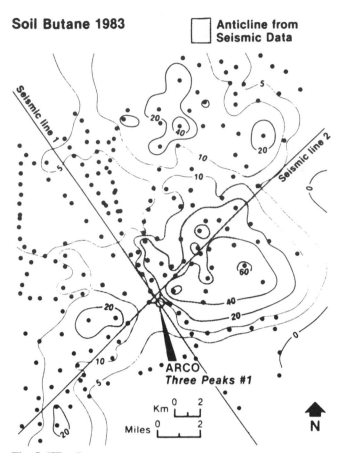

Fig. 5-67D Butane (ppm), second survey, Iron Springs laccolith, southwestern Utah (reprinted with permission of the Geological Society of America).

tempered, and thus the direct methods can be powerful exploration tools. The success of these methods is also a function of addressing possible soil factors that could mislead the explorationist. These factors can be minimized by carefully determining a viable technique for the known soil conditions and then statistically analyzing the data. If these methods cannot be used in a specific area, then they should be discarded without hesitation and an applicable indirect method used.

Selection of the right technique for onshore exploration is critical to implementing a successful survey. The most common methods are free-air and head-space. These methods are the most inexpensive, mobile, and developed techniques available. They are also used in environmental and mining geology, and thus the technology is advancing continually. Other methods, such as fluorescence, acid extraction, deabsorption, K-V fingerprint, and time-delay, seem to have limited application because of costs and time required for sampling. Their acceptance and applicability may improve with increases in data bases and decreases in costs. Other methods such as airborne, water-well, and surface-drainage surveys have little application in the continental United States because of the intense drilling that has already occurred in most areas.

However, they should be useful in remote and unexplored regions. Increasingly, offshore techniques are using bottom-sediment sampling, especially in mature areas. The use of sniffer surveys will continue to decline as all the major and minor seeps are found and drilled.

Interpretation of the data is critical to the success of any surface geochemical program. The interpretation cannot be limited to simple statistical methods or contouring volumes of gas. Instead, the use of ratios both for prospect areas and models needs to be implemented to increase confidence in interpretation.

As discussed in later chapters, the direct methods are more effective when used as a follow-up to indirect methods. To counter the problems of soil conditions and to minimize their effects and the time required to analyze them, direct methods should be limited to prospect detailing. Direct methods cost more than indirect methods. Therefore, direct methods will function more effectively when there is a specific target to delineate because they will be evaluating, and be affected by, only a limited number of influences over a small area. The impact of the soil environment will be minimized, and costs will be less. Reconnaissance surveys using direct methods that have been presented in the literature are difficult to interpret and are very unreliable.

References

Adams, C. W. (1990). Jace and Moore-Johnson Fields, in *Morrow Sandstone of Southeast Colorado and Adjacent Areas,* S. A. Sonnenberg et al. eds., Rocky Mountain Association of Geologists, pp. 157–166.

Arp, G. K. (1992). An integrated interpretation for the origin of the Patrick Draw Oil Field sage anomaly, *American Association of Petroleum Geologists Bulletin,* Vol. 76, No. 3, pp. 301–306.

Arshad, M. and W. T. Frankenberger Jr. (1991). Effects of soil properties and trace elements on ethylene production in soils, *Soil Science,* Vol. 151, No. 5, pp. 377–386.

Barringer, A. R. (1981). Airborne geochemical reconnaissance methods for oil and gas explorers, in *Unconventional Methods of Exploration II,* Southern Methodist University, Dallas, TX, pp. 219–239.

Blanchette, P. L., (1989). Effects of soil characteristics on adsorbed hydrocarbon data, *Association of Petroleum Geochemical Explorationists Bulletin,* Vol. 5, No. 1, pp. 116–138.

Calhoun, G. G. and R. Burrows (1992). Surface fluorescence method can identify potential oil pay zones in Permian Basin, *Oil and Gas Journal,* Pennwell Publications, Sept. 28, pp. 96–100.

Chakhmakhev, V. A. and T. L. Vinogradov (1980). Geochemical methods of predicting oil and gas at depth, *Neftegazovaya Geologiya i Geofizika,* No. 6, pp. 8–12.

Davis, J. B., 1967. *Petroleum Microbiology,* Elsevier Publishing Co., New York.

Debnam, A. H. (1969). Geochemical prospecting for petroleum and natural gas in Canada, *Geological Survey of Canada Bulletin,* No. 177.

Devine, S. B. and H. W. Sears (1975). An experiment in soil geochemical prospecting for petroleum. Della Gas Field, Cooper Basin, Australia *Petroleum Exploration Association Journal,* Vol. 15, pp. 103–110.

Devine, S. B. and H. W. Sears (1977). Soil hydrocarbon geochemistry, a potential petroleum tool in the Cooper Basin, Australia, *Journal of Geochemical Exploration,* Vol. 8, pp. 397–414.

Dickinson, R. G. and M. D. Matthews (1991). A regional microseep survey of the Wyoming-Utah Overthrust Belt (Abstract), *American Association of Petroleum Geologists Bulletin,* Vol. 75, pp. 562.

Dunlap, H. F., J. S. Bradley, and T. F. Moore, (1960). Marine seep detection—A new reconnaissance exploration method, *Geophysics,* Vol. 25, pp. 275–282.

Dunlap, H. F. and C. A. Hutchinson (1961). Marine seep detection, *Offshore,* Vol. 21, pp. 11–12.

Faber, E. and W. Stahl (1984). Geochemical surface exploration for hydrocarbons in North Sea, *American Association of Petroleum Geologists Bulletin,* Vol. 68, pp. 363–386.

Fausnaugh, J. M. (1989). The effect of high soil conductivity on headspace gas sampling techniques, *Association of Petroleum Geochemical Explorationists Bulletin,* Vol. 5, No. 1, pp. 96–115.

Filatov, S. S., V. P. Beloglazov, S. V. Vil'tovskaya, and I. B. Stepanenko (1982). Gas geochemical studies of stream sediments in oil-gas exploration on the Siberian Craton, in *Problemy geologii nefti i gaza Sibirskoy, VNIGRI,* Leningrad, pp. 91–100. Translation in *Petroleum Geology,* McLean, VA, 372–375.

Fisher, J. L. (1986). Hydrocarbon distribution patterns over uranium deposits: Implications for surface petroleum geochemical prospecting techniques, *Association of Petroleum Geochemical Exploration,* Vol. 2, pp. 21–36.

Fontana, J. (1988). The time delayed free (interstitial) soil gas method, *Association of Petroleum Geochemical Exploration Newsletter* No. 3, Sept. pp. 4–6.

Hebert, C. F. (1984). Geochemical prospecting for oil and gas, using hydrocarbon fluorescence techniques, in *Unconventional Methods in Exploration for Petroleum and Natural Gas III,* Southern Methodist University, Dallas, TX, pp. 40–58.

Hebert, C. F. (1988). Ballina Plantation Prospect, Louisiana: A case study in fluorescence geochemical exploration, *Association of Petroleum Geochemical Explorationists Bulletin,* Vol. 4, No. 1, pp. 102–119.

Heemstra, R. J., R. M. Ray, T. C. Wesson, J. R. Abrams, and G. A. Moore (1979). *A Critical Laboratory and Field Evaluation of Selected Surface Prospecting Techniques for Locating Oil and Natural Gas,* BETC/RI-78/1B, U.S. Dept. of Energy.

Horvitz, L. (1969). Hydrocarbon geochemical prospecting after 30 years, in *Unconventional Methods in Exploration for Petroleum and Natural Gas,* Southern Methodist University, Dallas, TX, pp. 205–218.

Horvitz, L. (1972). Vegetation and geochemical prospecting for petroleum, *American Association of Petroleum Geologists Bulletin,* Vol. 56, pp. 925–940.

Horvitz, L. (1981). Hydrocarbon geochemical prospecting after forty years, in *Unconventional Methods in Exploration for Petroleum and Natural Gas II,* Southern Methodist University, Dallas, TX, pp. 83–95.

Horvitz, L., (1985). Geochemical exploration for petroleum, *Science,* Vol. 229, pp. 821–827.

Horvitz, E. P. and S. Ma. (1988) Hydrocarbons in near-surface sand, a geochemical survey of the Dolphin Field in North Dakota, *Association of Petroleum Geochemical Explorationists Bulletin,* Vol. 4, No. 1, pp. 30–61.

Hulen, J. B., S. R. Bereskin, and L. C. Bortz (1990). High temperature hydrothermal origin for fractured carbonate reservoirs in the Blackburn Oil Field, Nevada, *American Association of Petroleum Geologists Bulletin,* Vol. 74, pp. 1262–1272.

Jaacks, J. (1991). Using pyrolysis mass spectrometry in mineral exploration, presented at the December meeting of the Geochemical Exploration Discussion Group, Golden, Co.

Jones, V. T. and R. J. Drozd, (1983). Prediction of oil and gas potential by near-surface geochemistry, *American Association of Petroleum Geologists Bulletin,* Vol. 67, pp. 932–952.

Kartsev, A. A., Z. A. Tabasaranskii, M. I. Subbota, and G. A. Mogilevskii (1959). *Geochemical Methods of Prospecting and Exploration for Natural Gas,* trans., P. A. Witherspoon and W. D. Romey, University of California Press, Berkeley, CA.

Klusman, R. W., K. J. Voorhees, J. C. Hickey, and M. J. Malley (1986). Application of the K-V fingerprint technique for petroleum exploration, in *Unconventional Methods in Exploration for Petroleum and Natural Gas IV*, M. J. Davidson ed., Southern Methodist University Press, Dallas, TX, pp. 219–243.

Landrum, J. H., D. M. Richers, L. E. Maxwell, and W. Fallgater (1989). A comparative hydrocarbon soil gas study of the Flathead Valley and Townsend Valley areas, Montana, *Journal Geochemical Exploration*, Vol. 34, pp. 303–335.

Malmquist, L. (1978). An interactive regression analysis procedure for numerical interpretation of regional exploration geochemical data, *Journal of Mathematical Geology*, Vol. 10, pp. 23–42.

McCrossan, R. G., N. L. Ball, and L. R. Snowden (1972). An evaluation of surface geochemical prospecting for petroleum, Olds-Carolina Area, *Alberta Geological Survey of Canada*, Dept. of Energy, Mines & Resources, Paper 71-31.

Pollard, N. (1989). Alabama Ferry (Glen Rose) field, in *Occurrence of Oil and Gas in Northeast Texas*, P. W. Shoemaker, ed., East Texas Geological Society, Tyler, TX, pp. 1–14.

Price, L. (1976). A critical overview and proposed working model of surface geochemical exploration, in *Unconventional Methods in Exploration for Petroleum and Natural Gas IV*, M. J. Davidson, ed., Southern Methodist University Press, Dallas, TX, pp. 245–304.

Richers, D. M., V. T. Jones, M. D. Matthews, J. Maciolek, R. J. Pirkle, and W. C. Sidle (1986). The 1983 Landsat soil-gas geochemical survey of Patrick Draw area, Sweetwater County, Wyoming, *American Association of Petroleum Geologists Bulletin*, Vol. 70, pp. 869–887.

Richers, D. M. and L. E. Maxwell (1991). Application and theory of soil gas geochemistry in petroleum exploration, in *Source and Migration Processes and Evaluation Techniques*, N. H. Foster and E. A. Beaumont, eds., *American Association of Petroleum Geologists*, Tulsa, OK, pp. 141–158.

Richers, D. M., R. J. Reed, K. C. Horstman, G. D. Michels, R. N. Baker, L. Lundell, and R. W. Marrs (1982). Landsat and soil-gas geochemical study of Patrick Draw oil field, Sweetwater County, Wyoming, *American Association of Petroleum Geologists*, Vol. 66, pp. 903–922.

Richers, D. M. and C. Weatherby (1985). A continued study of the Patrick Draw test site, Sweetwater County, Wyoming: *4th International Symposium on Remote Sensing of the Environment*, Proceedings, San Francisco, CA.

Sackett, W. M., (1977). Use of hydrocarbon sniffing in offshore exploration, *Journal of Geochemical Exploration*, Vol. 7, pp. 243–254.

Saenz, G., H. Pannell, and N. E. Pingitore (1991). Recent advances in near-surface organic geochemical prospecting for oil and gas, in *Source and Migration Processes and Evaluation Techniques*, N. H. Foster and E. A. Beaumont, eds., *American Association of Petroleum Geologists*, Tulsa, OK, pp. 135–140.

Saunders, D. F., K. R. Burson, and C. K. Thompson (1991). Observed relation of soil magnetic susceptibility and soil gas hydrocarbon analysis to subsurface accumulation, *AAPG*, Vol. 75, pp. 389–408.

Schumacher, D. (1992). Surface exploration for oil and gas: Advances of the eighties, applications for the nineties, Short Course sponsored by the Rocky Mountain Association of Geologists, Denver CO., Jan. 27.

Scott, C. and A. K. Chamberlain (1988). Blackburn Field, Nevada, a case history, in *Occurrence and Petrophysical Properties of Carbonate Reservoirs in the Rocky Mountain Region*, S. M. Goolsby and M. W. Longman, eds. Rocky Mountain Association of Geologists, pp. 241–250.

Scott, L. F., R. M. McCoy, and L. H. Wullstein (1990). Anomaly may not reflect hydrocarbon seepage: Patrick Draw Field, Wyoming Revisited, *American Association of Petroleum Geologists Bulletin*, Vol. 73, pp. 925–934.

Sweeney, R. E., (1988). Petroleum-related hydrocarbon seepage in a recent North Sea sediment, *Chemical Geology*, Vol. 71, pp. 53–64.

Tainter, P. A. (1984). Stratigraphic and paleostructural controls on hydrocarbon migration in Cretaceous D and J sandstones of the Denver Basin, In (eds.), J. F. Woodward, F. Meissner, and J. Clayton *Hydrocarbon Source Rocks of the Greater Rocky Mountain Region*, Rocky Mountain Association of Geologists, Denver, CO, pp. 339–354.

Ullom, W. L. (1988). Ethylene and propylene in soil gas: Occurrence, sources and impact on interpretation of exploration geochemical data, *Association of Petroleum Geochemical Explorationists Bulletin*, Vol. 4, No. 1, pp. 62–81.

Van Kooten, G. K. (1988). Structure and hydrocarbon potential beneath the Iron Springs laccolith, Southwestern Utah, *Geological Society of America Bulletin*, Vol. 100, pp. 1533–1540.

Weismann, R. (1980). Development of geochemistry and its contribution to hydrocarbon exploration, *Proceedings of the 10th World Petroleum Congress*, Heyden Publishers, London, pp. 369–385.

Weissenburger, K. S., (1991). Pitfalls and caveats in surface light hydrocarbon surveying (abstract), *American Association of Petroleum Geologists Bulletin*, Vol. 75, p. 691.

Xuejing, X. (1992). Local and regional surface geochemical exploration for oil and gas, *Journal of Geochemical Exploration*, Vol. 42, pp. 25–42.

Yasenev, B. P. (1986). New data on direct geochemical methods of prospecting for oil and gas, Vol. 3, *Petroleum Geology*.

6

Radiometrics

Introduction

The use of radiometrics predates the use of other geochemical methods. The technique dates back to at least 1926, when the Soviets observed anomalously low gamma counts over existing oil fields (Armstrong and Heemstra, 1972, 1973). The U.S. petroleum industry began applying radiometric methods about 1943. Since that time, there have been numerous articles and increasing curiosity and application by the industry, especially during widespread economic recession. When petroleum prices are low, the industry has sought inexpensive exploration methods, especially radiometrics. This method has proved that it can be an excellent tool for petroleum exploration, but environmental influences and a preponderance of primitive collection techniques have continued to cause confusion and have limited its acceptance. The consistent association between soil-gas anomalies and radiometric anomalies has never been clearly explained. Only a small percentage of all radiometric anomalies found have been confirmed by soil-gas techniques as being related to petroleum seepage.

Radiometrics detects changes in the total gamma-ray count at the surface. The changes result from the radioactive decay of the uranium-radium, potassium, and thorium-radium series. By definition, radiometrics is a geophysical measurement and technically should be addressed in that context. In practice, little work has been done by the geophysical community to relate these phenomena to microseepage or to explain the response associated with petroleum accumulations. Conversely, numerous articles discussing radiometrics have appeared under the heading of surface geochemistry. Radiometric anomalies have been intimately associated with surface geochemical anomalies caused by microseepage, and this relationship will be discussed.

In 1926, early Soviet investigators employed radiometrics for petroleum exploration by using an electrometer to measure the gamma emissions of the Maikopsky oil region. This area and the Ukhta oil-bearing region were explored with a new 2150-cm^3 electrometer (Bogoyavlenskiy, 1927). The Maikopsky Field lies in a tectonically undisturbed area and, in the 1920s, the land was forested and had no logging. The Ukhta Field is located in a great anticlinal fold. Bogoyavlenskiy observed increases in the intensity of gamma radiation above oil reservoirs containing both light and heavy crudes. Bogoyavlenskiy concluded that the oil had absorbed large amounts of radium and gave off "emanations" through the earth.

Following Bogoyavlenskiy's reports, other papers were published before 1956 on the "radioactivity halo" phenomenon. After 1956, instrumentation became increasingly sophisticated. Many investigators began addressing the impact of source and environmental influences on radiometrics. Alekseev (1957) outlined the state of the art at that time. A review of radon migration by Tanner (1960) included environmental and petroleum exploration radiometrics. Sikka (1963) discussed radiometric surveys and the many theories on radiation "halo" anomalies around oil and gas fields. General texts on radiometrics by Broda and Schonfeld (1966) and Durrance (1986) discussed the method for prospecting in their list of applications. In the Soviet Union, Polshkov (1965) recognized radiometrics as a possible addition to other tools for structural mapping. A comprehensive article by Pirson et al. (1966) on geochemical concepts contributed much to an understanding and awareness of this unconventional exploration method. Foote (1969B) and Weart and Heimberg (1981) presented updates, but there was little new information on advances in the technology since 1960s.

Geochemistry/Natural Radioactivity

Natural Radioactivity

The radioactive daughter products generating radiation emissions are generally short-lived but are continuously regenerated by their long-lived parent. The initial nuclide is called the *parent*, and the product of transformation is called the *daughter*. A parent radionuclide can produce a daughter product that will in itself undergo continual transformation until a stable daughter product is achieved. This chain of parent and daughter products is called a *decay series*. The occurrence of naturally unstable nuclei that undergo spontaneous transformation by decay is not common. Radionuclei are held together by strong forces, and it takes an excess of neutrons to cause the decay process to occur. The number of neutrons that can be in excess varies, and many radionuclei have minor excesses without undergoing decay. When decay occurs, it is because the nucleus becomes unstable as a result of neutron excess, and the electrostatic repulsion between the protons overcomes the binding forces. The binding energy is the energy that must be exceeded to decompose the radionuclide into its component protons and neutrons.

The radioactive emissions derived from the decay chain are divided into alpha, beta, and gamma radiation. The length of time during which decay occurs varies with the energy of the emitted particle in the decay process. Alpha particles tend to be large and to lose energy easily. Beta particles, which are much smaller, tend to ricochet off other objects and do not lose their energy as rapidly. Gamma radiation occurs after the emission by alpha- and beta-particle radiation has ceased because the energy remaining is insufficient to generate these types. Until the particle reaches a stable form, it will continue to emit gamma radiation.

Cosmic rays are charged particles that enter the earth's atmosphere and whose collisions cause a cascade of gamma radiation. They consist mainly of protons with minor amounts of alpha particles. These particles are undergoing decay, and they add to the radiation emissions measured at the earth's surface during a radiometric survey.

Natural Radioactivity Sources

The three principal sources of natural radioactivity in the earth are potassium 40, the uranium-radium series, and the thorium-radium series (Figs. 6-1–6-3). The migration and fixation of various members of these series are responsible for natural radioactivity originating from the earth. The relative contribution of each species to the total intensity of radioactivity is influenced by its half-life, the type of radioactive emissions, and the energy release of subsequent radia-

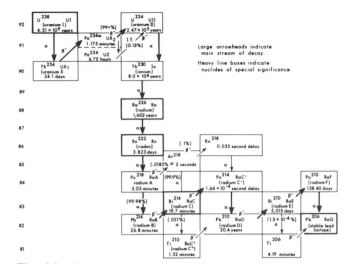

Fig. 6-2 Uranium-radium decay series. Large arrows indicate main decay routes, and heavily outlined boxes indicate nuclides of significance (Armstrong and Heemstra, 1073).

tion (i.e., intensity). The radiation energies, intensities, and half-lives for the uranium series are given in Table 6-1 and those for the thorium series in Table 6-2.

Uranium and thorium series nuclides are relatively large in molecular size compared to bismuth, thallium, and potassium nuclides. Potassium 40 generally constitutes 0.012% of the total potassium present. Potassium is widely distributed and is an essential element of most life-forms. The original concept for the existence of radiometric anomalies relied on microseepage that brought not only hydrocarbons to the surface but radioactive elements and their daughter products as well (Crews, 1959; Collins, 1963; Armstrong and Heemstra, 1973). The size of the uranium molecule alone would reject this concept. However, the close relation-

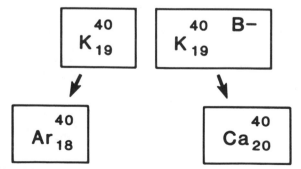

Fig. 6-1 Potassium decay series.

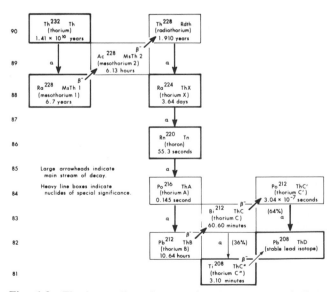

Fig. 6-3 Thorium-radium decay series. Large arrows indicate main decay, and heavily outlined boxes indicate nuclides significance (Armstrong and Heemstra, 1973).

Table 6-1 Uranium series: energy, intensities, and half-lives

Nuclide	Half-life	Decay	α Energy (MeV)	Amount (%)	β Energy (MeV)	Amount (%)
^{238}U	4.468×10^9 years	α	4.196	77		
	Spontaneous fission		4.149	23		
	8.19×10^{15} years					
^{234}Th	24.1 days	β^-			0.199	73
					0.104	21
					0.060	5
					0.022	1
^{234m}Pa	1.18 min	β^-			2.290	98
^{234}Pa	6.7 h	β^-			1.190	5
					0.680	19
					0.512	63
					0.280	12
^{234}U	2.48×10^5 years	α	4.774	72		
	Spontaneous fission		4.723	28		
	2.10^{16} years					
^{230}Th	7.52×10^4 years	α	4.689	76		
	Spontaneous fission		4.621	24		
	$>1.5 \times 10^{17}$ years					
^{226}Ra	1602 years	α	4.785	94		
			4.602	6		
			5.490	1006		
^{222}Rn	3.825 days	α				
^{218}Po	3.05 min	α				
		(99.98%)	6.002	100		
		and				
		β^-			0.330	100
^{214}Pb	26.8 min	β^-			1.030	100
^{218}At	2 s	α				
		(99.9%)	6.757	4		
			6.694	90		
			6.654	6		
		and				
		β^-			0.670	100
^{214}Bi	19.7 min	α				
		(0.04%)	5.512	40		
			5.448	54		
			5.268	6		
		and				
		β^-			3.260	19
					1.880	9
					1.510	40
					1.020	23
					0.420	9
^{214}Po	1.64×10^{-4} s	α	7.687	100		
^{210}Tl	1.32 min	β^-			2.340	19
					1.870	56
					1.320	25
^{210}Pb	22.3 years	α				
		(<1%)	3.72	—		
		β^-			0.610	19
					0.015	81

Continued

Table 6-1 *Continued*

Nuclide	Half-life	Decay	α Energy (MeV)	Amount (%)	β Energy (MeV)	Amount (%)
^{210}Bi	5. 02 days	α				
		(<1%)	4.687	40		
		and	4.650	60		
		β^-			1.161	100
^{210}Po	138.3 days	α	5.305	100		
^{206}Tl	4.19 min	β^-			1.527	100
^{235}U	7.13×10^8 years	α	4.397	57		
	Spontaneous fission		4.598	5		
	3.5×10^{17} years		4.557	4		
			4.503	1		
			4.416	4		
			4.397	57		
			4.367	18		
			4.345	2		
			4.324	3		
			4.217	6		
^{231}Th	25.64 h	β^-			0.302	52
					0.218	20
					0.138	22
					0.090	6
^{231}Pa	3.43×10^4 years	α	5.058	11		
			5.031	3		
			5.029	20		
			5.012	25		
			4.985	1		
			4.951	23		
			4.934	3		
			4.852	1		
			4.737	8		
			4.713	1		
			4.681	2		
^{227}Ac	22 years	α				
		(1.38%)	4.951	47		
			4.938	40		
			4.870	6		
			4.853	4		
			4.766	1		
					0.046	100
^{227}Th	18.17 days	α	6.038	25		
		and	6.009	3		
		β^-	5.978	23		
			5.960	3		
			5.867	2		
			5.808	1		
			5.755	21		
			5.713	5		
			5.709	8		
			5.701	4		
			5.693	2		
			5.668	2		
^{223}Fr	21 min	α				
		(<0.01%)	5.340	—		
		and				
		β^-			1.120	100

Continued

Table 6-1 *Continued*

Nuclide	Half-life	Decay	α Energy (MeV)	Amount (%)	β Energy (MeV)	Amount (%)
^{223}Ra	11.68 days	α	5.748	9		
			5.717	54		
			5.608	26		
			5.540	9		
			5.434	2		
^{219}At	0.9 min	α (97%) and β$^-$	6.28	100	—	—
^{219}Rn	3.92 s	α	6.819	81		
			6.553	11		
			6.425	8		
^{215}Bi	8 min	β$^-$			—	—
^{215}Po	1.83×10^{-3}s	α	7.386	100		
^{211}Pb	36.1 min	β$^-$			1.355	92
					0.525	6
					0.251	1
^{211}Bi	2.16 min	α (99.73%) and β$^-$	6.623	84	—	—
			6.279	16		
^{211}Po	0.52 s	α	7.450	99		
^{207}Tl	4.79 min	β$^-$			1.431	100

After Durrance, 1986, Ivanovich, 1982; Lederer and Shirley, 1978; reprinted with permission of John Wiley & Sons.

ship between uranium-thorium-potassium and their daughter products and petroleum has suggested this possibility. No empirical evidence currently exists to suggest that uranium in the soil is a result of vertical migration. If radioactive elements were migrating vertically, apical radiometric anomalies would be more likely expressions than the halos that are typically observed. Bogoyavlenskiy (1927) did find increases in emissions over fields rather than lows in the Maikopsky oil region in the Soviet Union.

The removal of uranium from solution to insoluble form occurs on the surface layer of the insoluble organic phase. While uranium is reduced in valency to U^{4+}, the organic matter involved (peat, wood, cellulose, oil shale kerogen) is oxidized, and new reactive groups are formed to complex with more uranium. Although most uranium-organic associations are considered unstable, the complexes of uranium with ethylene, propylene, butylene, and other organics are thought to be stable.

Berger and Deul (1956) outlined the importance of the organo-uranium association in the geochemical cycle of uranium. Although petroleum from nonuraniferous areas may contain very little uranium, petroleum still has the ability to transfer uranium from its original site to another. Analysis of the organic matrix of uranium ore points to a humic acid origin. When metals are present, they hydrolyze uranium and related daughter products easily into solution. Various authors have proposed that uranium is carried in a colloidal suspension by groundwater to the deposition site. The effects of radioactivity on organics are well documented, especially in relation to fatty acids. Numerous organic shales are highly radioactive, such as the Pennsylvanian black shales of the Atoka, Cherokee, and Marmaton groups in the Illinois and Western Interior basins. However, radioactive minerals are not always found in highly organic shales.

Petroleum will remove uranium from groundwater very efficiently, forming insoluble organometallic complexes. This process has been found to be irreversible. Whether in the uranous or uranyl ionic form, uranium is easily absorbed onto organic and inorganic materials because of its large size and high density. Absorption of uranium onto various clays, such as kaolinite, illite, and montmorillonite, is less common than on organic matter. Clay layers usually do not allow migration or diffusion of uranium-rich fluids through them because of the clay's ability to absorb most of the uranium present. Therefore, the likelihood of uranium migrating vertically from any depth is not strong because of the absorptive ability of clays and organic matter.

Radium and uranium typically form different organometal

Table 6-2   Thorium series: energy intensities and half-lives

Nuclide	Half-life	Decay	α Energy (MeV)	Amount (%)	β Energy (MeV)	Amount (%)
^{232}Th	1.39×10^{10} years	α	4.016	77		
	Spontaneous fission $>10^{21}$ years		3.957	23		
^{228}Ra	5.75 years	β^-			0.039	60
					0.015	40
^{228}Ac	6 .13h	β^-			1.850	10
					1.700	7
					1.110	53
					0.640	7
					0.450	13
^{228}Th	1.913 years	α	5.421	73		
			5.339	27		
^{224}Ra	3.64 days	α	5.686	95		
			5.449	5		
^{220}Rn	55.6 s	α	6.288	100		
^{216}Po	0.145 s	α	6.779	100		
^{212}Pb	10.64 h	β^-			0.569	12
					0.331	83
					0.154	5
^{212}Bi	60.5 min	α (36%)	6.090	27		
			6.051	70		
			5.769	2		
			5.607	1		
		and				
		β^-			2.270	63
					1.550	10
					0.930	8
					0.670	6
					0.450	8
					0.085	5
^{212}Po	3.04×10^{-7} s	α	8.785	100		
^{208}Tl	3.1 min	β^-			1.796	52
					1.521	22
					1.274	21
					0.640	5

After Durrance, 1986; Ivanovich, 1982; Lederer and Shirley, 1978; reprinted with permission from John Wiley & Sons.

complexes. As a result, one type of complex may be enriched and the other depleted, and both may show no relationship to any particular petroleum accumulation. However, it has been observed that an increase in radium exhibits a linear relationship with petroleum as a function of increasing age of the strata.

Mobility of Uranium and Thorium

Uranium is transported as colloids, as emulsoids, or through diffusion in groundwater (Boyle, 1982). The presence and overall mobility of metal complexes greatly in-

crease in the presence of uranium. Langmuir (1978) has proposed that uranium's mobility is controlled by the following factors:

1. The uranium content of source rocks, sediments, and soils and the ease with which they are leached

2. The distance of groundwater from the uranium source

3. The presence of highly sorptive compounds, such as petroleum; organic matter; oxhydroxides of Fe^{3+}, Mn, and Ti; and clays

4. The Eh, pH, salinity, temperature, pressure, and oxidation state of the water

5. Climatic effects, seasonal variations, and rate of water evaporation

6. The concentration of suitable complexing agents that lead to the formation of insoluble uranium minerals

7. The degree to which the water involved is isolated hydrodynamically from areas of groundwater mixing

Thorium and uranium form hydrated cations of UO^{2+} and Th^{4+} and are responsible for the solubility of these metals over a wide range of pH (Harmsen and de Haan, 1980). Several organic acids increase the solubility of thorium and uranium. However, uranium and thorium mobility is limited as a result of absorption on clays and organic matter and less soluble phosphates and oxides. Uranium is probably transported in the form of the uranyl ion. In the presence of silicon and pitchblende, potentially economic sedimentary deposits of U_3O_8 form. Pitchblende, which is a derivative of petroleum, accumulates radioactive elements (Brookins, 1988). However, a multitude of uranium compounds may be precipitated out, depending on Eh-pH conditions and the absence or presence of carbonates, iron sulfides, and iron oxides.

Thorium is generally absorbed onto clays or organic matter more effectively than uranium. Thorium usually forms ThO_2 under most Eh and pH conditions (Brookins, 1988). Other forms of Th will be present, such as $Th(OH)_4$, but they will typically age to ThO_2. When the pH falls below 3, ThO_2 will dissolve to form $ThSO^{4+}$. Thorium carbonates are not very common.

Reducing conditions are usually caused by the presence of hydrocarbons, hydrogen sulfide, or unstable sulfur compounds. Both Eh and pH are considered important factors in the transport of uranium by groundwater movement. Under reducing conditions, uranium is believed to migrate from water and be absorbed by petroleum moving through the rocks, which explains the very high gamma activity measured at the oil/water contact in producing wells. Because of the ability of thorium and uranium to substitute for each other, the above-mentioned factors apply to thorium as well. Most uranium minerals are precipitated under oxidizing conditions. Migrating hydrocarbons in the soil cause a reducing environment, which thus increases the solubility of uranium. This allows its removal in the near-surface by both surface and groundwater movement.

Radiation Anomalies

Positive radioactive anomalies near oil deposits were considered important prospecting clues in the outer zone of the East Carpathian Downwarp (Shchepak, 1964; Schchepak and Migovich, 1969). The higher ratios of radium content to salinity for a particular formation correlated well with waters from natural gas and petroleum deposits. No correlation was found with waters above or below these accumula-

tions. Mal'skaya (1965) disagreed with Shchepak (1964) on this correlation, except with respect to certain gas deposits. The increase in radium content of the groundwater horizon near the oil/water contact was credited to the high grade of metamorphism associated with the reducing environment of deep, stagnant waters in isolated horizons (Alekseev and Gottikh, 1965). Some surface waters showed an increase in radium and a decrease in uranium concentrations over the productive centers of the fields (Alekseev et al., 1960) but, generally, radium was not observed in the surface water within the oil and gas region.

The radiometric anomalies detected over a petroleum accumulation represent a phenomenon that is present only in the upper few millimeters of the soil. Whether these anomalies extend down to the petroleum accumulation has not been investigated. However, several researchers have documented increases in shale radioactivity in the strata overlying petroleum accumulations. These increases have probably been used as an exploration tool, but Pirson et al. (1966) reported the opposite—radiation lows in the shales immediately overlying petroleum deposits. The latter seems to be the exception and not the rule.

Radium activity anomalies (often misconstrued as radon anomalies) can inaccurately reflect geologic features under thick overburden because of displacement by groundwater movement unless the features are very large (Tanner, 1960). The radiation halos observed on the surface cannot have been produced solely by groundwater carrying dissolved radium salts up from the vicinity of the accumulated petroleum and associated brines. Rothe (1959) calculated that only 2% of the radium migrating from "X deposit" at a depth of 242 m reached the surface. However, all the uranium reached the surface.

Causes of Radiometric Anomalies

Association with Hydrocarbons

The history of radioactivity measurements associated with produced oil dates back to Bogoyavlenskiy (1927), who determined an inverse relationship of ash radioactivity to ash content in petroleum. The presence of uranium has been demonstrated to be genetically related to the organic matter in uranium-rich shales. Uranium enrichment increases under a low sedimentation rate, which allows more organic matter to accumulate.

High concentrations of uranium are found in association with asphalt and petroleum deposits. The relationship is not just chemical scavenging of uranium by petroleum. There is a strong link between radioactivity and the generation or sourcing of hydrocarbons. The association of uranium with petroleum can be categorized as follows (Bloch and Key, 1981):

1. Groundwater intrusion and water washing of a petroleum reservoir containing uranium may lead to precipitation there.

2. Microbiological oxidation of vertically migrating hydrocarbons results in the formation of solid bitumen. The bitumen becomes enriched in uranium and other metals.

3. Hydrogen sulfide generated from petroleum will reduce uranium in aqueous solution, causing precipitation, along with other metal compounds.

Several investigators have advanced the concept that radioactive anomalies form because of the direct presence of hydrocarbons in the soil or rock. Sikka (1962A, 1962B, 1963) theorized a linear system of fracture patterns beginning near the flanks of an anticlinal structure, progressing upward, and emerging at the surface away from the petroleum accumulation. These fractures, it was suggested, would provide easy outlets to rising hydrocarbons, groundwater, and radioactive elements and compounds.

Foote (1969A, 1969B) indicated that a radiation halo depends directly on the presence of the hydrocarbons in the soil and in the lower formations overlying the petroleum accumulations. Sikka (1962A, 1962B, 1963) and Merritt (1952) observed oil from wells drilled on the edge of a field and those drilled near the center, and they proposed temperature differences to explain the formation of the soil-gas halo anomalies and possibly also the radiometric anomaly.

Alekseev and Gottikh (1965) believed that the radiation anomaly was related to the progressively developing structure that may or may not have accumulated a petroleum deposit during the course of its formation. Their studies suggested that the radiation anomalies were not directly related to the presence of petroleum, and they demonstrated that this same halo is present around the very promising but, at that time, dry Schhelkovo structure. (Did they drill deep enough?) Alekseev indicated that changes in grain size were associated with the presence of the radiometric anomaly. Thus, Alekseev painted a convincing picture of the origin and explanation of the so-called oil field halo. However, this interpretation was not convincing enough for Pirson et al. (1966), who questioned the unproductiveness of the dry structure and its history of petroleum accumulations. They also questioned the actual grain size variation over the studied structures, which is the basis for Alekseev's explanation of the radiation anomaly.

Gregory (1956) maintained that no correlation exists between radioactivity anomalies and deeply buried oil reservoirs. Instead, a direct correlation of radioactivity to the soil types was pointed out. However, in early attempts to reconcile radiation data to geochemical data, today's sophisticated radiation counting techniques were not available.

Pirson et al. (1966) and Pirson (1969) viewed radiometric anomalies as originating from mineralization phenomena that occur during vertical water leakage from the formation.

The leakage is due to consolidation and compaction of sediments as the petroleum accumulates in geologic traps. In other words, a funnel of mineralization containing radioactive elements is formed by the vertical escape of compaction-expelled waters during oil formation and migration. These waters carry suspended hydrocarbons upward, bypassing the oil and gas accumulations and forming a chimney of reduced rocks extending to the earth's surface. These chemically reduced zones were identified by Pirson (1969) as the cause of certain telluric currents that can be used to help locate oil. Pirson felt that the radioactivity lows form during the actual formation of the trap and accumulation of the petroleum. These lows indicate the presence of primary hydrocarbon accumulations at depth but are not expected to be associated with secondary deposits where petroleum has migrated from the original location. Also, it has been observed that the radiometric halo does not disappear after the oil has left the region. According to Pirson's reasoning, the radiation anomaly should take a long time to form during the geologic history of the petroleum accumulations and should also take a long time to disappear.

Increases in radon, helium, argon, radium, uranium, and thorium tend to occur in association with petroleum deposits. No published investigations have looked at the strata below a petroleum reservoir to see if there is a "chimney" effect down to the basement. Extremely high amounts of radon and helium associated with natural gases have been found in the Panhandle Field, Texas (Pierce et al., 1956). These reservoirs also contain uraniferous asphalt, radium-bearing brines, and radioactivity anomalies in the rocks. The helium that is generated is believed to be derived from uranium and its daughter products, which are contained in the natural gas. Radon levels are easily lowered by absorption into petroleum, which is still present in the pore spaces of the gas reservoir.

An alternative concept for radiometric anomalies is that uranium and other radioactive products present in the soil in the area of petroleum microseepage are being subjected to reduction. An increase in moisture will enhance this mechanism. The stronger the reducing environment, the greater will be the increase in reduction of the uranium and thorium series to a soluble form in the soil. Radiometric methods detect emissions in only the upper few millimeters to several centimeters of the soil. In this part of the soil, reduced products are easily removed by water, wind, and biological activity. Oxidation of the uranium and thorium series minerals occurs outside the area of petroleum microseepage. The edges of the radiometric anomalies sometimes display spikes (if not faulting), suggesting that the accumulation of oxidized uranium and thorium series is occurring. The spikes thus represent an area of transition between oxidation and reduction zones in the soil. Well-defined radiometric lows are usually indicative of old fields. It is possible that, as a result of vertical migration, hydrocarbons are moving uranium from lower levels in the soil (not the bedrock) toward the surface. As depletion of radioactive elements occurs from the soil, there is less and less radioactivity, and subsequent lack of emissions causes a low over the field.

Durrance (1986) suggested that the presence of uranium in relation to petroleum deposits indicates that groundwater is enriched with radium 224, 226, and 228. Radium 226 is formed from the decay of uranium. Ternary diagrams of barium, radium 224, and radium 228 have been used by Bloch and Key (1981) and Durrance (1986) to distinguish between uranium deposits associated with oil field brines and naturally radium-rich brines.

Hydrocarbon deposits and coal seams (during and after peatification) tend to collect heavy metals, especially those accumulations that are more oxidizing. Uranium has been found in minable quantities in the ash from coal seams of the Fort Union Formation, Williston Basin, North Dakota. Armstrong and Heemstra (1973) established a clear link between petroleum, heavy metals, and radioactive isotopes.

Surface Expression of Radiometric Anomalies/Prediction of Production

Many articles on radiometrics have indicated results that, their authors feel, definitely outline existing fields. Other studies suggest that new productive areas have been found by using this method as a prospecting tool. Some early investigators reported that a radiometric high was a typical response to a petroleum accumulation. However, the majority of investigators have reported radiometric lows over, partially over, or offset from, a later discovered field. Lang (1950) reported that the Ceres Field, a shoestring sand stratigraphic trap in Noble County, Oklahoma, was drilled on the basis of a radiometric anomaly. He further asserted that the productivity of the field was directly related to the total radioactive count. Scherb (1953) reported on a radiometric survey of the Van Field, Van Zandt County, Texas, where the radiation low was displaced to the west of the center of production. Quantitative radiometric results outlined similar shoestring sandstones in Kansas and indicated the degree to which the productive sand is thinning and thickening (Sterrett, 1944).

Langford (1962) claimed that old fields show a more pronounced halo effect than new oil fields do. He concluded that the wells that had been drilled into old fields had allowed the release of radioactive molecules from the crest of the pool, causing a low-radiation zone.

A decrease in radioactivity related to uranium was demonstrated by Flerov (1961) over the Shkapovo Oil Field. This anomaly was a uranium low, but thorium did not exhibit the same depression. When soil and lithologic variations were taken into account, Flerov claimed a 60% to 70% success rate in finding productive wells.

Radiometric anomalies in the form of a depression above oil and gas reservoirs have been reported by many investigators. Activity profiles recorded with car-borne equipment include those of the Binagady-Khurdalan anticline in Azerbaijan by Zolotovitskaya (1965) and Yermakov and Shatsov (1966); the Kum-Dag and Kyzyl-Kum Oil Fields in western Turkmen in the Soviet Union by Yermakov and Shatsov (1959); and the Cambay and Ankleshwar areas in India by Aithal (1959). Us and Kripnevich (1966) measured anomalous lows with respect to background outside the productive area in the western Fore-Caucasus region that were 60% to 80% greater than inside the productive area.

Positive activity increases were shown by Stothart (1943, 1954) in radon measurements over the petroleum-producing fields of Darst Creek and other East Texas fields in Fort Bend, Fisher, and Scurry counties, Texas, and in Seminole County, Oklahoma. Stothart (1954) determined the productive limits of the reservoir to depths of several thousand feet by measuring the anomalous highs and he claimed a success of 70%. Stations were established at 100-m intervals and monitored for 15 min by portable equipment carried in a vehicle (Stothart, 1948). Stothart indicated seven positive radon anomalies, which resulted in four new fields.

Sikka (1963) found that many investigators had recorded various radiometric anomalies associated with oil fields. He concluded that the number of failures outweighed the number of successful predictions. He also showed that 28 investigators had claimed that negative radiometric anomalies occurred over oil reservoirs whereas 14 claimed that positive radiometric anomalies developed over reservoirs. Over one anticlinal field, the Ten Section, Oklahoma, both a low and a high were observed by different investigators. The reason for Kellog's (1957) high and Sikka's (1963) low over the field was that Kellog did not apply a correction for the soil in his data interpretations (Armstrong and Heemstra, 1973). Curry (1984) and Weart and Heimberg (1981) indicated that radiometric anomalies were 50% successful (if not over 70%) when predicting a dry hole or a producer. The surveys done by these authors suggest that, for the most part, there was a good correlation between production and the radiometric anomaly in a regional sense. The required level of confidence for indicating a dry hole or potential accumulation related to an anomaly can be considered subjective and rarely reliable. To date, drilling in several anomalies in the Weart and Heimberg investigations has resulted in dry holes. They did not correct their radiometric data for soil, topography, moisture, and other factors.

Aithal (1959) observed that anomalously high radiometric readings are the result of radium activity and that the lows in the center of the field are due to potassium 40. He further found that background levels varied from area to area. Alekseev and Gottikh's (1965) work showed that variations in the surface radiometric expression were related entirely to potassium 40.

Soil samples taken by Baranov et al. (1959) from a depth of 25 cm over the Gekcha Oil Field, western Turkmen, exhibited a pronounced radiometric anomaly. It was found that equal increases in concentrations of uranium, radium, thorium, and potassium caused the anomaly. Karasev (1970) also found at the Gekcha Field that the uranium was tightly bound to the soil and was in isotropic equilibrium with radium and with other members of the uranium family, specifically ^{234}U. The results of all this work indicated that the halo was not the product of radium precipitation around the periphery of the oil field. Lundberg (1956) supported this view and implied that the radium present was derived from uranium minerals in the near-surface. His calculations indicate that if radium migrated vertically, only 1% would remain after ascending 242 m (800 ft) at 2.5 cm (1 in.) per year.

Sikka (1963) discussed the different individual anomalies over petroleum trapped in a reef, an anticline, and a fault reservoir. He maintained that a reef shows scattered highs bordering a low over an oil reservoir. An anticline is similar, but the scattered highs are farther from the edge of the accumulation. A fault shows highs along its length.

Thomeer (1966) observed the similarity between hydrocarbon anomalies and radioactivity patterns at the surface. Miller (1961) suggested that anomalous radiation patterns agreed with the general locations of hydrocarbon increases in the soil. Sikka (1962B) also found a radiometric anomaly to coincide with a hydrocarbon anomaly over the Ten Section Oil Field, Kern County, California. Even a radiation anomaly pattern from alpha counts on the soil samples taken from drill holes in Alberta was reported to correspond to chloride and methane anomalies (Lundberg, 1956).

Many countries other than the United States and the former Soviet Union have participated in petroleum exploration programs using radiometrics. Bisir (1958) found that oil-productive areas in Romania were indicated by low-radioactivity anomalies from the uranium-radium family of nuclides. The direct correlation of radiometric anomalies to nine Romanian oil fields was summarized by Gohn and Bratasanu (1967). Four radiometric profiles defined the La Cira Oil Field in Colombia, South America (Trapp and Victoria, 1965). A general radiometric low over the Al-Alamein Oil Field in Egypt was described by El Shazly et al. (1970), and the Redwater Oil Field in Alberta, Canada, was studied by Pringle et al. (1953) and by Sikka (1962A).

Gregory (1956) maintained that radiation lows are only coincidental with petroleum reservoir outlines. Nevertheless, he associated the high gamma activity with radium-bearing waters draining from petroliferous formations and accumulating upslope to the alkaline areas of the soil.

According to Thomeer (1966), halo patterns from hydrocarbons, salt concentrations, and radioactivity can be explained by fluid expulsion and capillary barriers resulting from basin compaction. These patterns are so distorted, however, that they might be useful only as supplementary information for use with other methods.

Soils

Thorium and uranium minerals in the soil are generally complex as a result of different contributing factors. There is a tendency for uranium to accumulate primarily in the A or eluvial horizon because of its association with ferric oxyhydroxides. Thorium is usually concentrated in the illuvial or B horizon because it associates with organic matter. The C horizon typically has concentrations of thorium and uranium that mirror the composition of the parent material immediately below. The concentrations of uranium and thorium can be highly variable in transported materials. Radiometric sampling for petroleum exploration is typically limited to the surface and, thus, the A horizon is specifically targeted. Because this is an area of rapid changes in clays and organic matter, care must be taken in analyzing the results of a radiometric survey.

The concentrations of uranium in soils range from 0.1 to over 11 parts per million (ppm) but typically are less than 3 ppm. Uranium has concentrations of less than 0.5 to 4 ppm in sedimentary rocks. The uranium content of sandstones (0.45 to 0.59 ppm) is typically less than in shales (3.0 to 4.0 ppm). Thorium contents are typically less than 1.7 to over 12 ppm in sedimentary rocks. In soils, thorium ranges from 0.4 to 76 ppm, with an average of 3 to 13 ppm.

Different soils typically have different radiation emissions associated with them. Consequently, a radiometric profile traversing from a low to a highly organic soil shows a change in background emissions. If the soils are highly variable over an area, the changes may be too numerous to quantify and correct for. Sandy soils typically have low radiation emissions and, in some cases, there are insufficient quantities of clays and organic material to allow for the accumulation of uranium and thorium that indicates an anomalous condition.

Kilmer (1985) found that determining the percentage of clay in the survey soils and normalizing the data aided in the interpretation and helped eliminate anomalies that were note caused by a petroleum seep. Organic matter affects the accumulation of uranium and thorium minerals and needs to be evaluated in a survey area to determine variability. In areas where dunes are prevalent, radiometrics have been known to outline and identify the locations of these recent soil formations as depressions or lower gamma-ray counts.

Several inert gases, both radioactive and nonradioactive, are produced by the nondisintegration of uranium, thorium, and potassium 40 isotopes. They add to the radioactivity emissions being generated. These gases are chemically inert and are physically immobilized in the formations below the water table and thus dissolve into the groundwater. As the groundwater reaches the water table, the gases leave the water, go into a vapor phase, and enter the pores of the rock and soil. A natural breathing action of the soil causes the inert gases to diffuse into the air at the surface. Here their radioactive components are detected as diurnal variations of the radiation background. Fertilizers add potassium, thorium, and uranium to the soil but not in amounts sufficient to affect prospecting by radiometric methods. Durrance (1986) estimated that it would take at least 100 years of application to increase the uranium content of a soil by 1 ppm, and 200

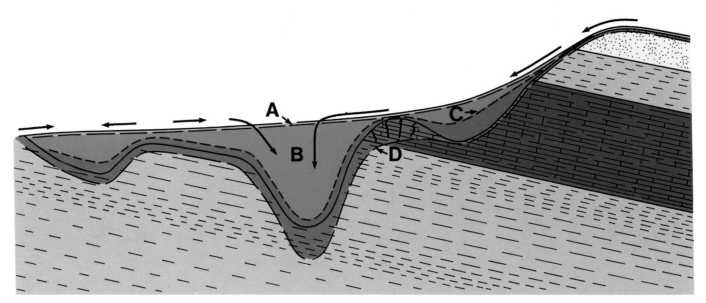

Fig. 3-3 Variations in topography cause variations in the weathering profile. This is a simplified example. A, B, C, and D refer to the A, B, C, and D soil horizons, respectively.

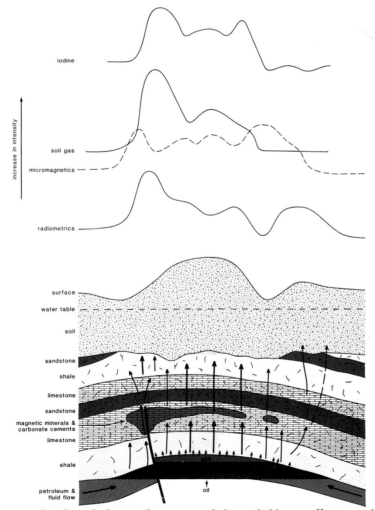

Fig. 4-1 A generalized cross section through the petroleum accumulation and chimney effect caused by migrating hydrocarbons. The surface expressions of various geochemical methods are indicated, but their responses are idealized and may be entirely different over other deposits or even absent.

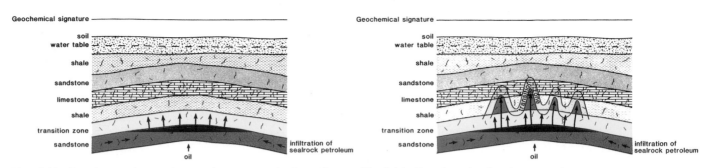

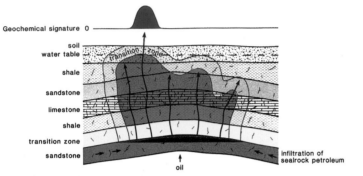

Fig. 4-13 Petroleum migrates (arrows that are near-horizontal) into the trap and concurrently begins to migrate in smaller amounts through the seal rock (vertical arrows) toward the surface. There is no surface expression at this stage.

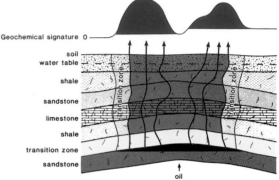

Fig. 4-14 Petroleum is migrating upward, establishing a transition zone that is still predominantly water-wet. As the transition zone continues to move upward, the lower part of the zone becomes increasingly saturated with petroleum and thus is hydrocarbon-wet.

Fig. 4-15 The migrating hydrocarbons begin reaching the soil substrate. The surface geochemical expression starts to develop. Zones of cementation and iron-rich minerals may develop in various stratigraphic zones.

Fig. 4-16 Vertical migration of petroleum is constantly reaching the surface, and minor diffusion is occurring in the seal rock. The accumulation is attempting to reach equilibrium with the overlying strata and is releasing a relatively constant flow of petroleum.

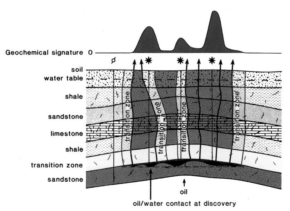

Fig. 4-17 Production has been established and has drawn down the reservoir pressure, causing deterioration of the geochemical anomaly. Numerous transition zones develop in response to areas of the reservoir that still have virgin or higher pressures. The overall amount of hydrocarbons migrating to the surface is decreasing with time.

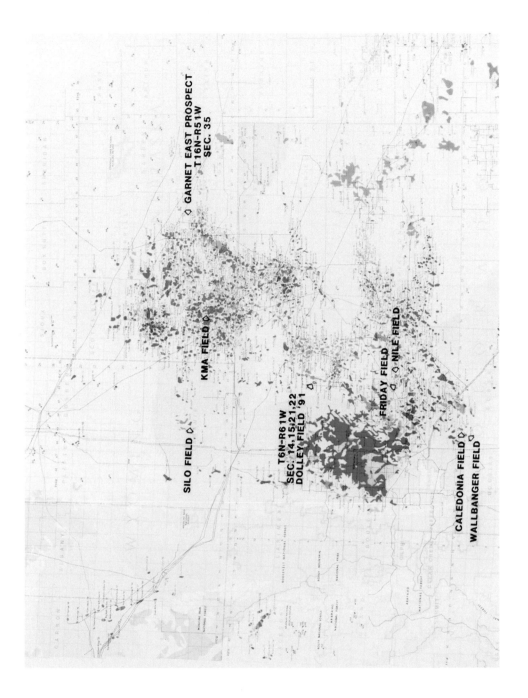

Fig. 5-57 Location map for the Silo, Garnet Test, KMA, Dolley, Friday and Niles Fields, Denver Basin.

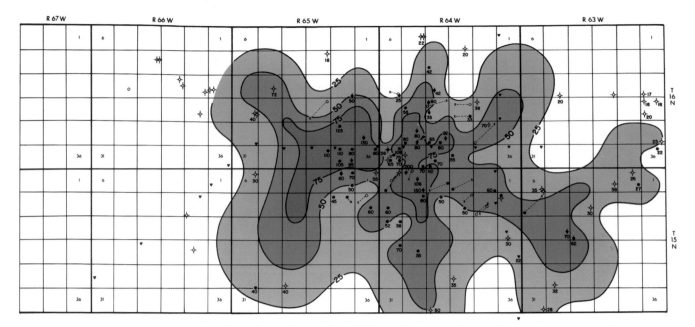

Fig. 5-59 Resistivity of the Niobrara Shale in and around the Silo Field, Laramie County, Denver Basin, Wyoming. Productive area is related to the resistivity of the Niobrara Shale (reprinted with permission from Scientific Geochemical Services).

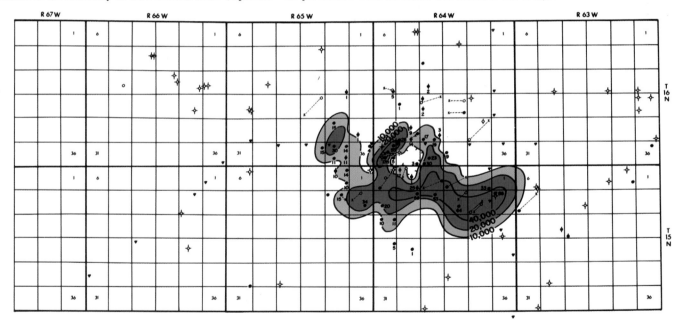

Fig. 5-60 Cumulative oil production for the Silo Field, Laramie County, Denver Basin, Wyoming (reprinted with permission from Scientific Geochemical Services).

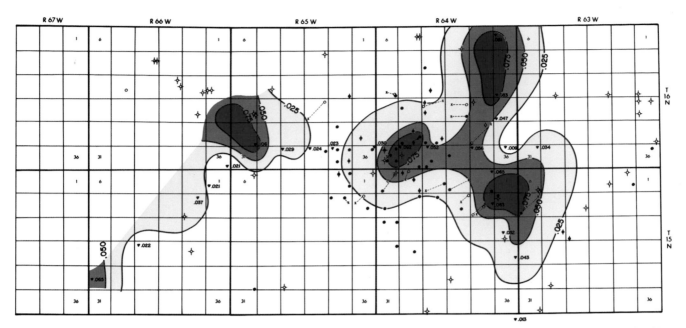

Fig. 5-61 Ethane map (in ppb) for the Silo Field, Laramie County, Denver Basin, Wyoming (reprinted with permission from Scientific Geochemical Services).

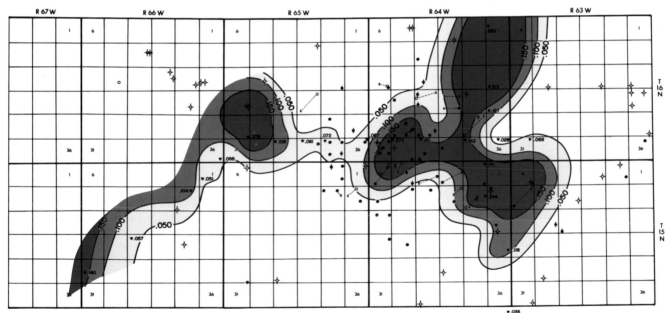

Fig. 5-62 Propane map (in ppb) for the Silo Field, Laramie County, Denver Basin, Wyoming (reprinted with permission from Scientific Geochemical Services).

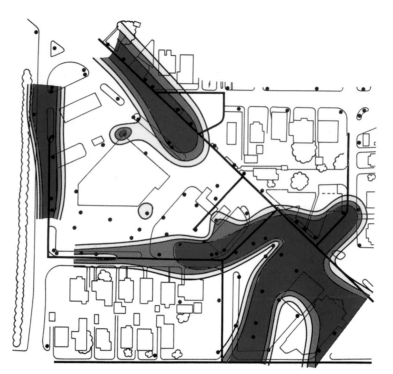

Fig. 5-63 Soil-gas survey to identify leakage from a series of pipelines, western United States (reprinted with permission from Scientific Geochemical Services).

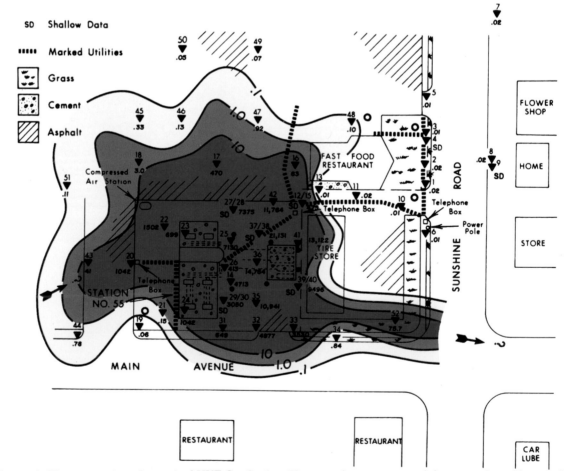

Fig. 5-64 *n*-Butane (*n*C₄) concentrations (in ppm) of XYZ Gas Station. The anomaly centers around the gas station with a trend toward the right side of the map (reprinted with permission from Scientific Geochemical Services).

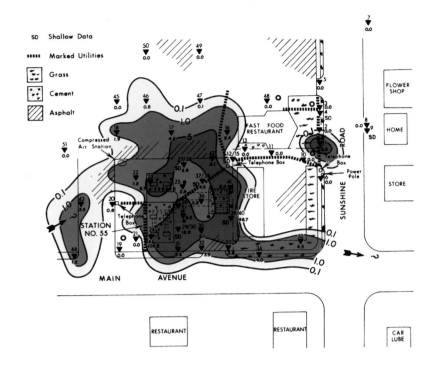

Fig. 5-65 Hexane (C$_5$) concentrations (in ppm) in XYZ Gas Station. Seepage is occurring toward the top and right side of the map (reprinted with permission from Scientific Geochemical Services).

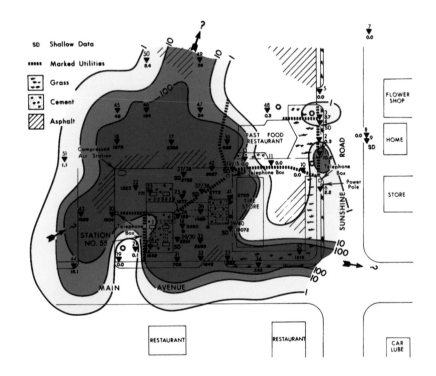

Fig. 5-66 *n*-Butane/benzene-toulene ratio from the XYZ Gas Station. The ratios increase toward the gas station (reprinted with permission from Scientific Geochemical Services).

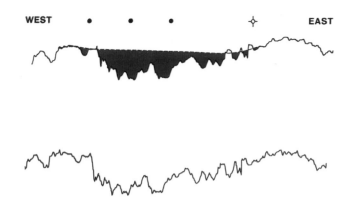

WEST   •   •   •   ✧   EAST

GOLDEN SPIKE OIL FIELD

Fig. 6-7 Continuous radiometrics profile across the Golden Spike Oil Field, Cheyenne County, Colorado, Las Animas Arch area, surveyed in 1989. See. Figure 5-32 for location of the field (reprinted with permission of CST Oil & Gas Corporation).

ALBION-SCIPIO, JACKSON CO., MICHIGAN

1973

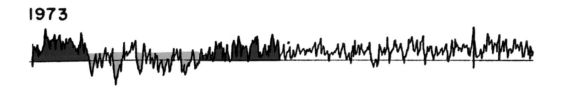

1978

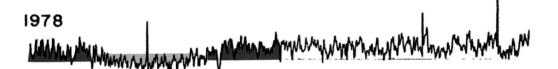

ALBION-SCIPIO, JACKSON CO., MICHIGAN

1973

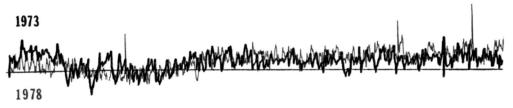

1978

Fig. 6-11 Continuous survey across the Albion-Scipio Field, Hillsdale and Jackson counties, Michigan, Michigan Basin. Surveys were done in 1973 and 1978 (Weart and Heimberg, 1981; reprinted with permission of the Institute for the Study of Earth and Man, Southern Methodist University).

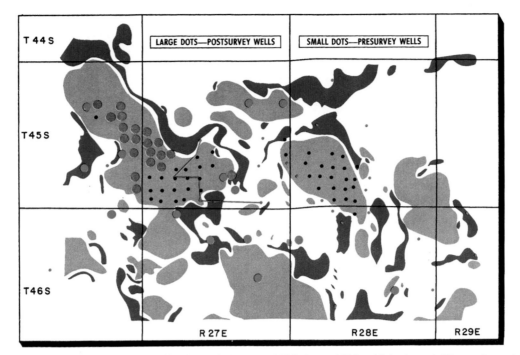

Fig. 6-17 Regional radiometric study of the Sunniland Trend (Weart and Heimberg, 1981) with both predrilling and postdrilling results. Production is from the Middle Sunniland limestone at 3484 m (11,500 ft). Blue dots are dry holes; green dots are productive. Results of 48 postsurvey wells indicate a correct call according to the authors. No data were presented to justify the basis for the anomalous highs and lows (reprinted with permission of the Institute for the Study of Earth and Man, Southern Methodist University).

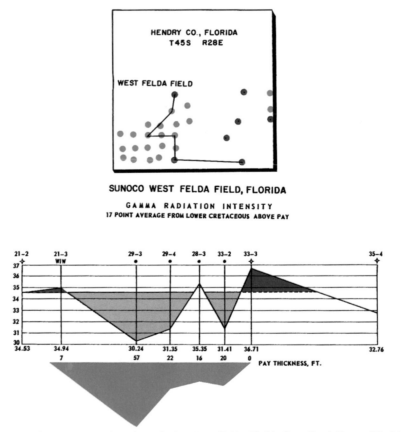

Fig. 6-18 Radiometrics and subsurface cross section through the West Felda Field, Sunniland Tract, Florida Platform, Florida. The radiometric low is depicted in yellow, the high in red, and thickness of the pay in green (Weart and Heimberg, 1981; reprinted with permission of the Institute for the Study of Earth and Man, Southern Methodist University).

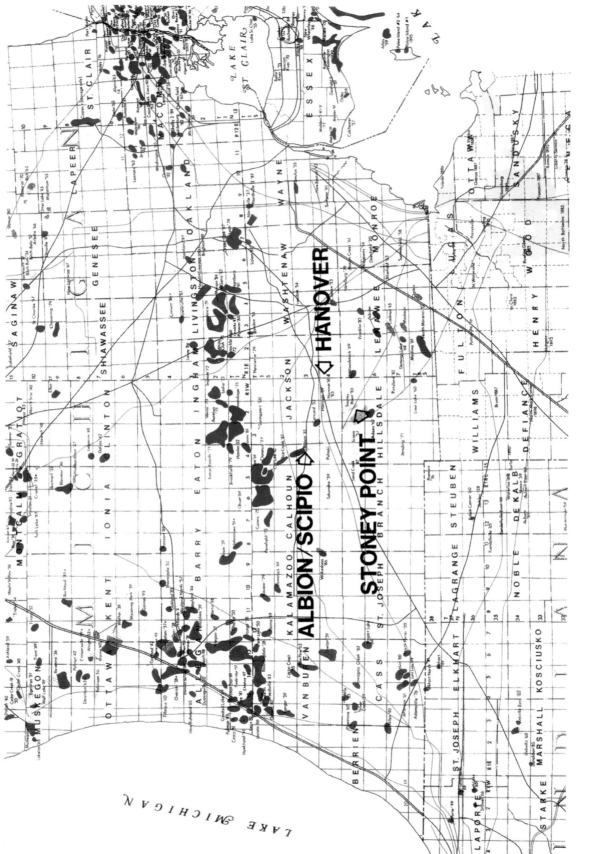

Fig. 6-21 Location of the Albion-Scipio Field, Hillsdale and Jackson counties, Michigan, Michigan Basin (reprinted with permission from Promap Corporation).

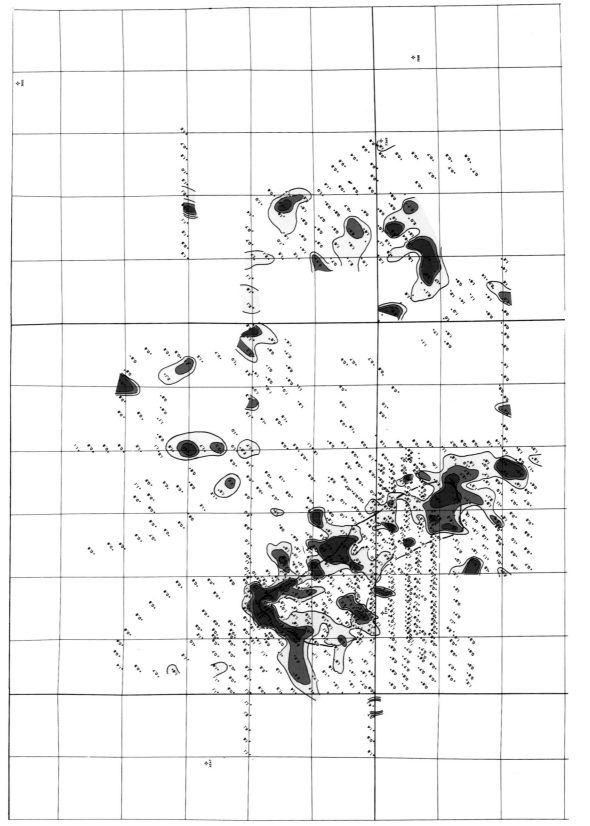

Fig. 7-1 A typical iodine survey in Lincoln County, Denver-Julesburg Basin, Colorado. Background values range from 0.6 to 1.0 ppm. Contour intervals are 1.6, 2.2, and 2.8 ppm. Soils are sandy, and a combination of rangeland and winter wheat farming is typical (reprinted with permission from Exeter Oil & Gas, Inc.).

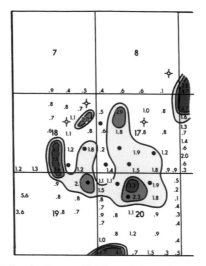

Fig. 7-8 Iodine survey, Niles Field, Township 1 South, Range 60 West, Adams County, Colorado. Surveyed in the summer of 1988, the field produced 2,996,386 bbls of oil and 3,629 MMcf of gas to the end of December 1991. Contour intervals are 1.5, 2.0, and 2.5 ppm (reprinted with permission from Atoka Exploration Corporation).

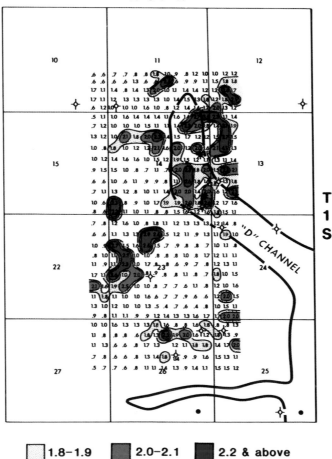

Fig. 7-9 Drilling post–iodine survey and reinterpretation of the D sand channel, Friday Field, Section 14, Township 1 South, Range 61 West, Adams County, Colorado. The structure is on top of the D sandstone; the green indicates the location of the channel. The northernmost well had 7.5 m (25 ft) of pay and the southern 1.7 m (5 ft) of pay. Second iodine survey, postdrilling, Friday Field, the survey was done on a 0.10-mile grid. Contour interval is 1.8 (light pattern), 2.0 medium pattern), and 2.2 (dark pattern) ppm. The anomaly is still exhibiting a halo form (reprinted with permission from Atoka Exploration Corporation)

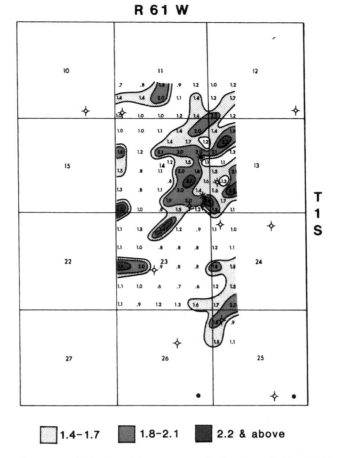

Fig. 7-10 The values from the second survey, which shared the same sample locations, Friday Field, Section 14, Township 1 South, Range 61 West, Adams County, Colorado. Contour intervals are 1.4, 2.0, and 2.4 ppm. The anomaly is not definable into an apical or halo form.

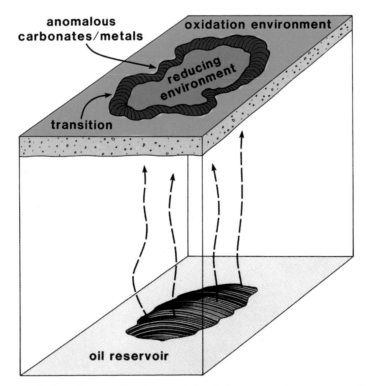

Fig. 8-1 Simplified proposal for accumulation of trace and major metals in relation to petroleum seepage. The seeping petroleum creates a reducing environment. Increasing solubility and movement of the elements away from the seepage areas occur. Redeposition takes place when conditions begin to change to an oxidizing state, usually outside the seepage areas.

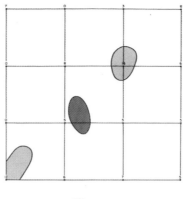

Fig. 13-4 Sample density of 1600 m attempting to detect a 400-m by 400-m target at an unspecified depth. Geochemical data are hexane in parts per billion and were acquired using the head-space method. Yellow is >10 ppb. The target is green.

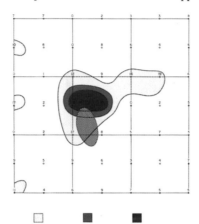

Fig. 13-5 Sample density of 800 m. The contour interval is 10 to 19 ppb (yellow), 20 to 29 ppb (orange), and >30 ppb (red). The presence of an anomaly is clearly indicated but does not clearly coincide with the target.

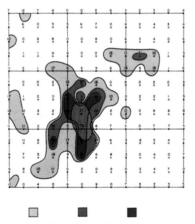

Fig. 13-6 Sample density of 400 m. The contour interval is 10 to 19 ppb (yellow), 20 to 29 ppb (orange), and >30 ppb (red). The target has become more clearly defined, and the presence of isolated anomalies is noted.

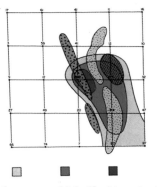

Fig. 13-7 Sample density of 1600 m in an area where there are multiple (five) targets of different size, length, extent, depth, and geologic age. Geochemical data are ethane in parts per billion and were taken using the head-space method. The blue is a structural high, the green is a series of narrow oolite bodies, and the brown represents channel sands deflected as a result of the structural high. The contour interval is 100, 200, and 300 ppb. The anomalous area is not definitive.

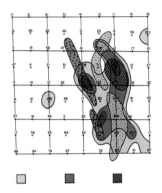

Fig. 13-8 Sample density of 800 m. The blue is a structural high, the green is a series of narrow oolite bodies, and the brown represents channel sands deflected as a result of the structural high. The contour interval is 100, 200, and 300 ppb. The anomalous area is starting to give some definition but is inconclusive.

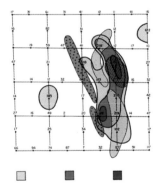

Fig. 13-9 Permutation of the square grid that will miss one of the targets. The blue is a structural high, the green is a series of narrow oolite bodies, and the brown represents channel sands deflected as a result of the structural high. The contour interval is 100, 200, and 300 ppb. The geochemical data do not detect the oolite trap at all.

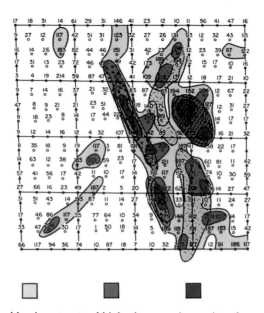

Fig. 13-10 Sample density of 400 m. The blue is a structural high, the green is a series of narrow oolite bodies, and the brown represents channel sands deflected as a result of the structural high. The contour interval is 100, 200, and 300 ppb. The different productive fields are better defined, and five anomalies are present. The oolite and structural high are clearly defined. The channel is somewhat erratic.

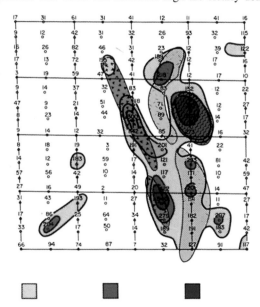

Fig. 13-11 Permutation of the sample grid from Fig. 13-10. The blue is a structural high, the green is a series of narrow oolite bodies, and the brown represents channel sands deflected as a result of the structural high. The contour interval is 100, 200, and 300 ppb. The northern oolite body is not as well defined. The channel system is not defined to any degree of confidence. The structure is clearly outlined.

years to increase potassium by a similar amount. Unless a fertilizer is applied in far greater than normal quantities, this source of interference is negligible.

Variations in radiation caused by changes in soil composition are the most common, and typically the most difficult, to detect. The radioactivity of the soil is usually determined by the parent material. This is not true in areas where the parent material is far from the soils. The location of parent materials and distance of transport may not be easily determined. Glacial deposits are an example. The total amount of radioisotopes in the soil is further affected by weathering processes, biological activity, and groundwater flow. Table 6-3, which summarizes the mean radiation values for soils, shows that variations in soil composition produce variations in radiation emissions.

Changes in the thickness of the total soil present or of individual beds or horizons can produce variation in radiation intensity. Organic matter and moisture content also cause variations in soil radiation intensity. Wetland areas present a particular and perhaps insurmountable problem. They accumulate radioisotopes in the organic matter and can exhibit rapid variations in radiation intensities over short distances. Areas with constant accumulation of surface water will show rapid changes in radiometric measurements, and areas that are poorly drained are natural sites for accumulation of radioisotope salts. This also occurs in arid areas that lack sufficient moisture and that have high evaporation rates and insufficient leaching. These conditions result in the development of alkaline soils and the concentration of radioactive isotopes in the soil. Areas of excellent drainage allow for the removal of salts and thus have low-intensity radiation. Areas with drainage systems that vary from poor to excellent have rapid variations in radiometric intensities.

Sikka (1963) found lower uranium contents in lime-rich sandy soils with pH of 7.5 to 8.5 and in acidic soils with pH of 2.5 to 5.0. Maximum uranium was removed from solution by organic matter with pH of 6.0 to 6.6. Positive radioactive anomalies have been found in areas with alkaline soil, probably because of the precipitation of radiation sulfate.

Moisture

Changes in moisture content of the soil affect gamma emissions. The greater the amount of moisture in the soil, the more likely that radiation emissions will be suppressed. Snow-covered areas indicate radiation profiles related to snow density and overall thickness on top of the soil. Soil does not dry uniformly over a given area. Therefore, variations in moisture content result in fluctuations in emissions as described before. Norwine et al. (1979) found that soil moisture variations related to changes in vegetation caused significant gamma-ray changes. These variations could be corrected by multiple linear regression statistics.

Table 6-3 Mean radiation values for soils in counts per second

Clay loam	327 cps
Loam	304 cps
Sandy loam	260 cps
Clayey sands	354 cps
Sands	227 cps

Sikka, 1962.

Topography

Topography represents a problem for radiometric data collection. Uranium removed from solution is redeposited downhill. This introduces two problems: (1) A uranium anomaly that developed on the top or flank of a topographic high is essentially removed, and (2) the uranium that is moved downhill is redeposited in another soil site. This creates anomalous areas that are related to reconcentration by surface and groundwater, and areas that are truly anomalous may not be detected. Heemstra et al. (1979) found that topography heavily influenced radiometrics and caused them to conclude that there was no correlation between soil-gas and gamma anomalies across producing fields. Typically seen in radiometric profiling using airborne and continuous sampling are changes in background intensities that mirror topographic variations. These changes may not always reflect the topography exactly, but they can mislead the interpretation.

Weather

Variations in barometric pressure due to changing weather can suppress or enhance radiation emissions from the soil. Thus, changing pressure can force recollection of all data across the survey area. Point radiometrics can avoid this problem by sampling at a base station through time, but extreme changes may render even this useless.

Seasonal Variations

Seasonal variations in radon, and subsequently changes in all radioactivity, were investigated by Klusman and Jaacks (1987), who determined that fluctuations in radon peaked in winter and were at a minimum in summer (Fig. 6-4). Mercury demonstrated an inverse relationship. Helium fluctuated the most, but essentially indicated higher volumes, in winter and the least in summer. Klusman and Jaacks concluded that fluctuations are controlled by changes in moisture, temperature, and pressure.

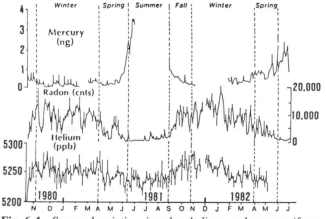

Fig. 6-4 Seasonal variations in radon, helium, and mercury (from Klusman and Jaacks 1987; reprinted with permission of Elsevier Scientific Publications).

Other Causes of Radiometric Anomalies

Radiometric anomalies are not uniquely associated with petroleum microseepage. Radiometrics have been extensively used in the search for uranium roll-front and Precambrian sedimentary and hydrothermal metal deposits. Precambrian sedimentary and hydrothermal deposits are not usually found in association with petroliferous basins (an exception is the Basin and Range Province of Nevada). Uranium roll-front deposits may occur in many petroleum-productive basins, such as Weld County, Denver Basin, Colorado, where hydrocarbons have been used in conjunction with radiometrics to map the roll-front deposits (Reimer and Otton, 1976; Reimer and Rice, 1977). Radiation leakage anomalies are common above many blind or exposed uranium bodies. The position of this leakage halo or any mineralization of the soil with respect to the buried ore deposit depends on the local structural tectonic conditions.

Some evidence indicates that groundwater tends to acquire and differentiate uranium from the daughter products when passing through uranium ore bodies. Other research demonstrates that colloidal uranium is absent in the groundwater around uranium deposits and that the uranium in the fluids is over 95% in the anion form. The uranium content of groundwater from a supergene uranium deposit indicates a nonlinear relationship with total salinity. The mobility of uranium is a function of CO_2 content as well. The $^{228}Ra/^{238}U$ solubility ratios can vary widely, especially in the presence of carbonates.

Radon and Earthquake Prediction

Radon is a short-lived decay product derived from either uranium or thorium. Radon 228 has a half-life of 3.82 days. Radon is the only gaseous element in the decay chain and is useful in soil-gas studies. The short life limits the distance

that radon can migrate or be transported and, thus, it can be used as an effective exploration tool. High concentrations of radon are essentially related to leaching of uranium in the soil substrate.

Radon surveys have been applied to areas with earthquake activity. Preliminary studies indicate that changes in radon and helium concentrations can possibly lead to the successful prediction of earthquakes. The relationship between He and Rn increases is not a constant but is related to the short-term, stress-release event known as the earthquake. As stresses build up in the earth, microfracturing increases which, in turn, creates enhanced zones of permeability. Both helium and radon use these pathways and, for a short period before an earthquake, amounts of helium and radon substantially higher than normal have been measured at the surface. After the earthquake has occurred, stresses are reduced, and microfracturing and the resulting permeability are also reduced. Subsequently, the amounts of helium and radon decline. Comparatively high radon concentrations are found in groundwaters that flow through acidic rock formations, and low concentrations are found in basic rocks.

Radon anomalies are found in association with faults or highly permeable strata. Because of its low diffusion coefficient compared to He, for example, radon migrates to the earth's surface slowly and not from great depths. Budde (1958) measured anomalous radon concentrations occurring from depths of 1 m in sand to 5 cm in clay. Radon diffusion in unconsolidated overburden is controlled by soil grain size and water content. Budde concluded that uranium deposits would be undetectable below 4 m if radon alone was used for exploration. Dyck (1968) set deeper limits, 6 to 9 m (20 to 30 ft), to detect buried uranium deposits by radon emissions. We can conclude that petroleum deposits would have to be close to the surface to cause any increase in the intensity of radon emissions.

Lang (1950) advocated the use of radon surveys for defining the presence of faults. Kondratenko (1957) assumed that radon emanation anomalies, with beta-activity anomalies lacking, were caused by mining operations. Subsurface faults located as deep as 1515 m (5000 ft) below unfaulted formations were thought to exist, but the data might also represent only massive leakage or breached traps, or both.

Foote (1969B) summarized what he felt were the most important factors affecting airborne radiometrics. These factors are listed below. Foote's discussion centered around bismuth 214, which can also include the daughter products because many radiometric anomalies can be a product of one or several specific daughter products. The specific daughter products causing the anomaly may vary from site to site.

1. For various radioactive elements, the diurnal variation resulting from environmental effects can amount to as much as a factor of 2. An airborne operation is required to survey a large area in a minimum amount of time to allow for diurnal corrections.

2. The contributions of bismuth 214 from atmospheric sources can account for a major portion of the bismuth 214 radiation measured at the surface and to an elevation of 400 ft above the surface.

3. The atmospheric concentrations of bismuth 214 must be measured every few miles. These concentrations must be properly removed to provide results representative of the surface concentrations.

4. The spectral energy shapes of uranium and thorium daughter product gamma radiation and of potassium 40 must be accurately known for the field geometry.

5. The cosmic ray energy spectrum must be known and its contribution removed from the data.

6. The radiation detector volume must be sufficient to give less than +5% statistical uncertainty in spectral results for a surface coverage between 90 m (300 ft) and 120 m (400 ft).

7. The correction for soil variations must be made for changes in soils before any anomalous relationships among uranium, thorium, and potassium concentrations are evaluated.

8. All data should be corrected to a constant surface elevation. This will correct for the topographic effects.

The majority of work has been with thorium, uranium, and potassium rather than bismuth, radium, and radon. Data collection should include evaluation of as many radioactive daughter products as possible.

Man-made sources have a great influence on radiation intensity, especially when surveys are adjacent to, or on, roads. The composition of the road material can cause variations in radiation intensity that are not related to natural changes. Dirt or gravel roads typically have a minimum impact on gamma-ray emissions and maximize background intensity. Bitumen and paved roads either nearly or completely absorb radiation so that intensity is at a minimum. The degree of compaction and the depth of the road material will also affect radiation intensity. The materials placed on roads in the winter, such as coal ash, brines, sand, and salt mixtures, will also impact the radiation intensity.

Detection and Measurement

The alpha-particle electroscope was one of the earliest means of measuring radioactivity in petroleum. It was used to measure radon over oil fields (Bogoyavlenskiy, 1929). Alpha-radiation surveying of surface soil samples was judged by MacElvain (1963A, 1963B) to be far superior to gamma surveying. A distinct halo pattern is produced by the presence of stable lead 206 and lead 208. MacElvain thought that these daughter products from radon and thorium may have been concentrated on the surface by ascending hydrocarbons.

Early gamma-radiation surveys over known anticlinal producing fields indicated increased total gamma activity at the crest of the structures (Lang, 1950). Sensitive Geiger–Müller detectors were placed at stations one-quarter mile apart, and gamma activity was recorded for six-month periods. Rothe (1959), walking and carrying his counting equipment, measured negative anomalies from background gamma radiation taken at 65-ft intervals across the old Fallstein Oil Field and the Diesdorf-hadde-Kath Field in what was then East Germany. Miller (1961) preferred a high-sensitivity, 10,000-in.^{3} ionization chamber (18 in. in diameter by 10 ft in length) instead of a scintillator for gamma detection because of its sensitivity and its slow but steady drift.

Drill holes have been used to monitor gamma radiation profiled across oil and gas fields (Kopia, 1967; Stothart, 1954). Gamma-radiation well log records were used by Omes et al. (1965) and Pirson et al. (1966) to outline radioactive oil field anomalies.

A scintillation counter was first used for geologic field work by Pringle in Canada in 1949. Both gamma-ray spectra and isoradiation plots of anomalous radiometric data were involved in this early work. The state of the art of detection in the 1940s was an extremely sensitive electroscope used by Stothart (1943, 1948). The scintillation counters employed by Trapp and Victoria (1965) were typical of the 1960s.

The detection of high-frequency electromagnetic radiation from uranium-radium, radium-thorium, and potassium series is based on the interaction of these emissions with the detector material. The detector interaction with the source causes the generation of a signal. Detectors typically can discern different types of charged particles and can isolate time-coincident events from a background of noncoincident events. The detection of radioactive emissions is a statistical time-dependent function of the radioactive decay process. The shorter the period in which counting of radioactive emissions occurs, the greater is the likelihood of statistical error. The statistical error varies inversely with the number of counts and can be expressed by the equation:

$$\gamma = 100/N^2$$

where γ equals the statistical error and N equals the number of counts. There is always a compromise between the statistical error and the number of counts. This can be compensated for by using detectors with a large crystal volume. A minimum number of 500 counts for continuous sampling or a time interval of 90 or greater per sample station is usually sufficient to achieve a 5% error.

A detector can be defined in terms of five response characteristics:

1. Intrinsic efficiency, the proportion of the emissions actually reaching the detector. The less sophisticated and the smaller the crystal size of the detector, the less confidence there will be in the data.

2. Total efficiency, the ratio of the detected emissions to those generated by the source.

3. Geometry, the solid angle around the detector where the emissions can actually be detected.

4. Range, the frequency or intensity spectrum limitations of the detector.

5. Resolution, the ability to distinguish between emissions of similar energies.

There are several different types of radiation detectors: gas ionization, proportional counters, Geiger–Müller counters, scintillation detectors, and semiconductor detectors. Some of these are not generally used in petroleum exploration because of their size and energy requirements.

Scintillation detectors are the most widely used form of radiometric equipment in petroleum exploration. The scintillator relies on the detection of radiation by materials known as *phosphors*. The phosphors are excited when bombarded by radiation, and they then emit light. The majority of phosphors are gases (especially the rare gases), organic molecules, and solids such as ZnS and NaI. The suitability of different phosphors to detect radiation is linked to their densities. For example, ZnS is best for detecting alpha particles and NaI for detecting gamma rays.

Airborne Method

An airborne instrument system has been developed for radiometric surveying. However, up to 75% of the detected radium decay gamma radiation will be the result of airborne radon decay (Foote, 1969B). This radon decay activity along with radioactivity from other background sources must be subtracted from the total gamma count (Balyasny et al., 1964) to determine the radiation response that originates from the earth's surface only. Anomalies containing as little as 6-ppm thorium and 3-ppm uranium have been detected by using airborne equipment (Adams, 1969). Small amounts of radioactive elements in the soil can affect gamma intensity at the surface. Weak radioactivity anomalies have been caused by thin mudstone beds containing as little as 0.002% U_{308}.

Airborne methods have been more aggressively used in the search for uranium than in petroleum exploration. In the 1970s, the United States pursued the National Uranium Resource Evaluation (NURE) program in order to evaluate and possibly quantify this specific resource. The survey was flown over most of the United States. Flight lines varied from 1 by 2 to 6 by 12 miles. The equipment was a scintillator detector of the type used in the search for petroleum. Because the radiometric response is similar to that found by using surface techniques, these regional methods have had classic (low) responses in association with existing fields. Therefore, the NURE data have been used to evaluate petroleum

potential in several basins and large areas on a regional basis. The flight line spacing is not conducive to detailed mapping or prospect generation but can provide leads for a more detailed program.

Lundberg (1956) was one of the first oil seekers to use the airborne scintillation detector. In 1952, he recorded radioactivity on flight lines over uranium deposits and oil fields in Texas and Canada. This airborne method has since been applied to petroleum prospecting by many investigators, and many successes have been reported.

MacFadyen and Guedes (1956) studied the effect of height, spacing, and "atmospheric and soil humidity." They found that 100- to 150-m altitudes were optimum. Resolution decreased significantly above 200 m. After a rain, three days were required before a single radiation profile could be discerned.

The Nuclear Regulatory Commission has developed equipment called ARMS (Aerial Radiological Measuring Survey). It will produce 360 counts/s from a 152-m (500-ft) altitude over an iodine 131 source concentration of 1 μCi/m^2. Six NaI crystals that were 4 in. by 2 in. in diameter with photomultipliers were used, which gave natural gamma-ray background ranging from 100 to 1000 counts/s. With this equipment, a geologic correlation was found to be similar to that calculated from the use of gamma-ray well logs.

Boyle (1955) noted that emissions of gamma rays are best detected at less than 1 ft from ground level. He felt that airborne prospecting for uranium was an adjunct to, not a substitute for, prospecting on the ground. It was determined that flying altitudes of 33 to 66 m (100 to 200 ft) still allowed a zone of detection forming a cone of 33 to 66 m (100 to 200 ft) around the flight path using a 3.5-in. NaI crystal.

Airborne surveying must be reduced to a common level, usually the earth's surface. One of the tools for reducing the surface radiation to a lithologically normal rad value of the earth's surface is the gamma-ray response of ^{40}K. This can be done only if the detector differentiates the different uranium daughter products.

In more recent times, airborne methods have been restricted to evaluating remote areas of the world prior to, or in conjunction with, surface exploration. Airborne methods are affected by the speed of the transport medium itself and the rate at which the detector collects emissions. Therefore, potential anomalies may be missed if the plane is flying too fast. The distance of the flight line from the surface may also affect detection of a radiation anomaly. Moisture, biomass variations, and the presence of stream sediments can be critical considerations in the evaluation of these data (Norwine et al., 1979).

Figure 6-5 is an example of NURE data across two existing oil fields and a prospect in eastern Colorado. The fields demonstrate a typical radiometric low. Morse and Alewine (1983) presented a similar interpretation of NURE data across the Patrick Draw Field, Wyoming. However, caution is necessary if using only NURE data for exploration

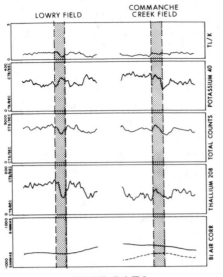

NURE DATA

Fig. 6-5 Examples of NURE data across two fields and a prospect in the Denver Basin, Colorado.

because some of the fields (of various areal extent and total petroleum in place) that were crossed by these airborne traverses did not display any type of anomaly.

Point Method

Point radiometrics is the simplest form of radiometric method. The equipment is only one step above a Geiger counter and, in some cases, is just an adaptation of one. A difficulty with radiometrics is repeatability, and this method, even though it is the simplest, seemingly can overcome that problem.

The radiometric equipment for point surveys is placed on the ground at the sample site. The detector is turned on and begins counting for a set period of time, preferably 1 to 3 min. This procedure is repeated three times at the sample site. The data generally are recorded by hand on field sheets (Fig. 6-6). The three readings are averaged, and the resulting number represents the data value for that sample site. If a wide variation in the count occurs between the three values, the operator keeps repeating the procedure until at least three data values are in close agreement.

One of the critical advantages of the point method is the use of a base station. A particular sample site (or sites) is chosen as a base station for the entire survey. Here a count is taken at the beginning of the day, at midday, and at the end of the day to determine the changes caused by diurnal effects. The survey data are plotted on graph paper, and the base station data are used to correct the results for diurnal variation. If major weather changes occur, the operator must immediately resample the base station site. Minor weather changes are discernible by normal sampling at the base station during the day. Soil changes should also be noted as sampling occurs; any distinct changes in composition, moisture, and man-made features should be noted on the operator's data sheet.

The disadvantage of this method lies in the difficulty in evaluating soils and other sources that could cause or suppress a radiometric anomaly associated with a petroleum accumulation.

Continuous Method

The published literature suggests that continuous radiometrics is the most common form of radiometric measurement in current use. The radiometric equipment is usually mounted on a vehicle. Some systems require a connection with the vehicle's odometer. This is recommended for greater accuracy and for maintaining a uniform speed during surveying. Figure 6-7 (see color plate) is an example of a typical continuous profile, which is over the Golden Spike Field, Township 16 South, Range 45 West, Cheyenne County, Las Animas Arch area, Colorado. The field is near the end of its productive life.

Profiles can be acquired along any form of grid. The best data acquisition occurs when the target size is known and traverses are perpendicular to strike. It has been noted by

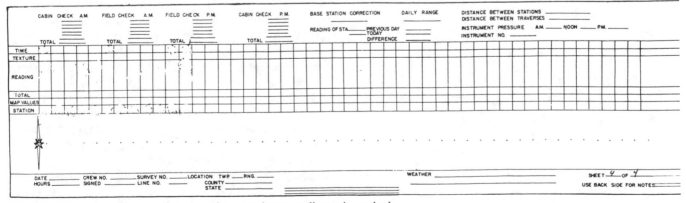

Fig. 6-6 Examples of a recording sheet for a continuous radiometric method.

many investigators that lines recorded in different directions across a target may not show comparable data. Usually, if the direction of surveying is consistent, the data are very usable. This method is often used along roads and barrow ditches that frequently contain a wide assortment of foreign materials, which can cause rapid and dramatic variations in radiation levels.

An advantage of this method is the real-time ability to repeat measurements to delineate and detail an anomaly that is found during the survey. The mobility of the method allows quick evaluation of a large area. The disadvantage is that it requires the operator to stop the vehicle and evaluate the soils for verification of the authority of the anomaly. Also, because of diurnal variations in intensity, the anomaly may not be repeated a short time later.

Integrated Radon Method

An integration method called *track etch* (Gingrich and Fisher 1976; Gingrich et al., 1980) is similar to the K-V fingerprint method for soil gas. A hole is drilled from 0.3 to 0.9 m (1 to 3 ft), and a plastic cup with a special film is placed upside down at the bottom (Fig. 6-8). The height of the cup is designed so that alpha-decay emissions from the soil do not penetrate too far into the air and reach the film. The alpha-decay pattern of radon (which is in the gas phase) impacts on the film and leaves a damage tract. A protective

lid over the hole prevents some damage by biological activity. The cup is left in place for three to six weeks and then is retrieved. The film is placed under a microscope, and the density of the tracks is determined. The number of etches is proportional to the radon gas activity in the atmosphere of the cup during the time interval it was in place. To eliminate radon 220 emissions from thorium decay and specifically target those from uranium, a plastic filter was placed at the bottom of the cup.

One of the concerns for this method is the source of the radon. Gingrich (1984) left the cups in the ground for a full winter, which compensated for temperature changes and provided clean contacts between anomalies and background. To eliminate areas of uranium, a portable scintillometer was used. Of course, this was used to define a nonpetroleum deposit, but its application should be the same for petroleum.

This method has the same problem as the K-V fingerprint technique (Chapter 5) in terms of groundwater table fluctuations and biological activity. In addition to that problem, what is the source of the radon?

Free-Air Radon Collection Method

This technique requires driving a hollow steel probe 1 to 3 m into the soil. The hollow portion of the probe has a measured volume. When the probe reaches a predetermined depth, a syringe is inserted into the probe, and the hollow portion is purged of atmospheric gases. A second sample is taken using a sample container whose interior is a vacuum. Theoretically, the gases entering the evacuated chamber are from the soil itself. The containers used for collecting the gas at the surface vary from simple syringes to double-sealed containers. The sample is then taken to the laboratory for analysis and determination of the radon content of the gas.

Interpretation of Data

Interpretation procedures vary for each form of radiometric method. Point radiometrics is fundamentally the simplest method in that normalization of the data occurs with respect to a base station. The base station, which is measured at specified intervals during the day, allows for correction of the data. This will generally eliminate the effects of the weather changes. The resulting data are plotted on a map and contoured.

The typical surface expression of radiometric anomalies is in the form of a halo and is seldom apical in nature. The halo form has led many to suggest that uranium and thorium increases indicate that they have migrated with upward-moving petroleum. The halo is defined by an abrupt increase in radiometric count at the edges of the accumulation, followed by rapid decreases inside the anomaly. In most cases, the radiometric drop across the center of the field is lower

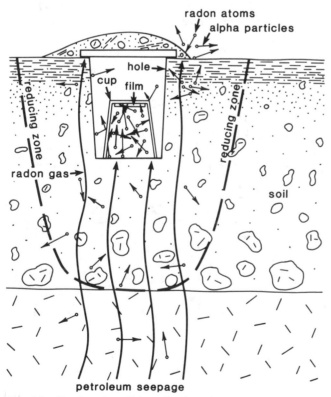

Fig. 6-8   Track etch technique for detecting radon.

in total counts than it is in the background areas. This is usually modified by clay, moisture, and organic contents of the soil. A radiometric low increases in intensity as production drains a reservoir over time. The concept of a fossil anomaly may have originated with the radiometric method.

Radioactivity mapping with a single gamma spectral line from radium, referred to as *radium metallometry* (Lauterbach, 1970), is considered a less ambiguous indication of hydrocarbons in the ground than the total gamma-ray activity (Zeuch, 1965). Many explorationists, using pulse height analyzers, have concentrated their measurements on specific nuclides in the gamma spectrum (Foote, 1968; 1969B; Serdyukova and Doshchechkin, 1970). The quantitative separation of gamma-ray spectra is accomplished by using certain suggested parameters. These parameters are radium decay (represented by the 1.12- and 1.76-MeV levels of RaC), thorium decay (represented by the 2.614-MeV level of ThC and 0.908- and 0.966-MeV levels of $MsTh_2$), and the potassium 40 presence (represented by the 1.46-MeV level).

Faults and fracture systems, if they are acting as conduits for migrating petroleum fluids, are typically seen as spikes in the radiometric data. Figure 6-9 is an example of a fault that was detected by a continuous profile.

Continuous ground or airborne profile data are typically the easiest to collect, and this method has the ability to acquire tremendous amounts of data in a single day compared to all other surface geochemical tools. However, the data are the most difficult to interpret. The complexity of the detector and its output determine the level of interpretation. Simple detectors generate one line of data, and it is impossible to determine what daughter products are causing variations in the data that change both the anomalous and background areas. Data are usually recorded on continuous graph paper and, thus, there can be no computer manipulation of the data. More sophisticated equipment records digitally so that the various profiles for each daughter product can be reviewed individually for analysis and interpretation.

Figure 6-7 is an example from the Golden Spike Field, Township 16 South, Range 45 West, Colorado. The field produces from the Morrow sandstone (Pennsylvanian) and

Mississippian carbonates and was discovered in 1969. The anomaly generally begins and ends at the known limits of the field, even though the wells in this area have been plugged and abandoned. Another example is from the Barada field, Township 2 and 3 North, Range 17 East, Nebraska (Fig. 6-10). It was discovered in 1941, produces from Hunton carbonates at 742 m (2450 ft), and has yielded 3+ MMbbl. Although the field is essentially depleted, a traverse across the eastern end still indicates an anomaly. Figure 6-11 (see color plate) is from Weart and Heimberg (1981) and indicates a profile across the Albion-Scipio Field, Michigan Basin, Michigan. The profile was acquired a considerable time after the field was discovered. The field produces from a dolomitized chimney in the Trenton and Black River formations (Ordovician) from 1220 m (4000 ft) and has produced 150 MMbbls of oil equivalent. The two radiometric surveys (in 1973 and 1978) both indicate anomalies corresponding to the field. However, the intensity of the anomalies is not as distinctive as at the Golden Spike and Barada fields. Background and anomalous areas are not as well defined, and it is questionable whether the prospect would have been drilled based on this data.

In summary, the radiometric profile is analyzed based on variations in the slope from one response to the next. Soil variations must be taken into consideration. The anomalous slopes of lines, if not caused by soil changes, man-made features, weather changes, or other effects, may be related to a petroleum accumulation at depth. Most of these effects can be eliminated or moderated if observed and understood before the data are interpreted.

Case Histories

Very few radiometric case histories record anomalies that were detected prior to drilling.

Powder River Basin, Wyoming

The Heldt Draw and Spearhead Ranch fields are located in the Powder River Basin in the eastern half of Wyoming,

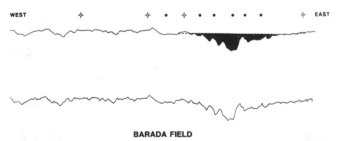

BARADA FIELD

Fig. 6-10 Continuous radiometrics profile across the Barada Oil Field, Richardson County, Nebraska, Forest City Basin. Surveyed in 1989 (reprinted with permission of Trinity Oil and Illuminating Corporation).

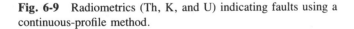

Fig. 6-9 Radiometrics (Th, K, and U) indicating faults using a continuous-profile method.

in Townships 45 and 46 North, Ranges 76 and 77 West, and Townships 38 and 39 North, Ranges 74 and 75 West respectively (Fig. 6-12). They were surveyed about the time they were both discovered (Curry, 1984). A truck-mounted continuous radiometric method was used.

The Heldt Draw Field was discovered in 1973 and produces from a 17-km (10-mile) northwest-trending Shannon sandstone reservoir (Fig. 6-13). The radiometric survey was conducted when there were four producing wells. The radiometric anomaly was the only data presented. Figure 6-14 indicates the drilling as of December 1991. The field produced a total of 4.8 MMbbls of oil and 4.3 BCF of gas. The correlation of the anomaly to the productive part of the field is excellent.

The Spearhead Ranch Field produces from the Frontier, Muddy, and Dakota sandstones. It was discovered in 1973 and has produced 6.5 MMbbls of oil and 56.5 BCF of gas. Figure 6-15 indicates the continuous radiometric anomaly in 1974 and the extent of drilling to that time. Figure 6-16 indicates the drilling up to December 1991, and the anomaly shows a good correlation with production.

Curry used several different types of radiometric equip-

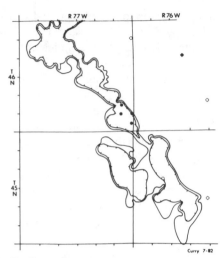

Fig. 6-13 Radiometric anomaly in association with Heldt Draw Field drilling in 1974 (after Curry, 1984; reprinted with permission of the Institute for the Study of Earth and Man, Southern Methodist University).

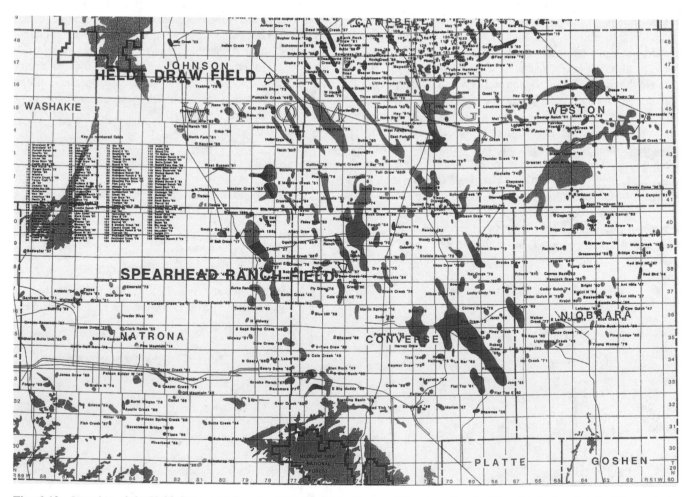

Fig. 6-12 Location of the Heldt Draw and Spearhead Ranch fields, Eastern Wyoming, Powder River Basin (reprinted with permission of Promap Corp.).

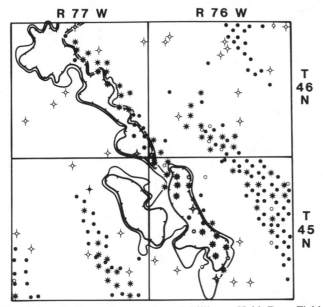

Fig. 6-14 Radiometric anomaly and drilling at Heldt Draw Field as of December 1991 (Curry, 1984). Note the lack of in-fill drilling compared to drilling in 1974 and the large size of the anomaly.

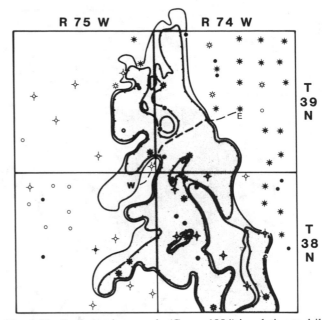

Fig. 6-16 Radiometric anomaly (Curry, 1984) in relation to drilling at Spearhead Ranch Field as of December 1991. Note the undrilled portion of the anomaly in Township 45 North, Range 77 West, and the fact that the main anomaly becomes offset from the productive area in Township 4 North, Range 77 West.

were presented that would allow the reader to scrutinize the results more carefully.

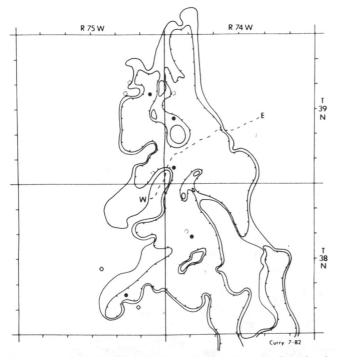

Fig. 6-15 Radiometric anomaly in association with Spearhead Ranch Field drilling in 1974 (Curry, 1984; reprinted with permission of the Institute for the Study of Earth and Man, Southern Methodist University).

ment. All showed the same general correlation of the location with the intensity of the radiometric anomaly. There was no attempt to discriminate between the different daughter elements, except for a mention of potassium 40. All the profiles relied on total gamma count. The size of the field may have made that effort unnecessary. No profiles or data

Sunniland Trend, Florida

Weart and Heimberg (1981) surveyed several different large field areas in several basins. The radiometric survey along the Sunniland Trend, specifically the West Felda Field, Florida Platform, Florida, was done because of the difficulty in utilizing seismic methods. Figure 6-17 (see color plate) indicates the radiometric anomalies, their association with production at the time of the survey, and the drilling after the survey. The anomalies had a general correlation with production, but the survey could not be considered site-specific.

Figure 6-18 (see color plate) is a cross section with the correlated radiometric anomaly. The low is not over the full extent of the field, and no explanation was given for the positive area in the field.

Aneth Field, San Juan County, Utah

The Aneth field is a reef complex located in San Juan County, Utah, Paradox Basin. A point radiometrics survey was carried out shortly after the field was discovered in 1956 (Fig. 6-19). The climate is desert, with little precipitation.

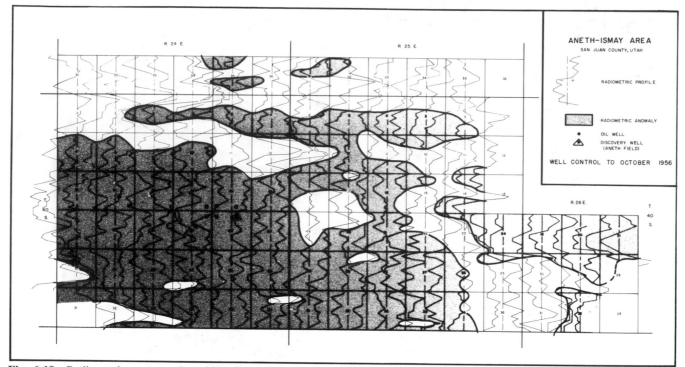

Fig. 6-19 Radiometric survey a short time after the Aneth Field was discovered (reprinted with permission from Delbert Long).

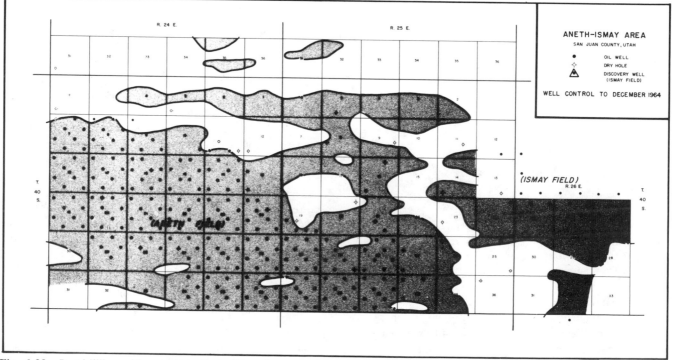

Fig. 6-20 Postdrillling with respect to the radiometric survey for the Aneth Field, the anomaly is essentially filled in with producing wells (reprinted with permission from Delbert Long).

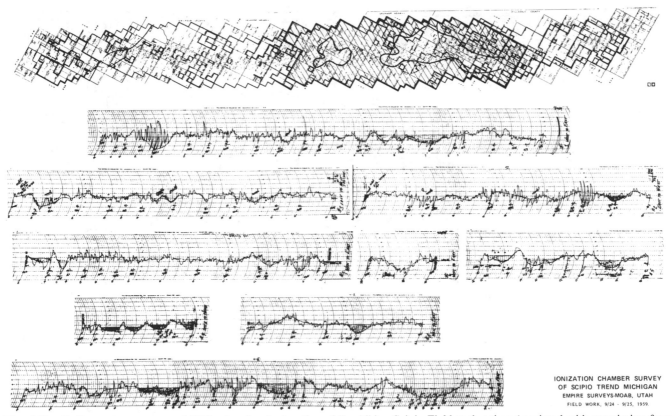

Fig. 6-22 Radiometric survey as of 1959, showing the location of the Albion-Scipio Field to that time (reprinted with permission from Delbert Long).

Drilling through the years has essentially remained within the outline of the radiometric anomaly (Fig. 6-20).

Albion-Scipio Field, Hillsdale and Jackson Counties, Michigan

The Albion-Scipio Field (Fig. 6-21; see color plate) was discovered in 1956 and produces from a dolomitized chimney in the Trenton-Black River formations (Ordovician). Soils are highly variable and range from clayey sands to highly organic clays. A continuous-profile survey was presented earlier in the text (Fig. 6-11) from Weart and Heimberg (1981). A point radiometric survey along roads was carried out in a wide grid that covered the existing production and the undrilled area to the east (Fig. 6-22). Subsequent drilling has not changed the outline of the field, and the radiometric anomalies cover only part of the production (Fig. 6-23). However, what is interesting is that two fields discovered later, Stoney Point and Hanover SE, lie in the survey area. The Stoney Point produces from the Trenton-Black River and the Hanover SE from the Traverse limestone (Devonian). These fields were discovered in 1985 and 1983, respectively. Stoney Point, which is the larger of the two fields, did not have an identifiable radiometric anomaly, but the smaller Hanover SE Field did have one.

Summary

The use of radiometric methods predates other surface geochemical methods, and the amount of literature available is substantial. Radiometric (including radon) methods are viable exploration tools for oil and gas exploration. A major problem is that numerous other radiometric sources may be present, and they must be identified and removed from the data. Corrections must be made for nonradiometric factors such as topography, weather, and moisture. All these corrective measures can be time-consuming, and complete success is never achieved. Experience in interpreting the results is essential. There are specific areas in which radiometrics is *not* affected by a multitude of outside sources but, in other areas, the amount of interference prevents a high degree of confidence in the results. Any radiometric anomaly should be verified by other surface geochemical or geophysical methods, regardless of the confidence that all the extraneous sources have been eliminated. As exploration tools, radiometric methods are best used for reconnaissance rather than for defining a prospect or a well site location exactly. The relatively low cost makes the technique a good first-pass method.

References

Adams, J. A. S. (1969). Total and spectrometric gamma-ray surveys from helicopters and vehicles, *Proceedings of the Sympo-*

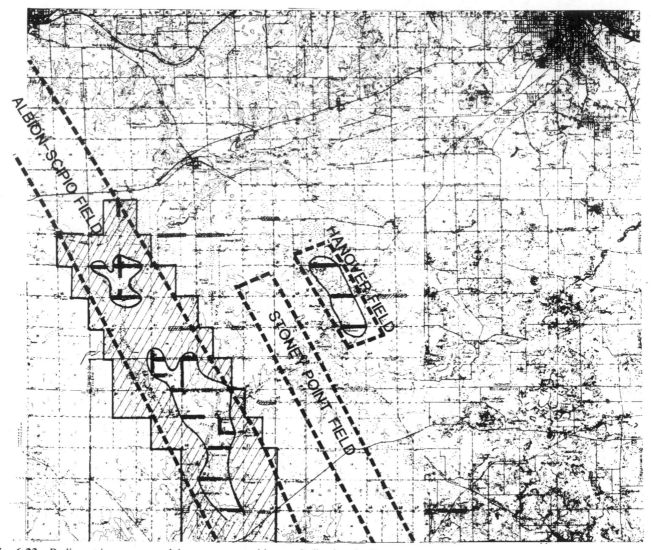

Fig. 6-23 Radiometric survey overlain on a topographic map indicating the Hanover SE and Stoney Point fields. Note that the Hanover Field has an associated anomaly but that Stoney Point does not (reprinted with permission from Delbert Long).

sium on Nuclear Techniques and Mineral Resources (Buenos Aires), International Atomic Energy Agency, Vienna, pp. 147–161.

Aithal, V. S. (1959). A note on oil exploration by radiometric survey, *Geophysical Exploration, a Symposium*, Baroda, India, pp. 128–134.

Alekseev, F. A. (1957). Radiometric method of prospecting for oil; the state and development of the method; and experience in its application, in *Trudy Vsesoyuznoi Nauchno-Tekhnicheskoi Konferentsii po Primeneniyu Radioaktivnykh i Stabil' nykh Izotopov i Izlucheniy v Narodnon Khozyaistve i Nauke*, April 4–12, *Rozvedka i Razrabotka Poleznykh Iskopaemykh.* (*Transactions in All Union Scientific Technical Conference on the Use of Radioactive and Stable Isotopes and Radiations in the National Economy and in Science*, April 4–12, 1957, *Prospecting and Development of Useful Minerals*), Gostoptekhizdat, Moscow, 1958, pp. 51–56: translation available as Atomic Energy Com-

mission (now the Nuclear Regulatory Commission) Report TR-4475, 1961, pp. 52–58.

Alekseev, F. A. and R. P. Gottikh (1965). Concerning the mechanism of formation of radiometric anomalies above petroleum deposits, *Sovetskaya Geologiya*, No. 12, Dec., pp. 100–120. Translation available in *International Geologic Review* (1966), Vol. 8, No. 10: pp. 1157–1171.

Alekseev, F. A., V. I. Yermakov, and V. A. Filonov (1960). Problems of the water content of radioactive elements in the waters of oil fields, in *Geophysical Abstracts* No. 183; p. 604.

Armstrong, F. E. and R. J. Heemstra (1972). Radiometrics proposed for exploration, *Oil and Gas Journal*, June 12, pp. 152–161.

Armstrong, F. E. and R. J. Heemstra (1973). Radiation halos and hydrocarbon reservoirs: A Review, Information Circular 8579, U.S. Bureau of Mines.

Balyasny N. D., A. V. Dmitiyev, and V. A. Ionov (1964). Device for subtraction of background noise, *Geofizicheseskoye Pribor-ostroyeniye*, No. 18, pp. 17–25; translation available in Foreign Technology Division Publication FTD-HT-23-885-67, Wright-Patterson Air Force Base, OH.

Baranov, V. I., N. G. Morozova, K. G. Kunasheva, E. V. Babi-cheva, and B. V. Karasev (1959). The radiometric method of exploration for petroleum and gas deposits, *Geochemistry*, No. 6, pp. 643–652.

Berger, I. A. and M. Deul (1956). The organic geochemistry of uranium, U.S. Geologic Survey Professional Paper 300.

Bisir, D. P. (1958). Discussion of some results in radioactive prospecting for petroleum and natural gas in Romania, *Proceedings of the 2nd United Nations International Conference on Peaceful Uses of Atomic Energy*, Vol. 2, United Nations Publications, New York, pp. 837–839.

Bloch, S., and R. M. Key (1981). Modes of formation of anomalous high radioactivity in oil-field brines, *American Association of Petroleum Geologists Bulletin*, Vol. 65, p. 154.

Bogoyavlenskiy, L. N. (1927). Radiometric exploration for oil, *Izvestiya Inst. Prikladnoy Geofiziki*, No. 3, pp. 113–123. Translation available in U.S. Bureau of Mines Information Circular 6072, 1928, pp. 13–18.

Bogoyavlenskiy, L. N. (1929). Radioactivity of ashes of some rock oils, U.S. Bureau of Mines Information Circular 6224, p. 22.

Boyle, R. W. (1982). *Geochemical Prospecting for Thorium and Uranium Deposits*, Elsevier Scientific Publishing, New York.

Boyle, T. L. (1955). Airborne radiometric surveying, *Proceedings of the International Conference on the Peaceful uses of Atomic Energy*, Geneva, Aug. 8–20, United Nations Publications, New York, pp. 744–747.

Broda, E. and T. Schonfeld (1966). *The Technical Applications of Radioactivity*, Pergamon Press, Oxford, Vol. 1, pp. 126–127.

Brookins, D. G. (1988). *Eh-pH Diagrams for Geochemistry*, Springer-Verlag, New York.

Budde, E. (1958). Radon measurements as a geophysical method, *Geophysical Prospecting*, Vol. 6, No. 1, pp. 25–38.

Collins, W. (1963). Gamma radiation survey of the Gilbertown area, Alabama, Information Series No. 31, *Geological Survey of Alabama*.

Crews, W. D. (1959). Radioactivity in exploration, *Oil and Gas Journal*, Pennwell Publications, Tulsa, OK, No. 21, May 18, pp. 391–397.

Curry, W. H. (1984). Evaluation of surface gamma radiation surveys for petroleum exploration in the deep Powder River Basin, Wyoming, in eds., M. J. Davidson and B. M. Gottlieb, *Unconventional Methods in Exploration for Petroleum and Natural Gas III*, Southern Methodist University, Dallas, TX, pp. 25–39.

Durrance, E. M. (1986). *Radioactivity in Geology: Principles and Applications*, Ellis Hoorwood, West Sussex, England.

Dyck, W. (1968). Radon-222 emanations from a uranium deposit, *Economic Geology*, Vol. 63, No. 3, pp. 288–289.

El Shazly, E. M., W. M. Meshref, K. M. Fouad, A. A. Ammar, and M. L. Meleik (1970). Aerial radiometry of El Alamein Oil Field, Egypt, *Geophysical Prospecting*, Vol. 17, No. 3, pp. 336–343.

Flerov, G. N. (1961). Use of nuclear physics in surveying and exploiting oil and gas deposits, *Proceedings of the Conference on Radioisotopes in the Physical Sciences and Industries*, Copenhagen, AEC Report TR-5404, pp. 1–5.

Foote, R. S. (1968). Application of airborne gamma-radiation measurements to pedologic mapping, *Proceedings of the 5th Symposium on Remote Sensing of the Environment*, University of Michigan, Ann Arbor, MI, pp. 855–875.

Foote, R. S. (1969A). Improvement in airborne gamma-radiation data analyses for anomalous radiation by removal of environmental and pedologic radiation changes, *Proceedings of the Symposium on Nuclear Techniques and Mineral Resources*, Buenos Aires, International Atomic Energy Agency, Vienna, pp. 187–196.

Foote, R. S. (1969B). Review of radiometric techniques, in *Unconventional Methods in Exploration for Petroleum and Natural Gas*, ed., W. E. Heroy, Southern Methodist University, Dallas, TX, pp. 43–55.

Gingrich, J. E. (1984). Radon as a geochemical exploration tool, *Journal of Geochemical Exploration*, Vol. 21, pp. 19–39.

Gingrich, J. E., H. W. Alter, and J. C. Fisher (1980). Recent advances in track etch radon detection system for uranium exploration, presented at the 8th International Geochemical Symposium, Hanover, Germany.

Gingrich, J. E. and Fisher, J. C. (1976). Uranium exploration using the track-etch method, *Exploration of Uranium Ore Deposits, International Atomic Energy Agency Proceedings*, International Atomic Energy Agency, Vienna, pp. 213–227.

Gohn, E. and E. Bratasanu (1967). On the possibilities of application of radiometry to direct prospecting for hydrocarbons, in *Geophysical Abstracts*, No. 247, p. 846.

Gregory, A. F. (1956). Analysis of radioactive sources in aeroradiometric surveys over oil fields, *American Association of Petroleum Geologists Bulletin*, Vol. 40, pp. 2457–2474.

Harmsen, K. and F. A. M. de Haan (1980). Occurrence and behavior of uranium and thorium in soil and water, *Netherlands Journal of Agricultural Science*, Vol. 28, p. 40.

Heemstra, R. J., R. M. Ray, T. C. Wesson, J. R. Abrams, and G. A. Moore (1979). *A Critical Laboratory and Field Evaluation of Selected Surface Prospecting Techniques for Locating Oil and Natural Gas*, BETC/RI-78/1B, U.S. Department of Energy.

Karasev, B. V. (1970). Determination of the isotopic composition of uranium in the soil, *Geochemistry International*, Vol. 7, No. 1, p. 203.

Kellogg, W. C. (1957). Observations and interpretation of radioactive patterns over some California oil fields, *The Mines Magazine*, July, pp. 26–28.

Kilmer, C. (1985). Shale normalized radiation surveying for petroleum exploration in geochemical prospecting, *IGM/Inside Oil & Gas Proceedings*, Houston, TX.

Klusman, R. W. and J. A. Jaacks (1987). Environmental influences upon mercury, radon and helium concentrated in soil gases at a site near Denver, Colorado, *Journal of Geochemical Exploration*, Vol. 27, pp. 259–280.

Kondratenko, A. F. (1957). Use of radiometric tools for geologic mapping, Chapter in *Trudy Vesesoyuznoi Nauchno-Teknicheskoi Konferentsii po Primeneniyu Radioaktivnykh i Stabil' nykh Izotopov i Izluchenii v Narodnom Khozyaistve i Nauke*, Gostoptekhizdat, Moscow, pp. 69–75.

Kopia, H. (1967). Detection and contouring of oil and gas deposits by means of the relative effectiveness of gamma radiation, in *Geological Abstracts*, No. 201, p. 913.

Lang, B. (1950). Gammatron surveys, *World Oil*, Vol. 131, No. 6, Nov., pp. 86–88.

Langford, G. T. (1962). Radiation surveys aid oil search, *World Oil*, Vol. 154, No. 5, pp. 114–119.

Langmuir, D. (1978). Uranium solution—mineral equilibrium at low temperatures with application to sedimentary ore deposits, in ed., M. M. Kimberley, *Uranium Deposits, Their Mineralogy and Origin*, Mineral Association of Canada, Short Course Handbook 3.

Latuerbach, R. (1970). Radium metallometry to indicate hidden tectonic breaks, in *Geophysical Abstracts*, No. 282, p. 816.

Lundberg, H. (1956). What causes low radiation intensities over oil fields? *Oil and Gas Journal*, April 30, pp. 192–195.

MacElvain, R. C. (1963A). What do near surface signs really mean in oil finding? Part I (of two), *Oil and Gas Journal*, Vol. 61, No. 7, pp. 132–136.

MacElvain, R. C. (1963B). What do near surface signs really mean in oil finding? Part II (of two) *Oil and Gas Journal*, Vol. 61, No. 8, pp. 139–146.

MacFayden, D. A. and S. V. Guedes (1956). Air survey applied to the search for radioactive minerals in Brazil, *Proceedings of the International Conference on Peaceful Uses of Atomic Energy*, Vol. 6, United Nations Publications, New York, pp. 726–739.

Mal'skaya, R. V. (1965). The radioactivity of groundwaters in the Western Ukraine, in *Geochemistry International*, Vol. 2, No. 6, pp. 1088–1091.

Marton, P. and L. Stegna (1962). On the basic principals of geophysical radioactive measurements, *Geofisica Pura e Applicata*, Vol. 53, No. 3, pp. 55–64.

Merritt, J. W. (1952). Radioactive oil survey technique, *World Oil*, Vol. 134, No. 1, pp. 78–82.

Miller, G. H. (1961). A geologist declares: "Radiation surveys can find oil," *Oil and Gas Journal*, Feb. 12, pp. 124–127.

Morse, J. G. and J. W. Alewine (1983). Airborne radiometrics: Case study of Patrick Draw Field, Wyoming, *Oil and Gas Journal*, Nov. 28, pp. 145–147.

Norwine, J. R., D. J. Hansen, D. F. Saunders, and J. H. Galbraith (1979). Near-surface moisture and biomass influences on the reliability of aerial radiometric surveys as a measure of natural radioelement concentrations, Bendix Contract EY-76-C-13-1664, U.S. Department of Energy.

Omes, S. P., Y. V. Bondarenko, N. I. Zakharova, and F. P.

Borkov (1965). A study of gamma fields above oil and gas deposits, Consultants Bureau, New York, pp. 179–189.

Pierce, A. P., J. W. Mytton, and G. B. Gott (1956). Radioactive elements and their daughter products in the Texas Panhandle and other oil and gas fields in the United States, *Proceedings of the International Conference on the Peaceful Uses of Atomic Energy*, United Nations Publications, New York, Vol. 6, pp. 494–498.

Pirson, S. J. (1969). Geological, geophysical and chemical modifications of sediments in the environment of oil fields, in *Unconventional Methods in Exploration for Petroleum and Natural Gas*, Southern Methodist University, Dallas, TX, pp. 158–186.

Pirson, S. J., N. Alarpone, and A. Avadisian (1966). Implication of log derived radioactivity anomalies associated with oil and gas fields, in *Logging Symposium*, 7th Annual, Tulsa, OK, May 9–11, SPWLA, Vol. 25.

Polshkov, M. (1965). New developments in geophysical and geochemical methods of prospecting for oil and gas in the U.S.S.R., Third Symposium on the Development of the Petroleum Resources of Asia and the Far East, Tokyo, Nov., Item 7.

Pringle, R. W., K. I. Roulston, G. W. Brownell, and H. T. F. Lundberg (1953). The scintillation counter in the search for oil, *Mining Engineer*, Dec., pp. 1255–1261.

Reimer, G. M. and J. K. Otton (1976). Helium in soil gas and well water in the vicinity of a uranium deposit, Weld County, Colorado, Geologic Survey, Open File Report 76-699.

Reimer, G. M. and R. S. Rice (1977). Line traverse surveys of helium and radon in soil gas as a guide for uranium exploration, central Weld County, Colorado, U.S. Geologic Survey, Open File Report 77-589.

Rothe, K. (1959). Problems of radiometric measurements over oil structures, *Berichte der Geologischen Gesellschaft*, Vol. 4, No. 2–3, pp. 183–187.

Scherb, M. V. (1953). Radioactivity in geophysical oil research, *Oil Forum*, Vol. 7, pp. 89–92.

Serdyukova, A. S. and V. P. Doshchechkin (1970). Possible uses of gamma-spectroscopy in logging wells containing radon, in *Petroleum Abstracts*, Vol. 10, No. 38, p. 2670.

Shchepak, V. M. (1964). Distribution of radium in ground waters of the outer zone of the East Carpathian Downwarp, in *Geochemistry International*, Vol. 1, No. 2, pp. 232–237.

Shchepak, V. M. and V. I. Migovich (1969). Silica in ground waters of petroleum reservoirs in the Pre-Carpathian Downwarp, in *Geochemistry International*, Vol. 6, No. 6, pp. 1093–1100.

Sikka, D. B. (1962A). Aeroradiometric survey of Redwater oil field, Alberta, Canada, *Metals and Minerals Review*, June, pp. 5–51.

Sikka, D. B. (1962B). Radiometric survey of Ten Section Oil Field, California, U.S.A., *Research Bulletin of the Punjab University*, Vol. 13, Parts I–II, pp. 149–161.

Sikka, D. B. (1963). Possible modes of formation of radiometric anomalies, *Geophysical Abstracts*, No. 206–265, p. 227.

Sterrett, E. (1944). Radon emanations outline formations, *Oil Weekly*, Vol. 115, No. 4, pp. 29–32.

Stothart, R. A. (1943). Radioactivity determinations set production delimitations, *Oil Weekly*, Vol. 108, No. 5, pp. 19–21.

Stothart, R. A. (1948). Tracing wildcat trends with radon emanations, *World Oil*, Vol. 127, No. 10, pp. 78–79.

Stothart, R. A. (1954). Delineation of petroleum areas by radioactive emanations survey, *World Petroleum*, Vol. 25, No. 4, pp. 78–79.

Tanner, A. B. (1960). Usefulness of the emanation method in geologic exploration, Chapter in *Geological Survey Research 1960*, U.S. Geologic Survey, Professional Paper 400-B, pp. B111–112.

Thomeer, J. H. (1966). Exploration for oil and gas by emanometric methods, in *Geophysical Abstracts*, p. 466.

Trapp, G. and H. F. Victoria (1965). Superficial radioactivity in the La Cira Field and its possible relation to petroleum, in *Petroleum Abstracts*, No. 57, p. 543.

Us, E. M. and V. L. Kripnevich (1966). Some data on anomalous natural gamma-radiation fields above oil and gas deposits in the Western Fore-Caucasus, *Petroleum Abstracts*, Vol. 6, No. 19, p. 1065.

Voegl, E. (1970). The quantification of uncertainty and profitability in petroleum and natural gas exploration and production, *Petroleum Abstracts*, Vol. 10, No. 44, p. 3009.

Weart, R. C. and G. Heimberg (1981). Exploration radiometrics: post-survey drilling results in *Unconventional Methods in Exploration for Petroleum and Natural Gas II*, Southern Methodist University, Dallas, TX, pp. 116–123.

Williams, W. J. and P. J. Lorenz (1957). Detecting subsurface faults by radioactive measurements, *World Oil*, Vol. 144, No. 5, pp. 126–128.

Yermakov, V. I., and A. N. Shatsov (1959). Radiometric survey in oil-bearing regions of West Turkmenia, in *Geophysical Abstracts*, No. 178, p. 373.

Yermakov, V. I., and A. N. Shatsov (1966). On the radiometric anomalies in areas of the Binagady-Khurdalan Oil Field, in *Geophysical Abstracts*, No. 239, p. 1190.

Zeuch, R. (1965). On new possibilities for prospecting by means of radioactivity measurements, in *Geophysical Abstracts*, No. 218. p. 230.

Zolotovitskaya, T. A. (1965). On the radiometric anomalies in areas of the Binagady-Khurdalan Oil Field, *Izvestiya Akademii Nauk Azerbaidzh, SSR*, Seriya Geologicheskiia-Geogr. Nauk, No. 5, pp. 80–87.

7

Halogens

Introduction

The halogen group is composed of chlorine, fluorine, bromine, and iodine. Halogens are important components of many types of compounds, especially when bonded with hydrogen. Halogens closely associate themselves with petroleum-derived organics; iodine specifically has been used in the search for hydrocarbons. All the halogens have been employed to delineate ore deposits of zinc, cooper, and gold (Boyle, 1987; Levinson, 1980). However, only iodine has been used consistently to delineate petroleum deposits (Kudel'sky, 1977; Gallagher, 1984; Allexan et al., 1986; Singh et al., 1987; Gordon and Ikramuddin, 1988; and Tedesco and Goudge, 1989).

History

Iodine has a strong affinity with gold, petroleum, and naturally occurring organic compounds. It has been used in petroleum exploration for a number of years, especially by the Russians, later by the Canadians, and most recently by the United States.

Figure 7-1 (see color plate) is an example of a typical iodine survey in a petroleum area. The background values in this survey range from 0.6- to 1.1-ppm molecular iodine. The anomalous values are typically over 2.0-ppm molecular iodine. Measurements between background values and the anomalous values associated with the field represent minor anomalies or a transition zone.

Iodine's association with gold has been documented by Boyle (1987). Gold and petroleum deposits are not usually found together (except in areas such as Nevada's Basin and Range Province) so that the explorationist generally is not concerned about obtaining a misleading interpretation with relation to gold.

Chlorine and fluorine also seem to be associated with petroleum hydrocarbons, but there are no published data to establish their viability and effectiveness. Bromine was mentioned by Gallagher (1984) as being present in the soils over some, but not all, oil fields. In the same study, iodine was consistently in anomalous amounts at each of the fields surveyed. Therefore, this chapter is focused almost entirely on iodine.

Halogen Reactions

Halogens combine with organic compounds by halogenization or photochemical reactions. Halogenization is a process that is thought to occur primarily under synthetic or drastic environmental conditions and thus is not usually present under normal circumstances. Halogenization seems to occur naturally in algae. Dehalogenization reactions occur more readily in the environment as either hydrolysis or disproportionation reactions (Moore and Ramamoorthy, 1984). Halogen compounds are susceptible to hydrolysis because of a charge separation between the halogen and carbon atoms. Table 7-1 is a summary of the half-life of some halogenated compounds (Maybe and Mill, 1978).

Photochemical reactions are the more likely cause of structural changes that allow absorption of halogens by organic molecules. These reactions result from the effect of

Table 7-1 Hydrolytic half-lives of some halogenated compounds.

Compound	Half-Life
CH_3F	30 years
CH_3Cl	339 days
CH_3Br	20 days
CH_3I	110 days
$CH_3CHClCH_3$	38 days
$CH_3CH_2CH_2Br$	26 days
$C(CH_3)_3Cl$	23 s
CCl	7000 years
$CHBr_3$	686 years
CH_2Cl_2	704 years

Source: Maybe and Mill (1978).

electromagnetic radiation in the near-ultraviolet range (240–700 nm). Ionizing radiation is not in sufficiently concentrated form to cause molecular alterations. Therefore, concentrations of halogens are caused by either (1) transfer of electrons or energy by an intermediate photosensitizer or (2) deactivation reactions causing direct absorption by the molecule entering into an excited state.

Organic molecules that undergo these electronic transitions have a wavelength of maximum response (λ max) and molar extinction coefficients (e) that are the magnitude of the molecules' ability to absorb photons. Thus with higher λ max, the structure of any organic compound in the environment will determine the ability of the reaction to occur. Ultraviolet absorption is common in many unsaturated and aromatic petroleum compounds. An increase in the number of double bonds in a petroleum molecule will require less energy for the electronic transition.

Table 7-2 is a summary of a λ max and e required for the organic molecules to absorb various chromophores. Figure 7-2 summarizes the various energies of electromagnetic radiation in different wavelength regions and the energies required for dissociation of some typical diatomic chemical

Table 7-2 Summary of wavelength of maximum response (λ max) and molar extinction coefficients (e) for certain halogens

Chromophore	Functional Group	λ max	ε max
$-Cl$	CH_3Cl	1730	100
$-Br$	CH_3Br	2040	200
$-I$	Ch_3I	2580	378

Source: Tinsley (1979); reprinted with permission of John Wiley & Sons.

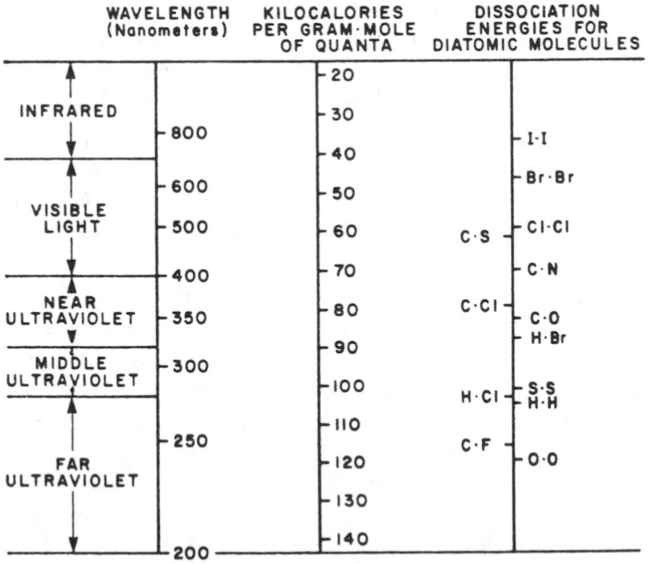

Fig. 7-2 Comparison of electromagnetic radiation and bond energies (Tinsley, 1979; reprinted with permission of John Wiley & Sons).

bonds. Both Table 7-2 and Fig. 7-2 suggest that iodine is more likely to bond with an organic molecule than are other halogens, followed by bromine and then chlorine. Fluorine is not present in this chart.

Process of Iodine Accumulation

Iodine's proposed surface relationship to petroleum accumulations at depth begins with the hydrocarbon gases reaching the soil/air interface. Under infrared and ultraviolet light, the hydrocarbons react with the iodine and form iodorganic compounds. Removing the hydrocarbon gases from the soil causes, over time, physical and biological processes that will reduce the amount of iodorganic compounds present.

The source of the iodine was originally thought to be predominantly sea spray or the atmosphere (Vinogradov, 1959; Chudecki, 1960; Karelina, 1961; Fuge, 1974). This suggested that background and anomalous values would decrease away from the coast and that samples taken near the coast would be difficult to interpret because of the large variations in the flux of the iodine. However, the sea spray concept presented conflicting results.

1. Normal soil iodine values in eastern Colorado in the central United States and iodine values from northwestern Turkey next to the Sea of Marmara have approximately the same background or normal soil values, 0.5- to 1.0-ppm molecular iodine (Allexan et al., 1986; Tedesco and Goudge, 1989). Eastern Colorado and northwestern Turkey have similar soil types. Central Texas in the southwestern United States, which is also relatively distant from a sea spray effect, has normal background values of 3.0- to 6.0-ppm molecular iodine (Allexan et al., 1986).

2. Iodine values initially increase with depth but then rapidly decrease below 10 ft (Runyon and Rankin, 1936; Hitchon et al., 1971). However, Collins and Egleson (1967), Collins (1969, 1975), Kudel'sky (1977), and Levinson (1980) documented iodine-rich brines in association with oil fields in specific basins. This suggests that infrared light may not be the catalyst and that other types of reactions and mechanisms are occurring.

The anomalous iodine compounds are formed by I^-, IO_4^-, and IO_3^- attaching to a vacant cation site or replacing another element that is weakly bonded. Sorption of I^- is only 30% of total iodine absorbed. IO_4^- is more strongly absorbed than IO_3^-, and both are absorbed readily in acidic solutions but less in alkaline condition. Even though absorption occurs onto iron oxides and aluminum oxides, the iodine is easily volatilized and removed. The compound IO_3^- is absorbed preferentially over I^- onto iron oxides. Figure 7-3 is an Eh-pH diagram for iodine, which suggests that, in most environments, the form of iodine is I^-. The point at which the form changes is well above the upper stability limits for

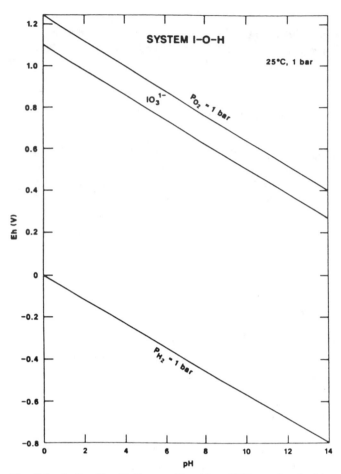

Fig. 7-3 Iodine Eh-pH diagram (Brookins, 1988; reprinted with permission of Springer-Verlag).

soils. This was confirmed by sampling in basins where coal-bed methane and biogenic gas are a recoverable resource. Current thought is that the iodine is associated with one or more of the heavier hydrocarbons. Very little research has been done on the form of the iodorganic compounds found in the soils above petroleum accumulations.

Sampling

Sampling for iodine is typically done within 2 in. of the surface. Organic material should be avoided if possible. Some authors have reported much deeper sampling of 6 in. to several feet. After collection, the sample is dried, and the clay fraction ($<$180 mesh) is sieved off for analysis by colormetric methods. The resulting value is represented as molecular iodine in parts per million (weight percent). Values are plotted and then contoured to determine if an anomaly (halo) exists. Typical background soil values can range from 0.5 to 6.0 ppm depending on the area but are usually below 2.0 ppm. Anomalous values are usually 150% to over 1000% higher than background values.

Factors Affecting Iodine Accumulation

Several different factors affect the accumulation of iodine in the soil substrate: rock types, organic matter, clays, Eh-pH, moisture content, and the general chemical composition of the soil.

water (Brookins, 1987). The form of all the iodine compounds related to petroleum accumulations has not been identified, but one form is possibly CH_3I (Moore and Rama-moorthy, 1984). This compound is reported to have a half-life of 55 days. However, the iodine compounds analyzed in the detection of petroleum accumulations seem to have a longer stability period. This is based on samples reanalyzed after several months of storage. The results were within -40% of the original values.

Iodine typically occurs as a minor component of various minerals but generally does not form separate minerals. Some of the compounds that iodine does form are: CuI, $Cu(OH)(IO_3)$, AgI, polyhalides, iodate, and periodates (Kabata-Pendias and Pendias, 1992). Iodine compounds are readily soluble. They are accumulated by organics, clays, and carbon in the soil depending on their sorption capacity. Prince and Calvert (1973) indicated that reducing environments contain higher amounts of iodine than oxidizing environments do.

Methane alone does not seem to cause increases in iodine. Normal decomposition of organic matter causes an increase in iodine but typically not in the higher amounts as when petroleum-generated hydrocarbons are introduced in the

Rock Types

Rocks are generally deficient in iodine relative to concentrations in soils and in brines and have a typical range from 0.2 to 0.8 ppm. The average concentration in the lithosphere is 0.3 ppm (Aubert and Pinta, 1977). Sedimentary rocks of marine origin are the exception, however, with values of 100 to 1000 ppm compared to metamorphic rocks (1–2 ppm), basic volcanic (0.5–0.8 ppm), and felsic or acidic volcanics ($<$0.5 ppm). Rivers and estuaries typically carry 0.1 to 40 μ/l, which is similar to amounts found in rainwater. Iodine in lake water is typically related to salinity. Iodine accumulation in lake bottom sediments undergoes a continuous process of reduction that releases much of the iodine into interstitial waters. Evaporite sequences of marine origin contain minor amounts of iodine, 0.005 to 0.2 ppm (Goldschmit, 1954). Evaporites of terrestrial origin have excessive iodine values of 200 to 1700 ppm. Typically these are formed under restrictive conditions with the absence of biological activity in closed desert basins such as the Chilean nitrate deposits. Soils are richer in iodine compared to their parent rock (Karelina, 1965; Aubert and Pinta, 1977; and Shacklette and Boergnegen, 1984). Concentrations vary from 0.2 to 30 ppm and depend on clay content, organic matter, and the amount of carbonates present.

pH

pH has also been said to affect iodine (Chudecki, 1960). However, in a later study (1963), Chudecki indicated that pH had no relationship to iodine accumulation. Kabata-Pendias and Pendias (1992) reported that soil acidity favors iodine sorption by organic matter, Fe and Al hydrous oxides, and illite clay, but alkaline soils accumulate iodine by a salinity process.

Organic Matter, Clay, and Humus

The majority of studies point to organic matter, humus, and clay as the main accumulators of iodine in soils (Kovda and Vasil'eyvskaya, 1958; Gallego and Oliver, 1959; Vil'-gusevich and Bulgakov, 1960; Zyrin and Bykova, 1960; Karelina, 1961; Zimovets and Zelenova, 1963; Glushenko et al., 1964; Tikomirov et al., 1980). As organic material or clay content, or both, increases, the amount of iodine in soil increases as well. Koch and Kay (1987) found that different types of organic matter absorbed iodine at different rates and in different amounts. Soils derived from carbonate parent material have also shown greater amounts of iodine present because of the carbonate absorption of halogens (Katalymov, 1964; Gallego and Oliver, 1959). However, the author has found that some carbonate-rich soils do not exhibit increases compared to noncarbonate soils in the same region. Most published studies indicate that samples were usually taken from the A horizon. The elluvial zone in one study indicated marked decreases in iodine content that are not necessarily related to organic matter (Zyrin and Bykova, 1960). It was not reported in any of the studies if the collected soil samples were near existing oil fields or in petroleum-producing basins. Therefore, it is difficult to determine if the wide range of values may be a result of petroleum that has migrated into the soil.

The amount of hydrocarbons migrating or seeping into the soil from below is generally measured in parts per billion and sometimes in parts per million. Organic matter is usually in parts per thousand to a percentage of the total soil. Small amounts of petroleum can increase the amount of iodine present from 1.5 to 50 times the amounts accumulated by the soils. Iodine enrichment of brines associated with petroleum accumulation has shown a correlation with short-chain aliphatic acid anions of marine origin, more commonly called *organic acids* (Means and Hubbard, 1985). When these hydrocarbons seep to the surface, iodine is preferentially absorbed.

Studies in oil field brines have shown dramatic increases in iodine (Runyon and Rankin, 1936; Collins and Egleson,

1967; Collins, 1969; Levinson, 1980). Organic matter that is altered, such as petroleum, is thus chemically more reactive in forming iodine compounds. However, organic matter and clay do play a role in the increases in iodine in the soil. The amount of organic matter and clay tends to control the amount of iodine that can be additionally accumulated by petroleum in the soil. Sandy soils, in general, will have values below 1 ppm of iodine, and iodine accumulations in sandy soils associated with oil fields will typically not exceed 4 ppm. But when the soils are rich in organics and clays and have iodine values of 4 to 8 ppm, the anomalous values will typically range from 10 to 40 ppm.

Soil Types

Iodine variations in soils have also been related to soil type and moisture content (Rao et al., 1971; Whitehead, 1978; Gallagher, 1984). Iodine has been reported to be closely associated with certain iron and aluminum oxides and organic matter under different pH conditions (Whitehead, 1978). Oxulated-soluble aluminum showed the closest correlation, and only when the pH fell below 4.8 did iodine show a closer association with iron oxides. Iodine also increases dramatically in organic-rich soils vs. lean or nonorganic soils. The study did not indicate whether any of the soils were sampled in or near a petroleum-productive basin. Some soils showed no variations in iodine with depth whereas others did. Rao et al. (1971) indicated that the clay mineral illite fixes iodide in greater quantities than montmorillonite and kaolinite. This suggests that the soil itself could be another source of the iodine. The hydrocarbons could scavenge iodine out of oxide coatings, minerals, and so forth, and redeposit as easily measured iodorganic compounds. Locally, exposed igneous rocks or evaporites could add iodine to soils. Therefore, the best results would come from an area where soil types are relatively uniform.

Figure 7-4 is a survey conducted in sandy soils having almost no visible organic matter. The background values range from 0.6- to 1.0-ppm molecular iodine. Figure 7-5 is a survey done in highly organic soils. The background values range from 1.5- to 2.8-ppm molecular iodine. However, in both cases, the variation between background values and anomalous values was significant and did not affect interpretation.

Moisture, Soil, and Lithology Variations

Moisture content is a factor that affects the measured iodine present in the soil. However, work by the author indicates that the moisture content in the soil causes less than a 10% variability in iodine values.

Iodine values can be affected if samples are taken from different horizons because the organic content will vary.

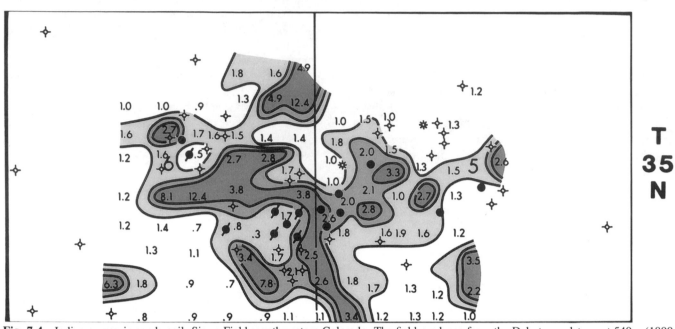

Fig. 7-4 Iodine survey in sandy soil, Sierra Field, southwestern Colorado. The field produces from the Dakota sandstone at 540m (1800 ft). The east side of the field is topographically higher than the west side (reprinted with permission from Atoka Exploration Corporation).

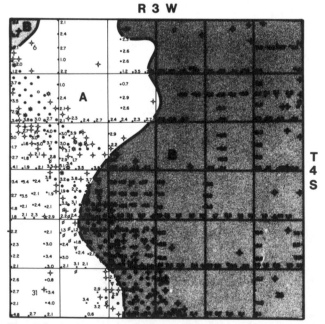

Fig. 7-5 Iodine survey in organic-rich soils. Albion-Scipio Oil Field, Michigan, Michigan Basin. (A) Deep, well-drained, and very poorly drained loamy, mucky soils formed in glaciofluvial deposits, glacial fill, or organic material (Boyer-Oshtemo-Houghton Association). (B) Deepened, moderately deep, well-drained, and somewhat excessively drained loamy soils that formed in glacial till material that weathered from sandstone or in glacial till over sandstone (Hillsdale-Eleva Riddles Association). (Reprinted with permission from Atoka Exploration Corporation.)

Consistently sampling the same soil layer results in data that are usable and comparable to samples taken right at the soil surface. Therefore, it is simpler to sample right at the surface to avoid a multidimensional problem and increased survey costs.

Investigators have reported that iodine values, both background and anomalous, do not change dramatically across widely divergent soils and changes in bedrock lithology (Fig. 7-6). However, the background and anomalous values may change when the bedrock changes from a low-organic to a high-organic lithology. Bedrock variations may require some normalization of the data but will not adversely affect interpretation as long as they are recognized.

Fertilizers

Fertilizers contain small quantities of iodine. However, when fertilizers are added to soils, they are distributed over large areas. The iodine that is part of the fertilizer becomes a small percentage of the total amount in the soil. Liming of soils does reduce the solubility of iodine and iodates and thus reduces iodine availability for biological processes.

Case Histories

Presented below are recent case histories using iodine as an exploration tool either alone or in conjunction with other nonsurface geochemical methods. The first four case histor-

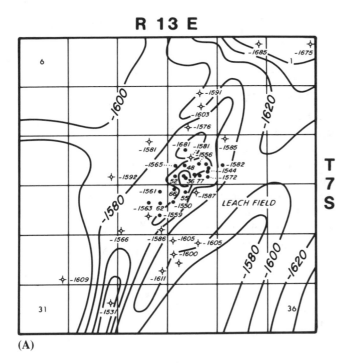

(A)

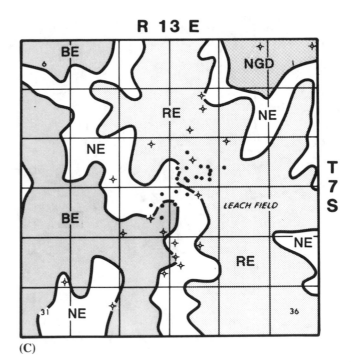

(C)

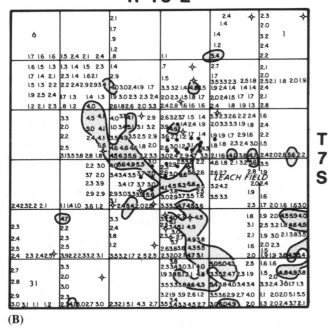

(B)

Fig. 7-6 Iodine survey with variable underlying bedrock topography and subcropping lithologies. The Leach Field is located in Township 7 South, Range 13 East, Jackson County, Forest City Basin, Kansas. (A). The field produces from the Hunton (Devonian-Silurian) and Viola (Ordovician) strata. The bedrock geology is highly variable. The iodine survey indicates an elongated, linear anomaly perpendicular to the field. Localized anomalies border the field. The large linear anomaly coincides with a suspected fault based on seismic. At the time of the survey, the field was near the end of its productive life (reprinted with permission from Farleigh Oil Properties).

ies all lie in the Denver Basin in the central United States (Fig. 5-57). The most productive reservoirs are the Lower Cretaceous J sandstone and the Upper Cretaceous D sandstone (Fig. 5-58). Both the D and J sandstones were deposited during two short-term transgression and regression sequences of the Cretaceous seas. Petroleum production is found in a 75-mile-wide fairway. Traps are porous channels located near, but not always in association with, structural highs such as the Garnet East Prospect and Friday Field. Other types of traps are permeability and porosity pinchouts and channel terminations such as the Dolley and KMA fields.

Friday Field, Adams County, Denver Basin, Colorado

The Friday Field, located in the Denver Basin, Section 14, Township 1 South, Range 61 West, Adams County, Colorado, was discovered in June 1988 using iodine analysis and subsurface geology. Prospect delineation was initially based on subsurface data from existing wells that outline a general D sandstone channel of Cretaceous age trending to the northwest (Fig. 7-7).

An iodine survey was conducted in January and February 1988 across the subsurface lead to determine if a petroleum

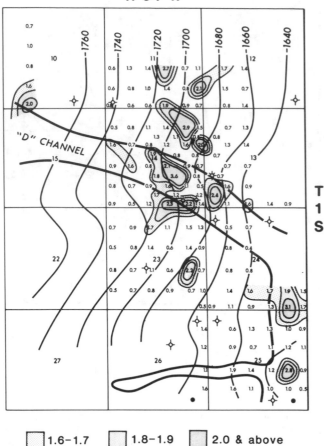

R 61 W

☐ 1.6–1.7   ☐ 1.8–1.9   ▨ 2.0 & above

Fig. 7-7 Geologic concept for D sand channel trend, Friday Prospect, Section 14, Township 1 South, Range 61 West, Adams County, Colorado. The structure is on top of the D sandstone. The solid lines indicate the projected location of the D channel. The first iodine survey across the Friday Prospect is indicated. Contour intervals are 1.6 (light pattern), 2.0 (medium pattern), and 2.4 ppm (dark pattern). Soils are sandy, variable organics and all are farmed. The anomaly suggest a halo or isolated series of apical anomalies (reprinted with permission from Atoka Exploration Corporation).

accumulation existed and to locate a possible drill site (Fig. 7-9; see color plate). The iodine survey indicated an anomaly trending to the north and creating a halo around the east half of Section 14. Comparison of the anomaly to a model of an existing field indicated similar values and patterns (Fig. 7-8; see color plate).

In the summer of 1988, the operator drilled the Friday No. 1 to the D sand and encountered 1.6 m of reservoir rock. The well was completed and flowed with a reported initial potential of 75 bbl of oil a day, 1.4 million ft³ (MMCF) of gas and no water. A short time later, another operator drilled 1200 m to the north and encountered 7.6 m of D sand. This well had an initial potential of 5 bbl of oil a day and 1 MMCF of gas. A reinterpretation of the subsurface data provides a much better match of the now defined chan-

nel to the iodine anomaly (Fig. 7-9; see color plate). The Friday Field produced 12,078 bbl of oil and 518,177 MMCF of gas as of December 1991.

A second iodine survey, using increased sample density, was carried out in July 1989 to further delineate the field (Fig. 7-9). This survey gives a clearer definition of the field and indicates little potential for finding additional production. This is supported by additional wells that were dry.

Figure 7-10 (see color plate) shows the values from the second survey at the same sampling points as the first survey. Both surveys indicate similar outlines of the field despite sampling in different years and in different seasons. The data suggest that iodine is not significantly affected by seasonal variations in soil geochemistry. Table 7-3 is the field's production to October 1991.

Dolley Field, Weld County, Denver Basin, Colorado

This series of iodine surveys takes the idea of repeatability and the lack of seasonal effects to a more refined level. The Dolley Field is located in sections 14, 15, 22, and 23 Township 6 North, Range 61 West in the Denver Basin, Weld County, Colorado (Fig. 5-57). It produces from the Cretaceous D sand. This field was pursued on the basis of a geologic concept for D sand development and on seismic's (propietary information) ability to find this event. Iodine was used as a secondary tool. Figure 7-11 is the interpreted subsurface geology prior to the iodine survey and drilling. Figure 7-11 is the iodine survey conducted in January 1991, which correlates closely with the known seismic anomaly and proposed geologic model. The first three wells drilled resulted in significant production from the D sand (Fig. 7-12). The geologic map now indicates the presence of a sand body in association with the iodine anomaly at the surface. The dry hole located to the north was drilled based on other methods, and the iodine results were not considered.

A second iodine survey was conducted in April 1991 (Fig. 7-13). Background and anomalous values increased slightly but did not change the interpretation. Subsequently, four additional wells were drilled. The two wells outside the anomaly had less than 1 ft of pay and had to be stimulated to obtain production. Figure 7-13 is the known geologic picture to this point.

The results of a third survey (Fig. 7-14) done in July 1991 show some significant differences from previous surveys. The background and anomalous values are quite elevated. During the three months prior to sampling, a significant amount of rainfall had occurred in this area.

Garnet East Test, Kimball County, Nebraska

The Garnet East Test is located in section 35, Townships 15 and 16 North and Range 51 West, Kimball County,

Table 7-3 Production from Friday Field, Township 1 South, Range 61 West, Adams County, Colorado.

	1988	1989	1990	1991	1992	Total
Well 1 Oil	2,228	3,627	874	246	0	6,975
Gas	69,795	25,266	2,044	0	0	97,105
Water	0	4,185	0	0	0	4,185
Well 2 Oil	743	3,026	1,321	756	146	5,249
Gas	66,790	320,253	18,712	15,317	6,746	427,818
Water	0	0	2,000	0	0	2,000

Nebraska. It was targeted to drill to the Cretaceous J sandstone (Fig. 5-57). The prospect was defined based on subsurface geology and detailed seismic data that indicated a structure in the J sandstone induced by deeper Permian Lyons dune buildup (Fig. 7-15). The iodine survey did not indicate anomalous values in the prospect area (Fig. 7-15), but there are some anomalous values within the Garnet Field. A well was drilled in 1991, the seismic structure was confirmed, and shows were encountered. The DST on the well yielded

only salt water, and the well was subsequently plugged. The results of the iodine survey were confirmed.

KMA Field Extension, Kimball County, Denver Basin, Nebraska

The KMA Field is located in Section 17, Township 14 North, Range 55 West, Kimball County, Nebraska. The field was discovered in 1983, and the initial well is located in the NE NE of Section 17 (Fig. 5-57). It was determined by engineering calculations that a larger reservoir was being

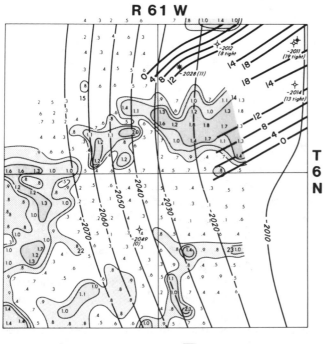

0.8-0.9   1.0-1.1   1.2 & above

Fig. 7-11 Subsurface structure map on top of the J sandstone prior to the discovery of the Dolley Field, Township 6 North, Range 61 West, Weld County, Colorado. Heavy lines represent an isopach of the D Sandstone. The contour interval is 2 ft. The contour interval for the structure map is 10 ft. The shaded area represents the iodine anomalies from the first survey. Iodine survey, January 1991, prior to the discovery of the Dolley Field. Contour intervals are 0.8, 1.6, and 1.2 ppm. Background values range from 0.4 to 0.8 ppm. Soils are sandy, with varying amounts of organics, and rangeland and winter wheat farming is typical (reprinted with permission from Atoka Exploration Corporation).

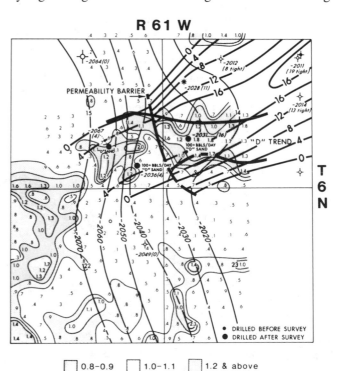

0.8-0.9   1.0-1.1   1.2 & above

Fig. 7-12 Subsurface structure map on top of the J Sandstone, post-discovery of the Dolley Field, Township 6 North, Ragne 61 West, Weld County, Colorado. Heavy lines represent an isopach of the D sandstone. The contour interval is 2 ft. The contour interval for the structure map is 10 ft. The shaded area represents the iodine anomalies from the first survey. Post-first iodine survey drilling of the Dolley Field. The location of the wells indicates that the iodine anomaly is apical. Pay zone thickness ranges from 1.3 m (4 ft) to 1.8 m (6 ft).

R 61 W

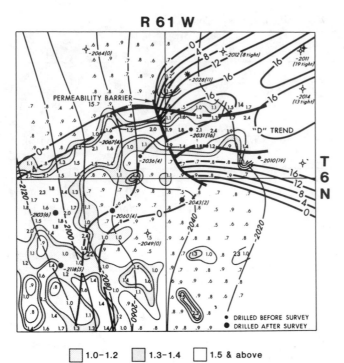

☐ 1.0-1.2   ☐ 1.3-1.4   ☐ 1.5 & above

Fig. 7-13 Post-second iodine survey drilling, Dolley Field, Township 6 North, Range 61 West, Weld County, Colorado. The wells located in the NW of section 23 and NW of the NE of section 22 had less than 1 ft of effective pay. The subsurface structure map is on top of the J sandstone. Heavy lines represent an isopach of the D sandstone, and the contour interval is 2 ft. The contour interval for the structure map is 10 ft. The shaded area represents the iodine anomalies from the first survey (reprinted with permission from Atoka Exploration Corporation).

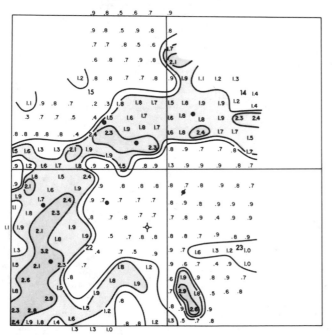

Fig. 7-14 Third iodine survey, July 1991, Dolley Field, Township 6 North, Range 61 West, Weld County, Colorado. Contour intervals are 1.0, 1.5, and 2.0 ppm of iodine. The field produced 95,245 bbl of oil and 613 MMcf by the end of December 1991 (reprinted with permission from Atoka Exploration Corporation).

R 51 W

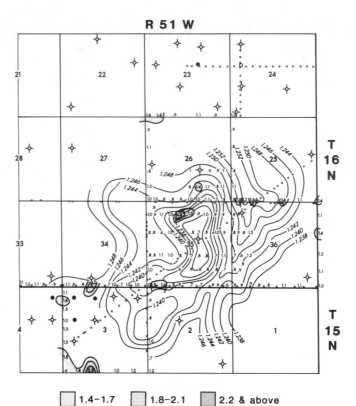

☐ 1.4-1.7   ☐ 1.8-2.1   ☐ 2.2 & above

Fig. 7-15 Structure on top of the Cretaceous J sandstone, iodine survey and postdrilling results, Garnet, East Prospect, Kimball County, Nebraska. Anomalous values are 1.4, 1.8, and 2.2 ppm. The structure was defined by subsurface geology and seismic and has a minor iodine anomaly associated with it. Subsequent drilling yielded shows, a DST produced all water, and the well was plugged (reprinted with permission from Atoka Exploration Corporation).

drained because there was no decline in the production for this one well over several years. Figure 7-16. indicates the geologic concept prior to the implementation of the iodine survey. It was thought that the southeast quarter of Section 8, the west half of the southwest of Section 9, the west half of the northwest of Section 16, and the northeast of Section 17 would be prospective. Figure 7-16 is the resulting iodine survey, which did not support the model. However, the decision was to drill, based on the geologic model. Subsequent drilling in sections 8, 9 and 16 yielded only one producing well. Drilling continued into Section 17, which had a large iodine anomaly covering most of the section. Four producing wells were completed in which each yielded approximately 30 bbls a day (Fig. 7-17). The iodine sampling density did not define production and, based on the results of the Dolley Field study above, this can be considered a reconnaissance survey.

Wehking Field, Atchison County, Forest City Basin, Kansas

The Wehking Field is located in Section 35, Township 6 South, and in sections 2 and 11, Township 7 South, Range

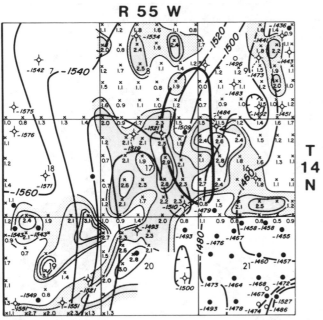

Fig. 7-16 Geologic concept for extending the KMA Field, Kimball County, Nebraska, which produces from the J sandstone. The projected porous J sandstone is outlined in green. The structure is on top of the J sandstone interval, and the contour interval is 10 ft. Iodine survey for the area around the KMA Field, Kimball County, Nebraska. The contour interval is 1.6 (light pattern), 2.0 (medium pattern), and 2.4 (dark pattern) ppm. Background values range from 0.8 to 1.2 ppm (reprinted with permission from CST Oil & Gas Corporation).

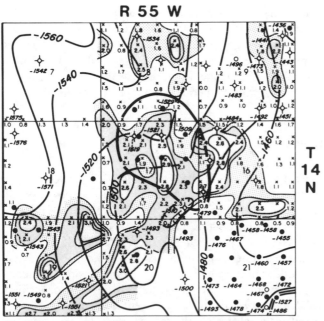

Fig. 7-17 Post–iodine survey drilling and geology for the KMA Field, Kimball County, Nebraska. The defined J sandstone porous section is outlined in green. Contour interval is 10 ft. The structure is on top of the J sandstone. The iodine anomaly and porous J sandstone subsequently correlate relatively well (reprinted with permission from CST Oil & Gas Corporation).

18 East, Atchison County, Kansas (Fig. 7-18). It was discovered in 1988. The field was drilled based on subsurface geology and a shallow core hole program. The Wehking No. 1 well was drilled to the Hunton (Devonian-Silurian) formation at 993m (2800 ft), but production was found in the Pennsylvanian McClouth sandstone (Fig. 7-19) at 580 m (1800 ft). The iodine survey was conducted after the discovery well was drilled but prior to field development and connection to a gas transmission line. The iodine anomaly is apical in shape. Subsequent drilling indicates a McClouth channel sand trending northeast-southwest, which dissects the Mississippian carbonate surface on a northwest plunging anticline. The gas produced is 96% to 98% methane, with minor amounts of heavy hydrocarbons. The best producing well lies in Section 35 and had an initial potential of 30 MMcf a day. All the wells, except two, do not seem to have any pressure communication.

Virburnum Trend, Forest City Basin, Missouri

Soil-gas and associated methods for petroleum exploration have recently been applied by the mining industry in select areas where ore deposits have accumulated appreciable amounts of petroleum. The Virburnum Trend is defined (Fig. 7-20) in the subsurface by a north-south dolomite body.

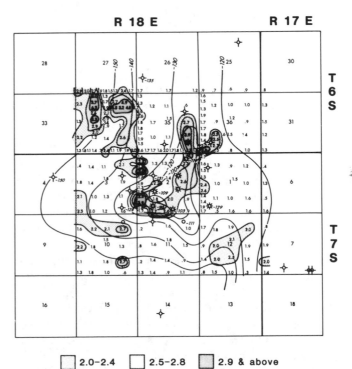

Fig. 7-18 Iodine survey and postdrilling geologic interpretation of the Wehking Gas Field, townships 6 and 7 South, Range 18 East, Atchison County, Kansas. Contour intervals are 2.0, 2.4, and 2.8 ppm. Background values range from 0.9 to 1.5 ppm. The anomaly is apical in form. The structure is on the base of the Ft. Scott limestone, and the contour interval is 10 ft (reprinted with permission from Atoka Exploration Corporation).

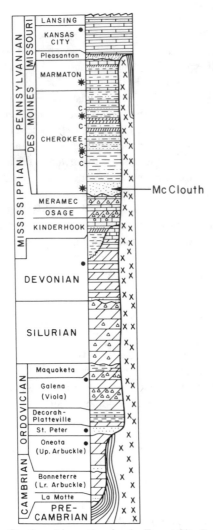

Fig. 7-19 Stratigraphic section for the Forest City Basin, Kansas and Missouri.

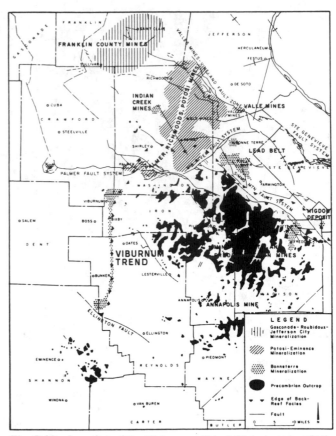

Fig. 7-20 Location of the Virburnum Trend (Hagni et al., 1986; reprinted with permission of the Geological Society of America).

The origin is similar to the petroleum-productive Albion-Scipio Field (which contains unusual amounts of sphalerite and galena) in the Michigan Basin. The petroleum associated with the Virburnum Trend is derived from Ordovician rocks deposited either during or after replacement of limestone by dolomite/Pb-Zn ore bodies. The petroleum has not been found in economic quantities but is consistently noted in association with the ore. Both the Virburnum and Albion-Scipio dolomite bodies seem to have formed under similar conditions but contain different commodities. Iodine and soil-gas reconnaissance surveys were carried out across a select part of the trend (Figs. 7-21–7-23). The iodine survey indicates a background of 1.2 to 1.9 ppm (Fig. 7-21). Where the deposits are located, the iodine is significantly elevated. One soil-gas survey line (Fig. 7-22) transected the main ore body, and the other line crossed supposedly barren ground. Figure 7-22 indicates the C_2, C_3, and ratios of C_2/C_3 across the first survey line. The results show increases in all three parameters that match the iodine data, indicating anomalous areas associated with the ore body. Figure 7-23 illustrates

the results across the second survey line using the same parameters. C_2 has three anomalous points at stations 16, 24, and 31. However, C_3 does not correspond with stations 16 and 24, but only with 31. The ratio of C_2/C_3 imitates the C_2 profile. The data indicate that C_3 reflects the iodine values. Therefore, C_3 and iodine reflect leakage to the surface in association with the petroleum trapped in the trend.

Belize, Central America

Using any form of surface geochemistry in tropical rain forests or swampy areas containing large volumes of organics has usually been considered prohibitive. However, recent research has proved this untrue (Mello et al., 1992). Figure 7-24 is a survey carried out across an identified seismic structure that lies in the northwestern corner of Belize, Central America. The area is part of the same basin that encompasses several large productive fields in the north of Mexico. Geology and structural history are assumed to be similar, based on the seismic interpretation (Fig. 7-24). The reservoir is probably the highly fractured Yalbac Formation. The iodine survey was collected along the existing seismic lines. Background iodine values ranged from 0.2 to 2.0 ppm. Profiles of the iodine data (Figs. 7-25–7-27) indicate a significant association with the general location of the seismic

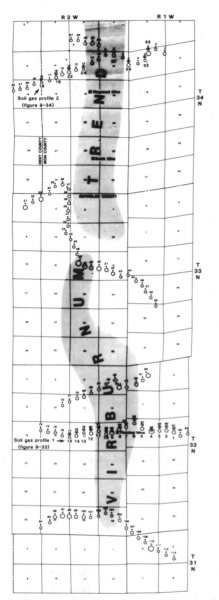

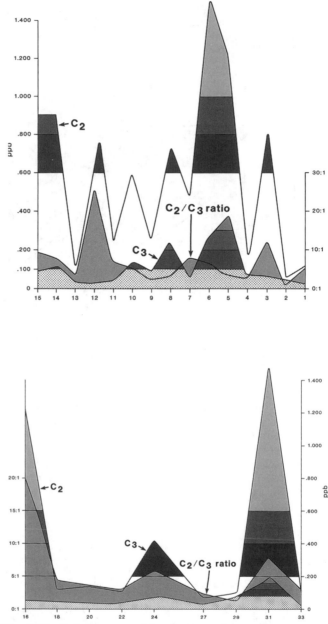

Fig. 7-21 Iodine reconnaissance survey across a select area of the Virburnum Trend, Forest City Basin, Missouri. The Virburnum Trend produces significant quantities of Pb and Zn from dolomite bodies. Contour intervals are 2.0, 3.0, and 4.0 ppm. The presence of iodine is due to petroleum leakage from the ore body. The leakage defines only the dolomite body, not the exact location of the Pb and Zn ore bodies within it (reprinted with permission from Atoka Exploration Corporation).

Fig. 7-23 Soil-gas survey, Line 2, reconnaissance-type survey, C_2, C_3, and ratio of C_2/C_3, profile in ppb (reprinted with permission from Atoka Exploration Corporation).

high. The presence of numerous faults suggests that they are probably acting as migration pathways and are thus the cause of the large areal extent of leakage. At the time of this printing, the structure was untested.

Conclusions

Of all the halogens, only iodine has been used successfully to delineate existing oil fields and to find new ones. There has been some resistance to accepting the iodine method,

even though the presence of ethane and heavier hydrocarbons in the soil seems to cause, or is closely associated with, increases in iodine. Other causes of increases or variations in iodine are usually of a lesser magnitude than petroleum. An iodine survey requires an extensive number of samples to establish background values before the presence of anomalous values can be considered. The cost differential of soil gas to iodine per sample is usually 5 to 20 to 1. It is much simpler to acquire, handle, and analyze iodine samples than to collect and measure soil gas. Interpretation is relatively easy if enough samples are obtained. Soil characteristics do not seem to be a factor as long as dramatic changes during the survey are noted and collection is restricted to the upper 2

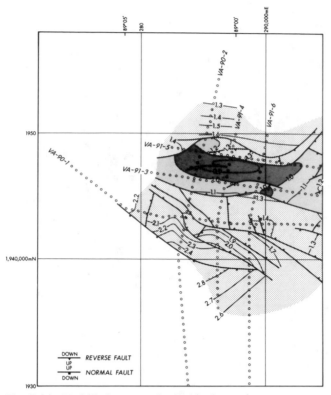

Fig. 7-24 Undrilled structure located in the northwestern corner of Belize, Central America. Structure mapping using the seismic data is on top of the Yalbac Formation. The contour interval is in 0.1 s of time. The top of the structure is denoted by the yellow color.

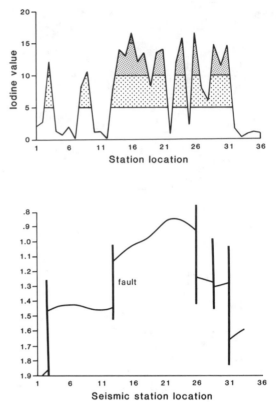

Fig. 7-26 Iodine profile along seismic line VA-91-4 with associated geophysical interpretation below. Contour intervals are 5.0 to 9.9 ppm (light pattern), 10.0 to 14.9 ppm (medium pattern), and >15.0 ppm (dark pattern) of iodine. Seismic is in 0.1 s of time (reprinted with permission of Exeter Oil and Gas, Inc.).

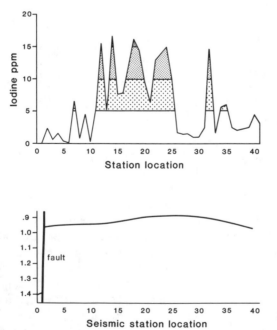

Fig. 7-25 Iodine profile along seismic line VA-91-5. Contour intervals are 5.0 to 9.9 ppm (light pattern), 10.0 to 14.9 ppm (medium pattern), and >15.0 ppm (dark pattern) of iodine. Seismic is in 0.1 s of time (reprinted with permission of Exeter Oil and Gas, Inc.).

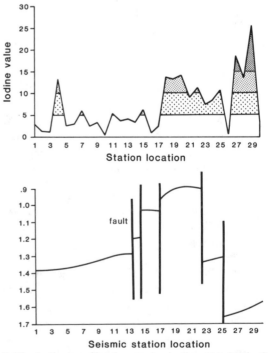

Fig. 7-27 Iodine profile along seismic line VA-91-6. Contour intervals are 5.0 to 9.9 ppm (light pattern), 10.9 to 14.9 ppm (medium pattern), and >15.0 ppm (dark pattern) of iodine. Seismic is in 0.1 s of time (reprinted with permission of Exeter Oil and Gas, Inc.).

in. of soils. An advantage of using iodine is its repeatability. Unlike gases or liquids, the iodorganic compounds are relatively stable; they are not easily affected by fluctuations in barometric pressures, rapid changes in quantities of gas present, wetting and drying of soils, or changes in the water table that easily affect vapor methods. Simplicity of sampling and analysis also reduces the possibility of error. The disadvantages of the iodine method are the large number of samples required for an accurate interpretation and the need to have a model for comparison.

References

Allexan, S., J. Fausnaugh, C. Goudge, and S. Tedesco (1986). The use of iodine in geochemical exploration for hydrocarbons, *Association of Petroleum Geochemical Explorationists Bulletin*, Vol. 2, No. 1, pp. 71–93.

Aubert, H. and M. Pinta (1977). *Trace Elements in Soils*, Elsevier Scientific Publishing Company, Amsterdam, pp. 248–255.

Boyle, R. W. (1987). *Gold, History and Genesis of Deposits*, Van Nostrand Reinhold, New York, p. 676.

Brookins, D. G. (1987). *Eh-pH Diagrams for Geochemistry*, Springer-Verlag, New York, p. 176.

Chudecki, Z. (1960). Investigations on the iodine content and distribution of soils of western Pomerania. *Roczn. gleboz.*, pp. 113–117 (in Polish).

Chudecki, Z. (1963). Some factors affecting the content of copper, iodine and zinc in the fundamental soil types of West Primor'e (Poland), *Zesz. Nauk Wyz. Szkol. rol. Szczecin*, Vol. 10, pp. 187–239 (in Polish).

Collins, A. G. (1969). Chemistry of some Anadarko Basin brines containing high concentrations of iodide, *Chemical Geology*, Vol. 4, pp. 169–187.

Collins, A. G. (1975). *Geochemistry of Oil Field Waters*, Elsevier Scientific Publishing Co., New York, pp. 164–166.

Collins, A. G. and Egleson G.C. (1967). Iodine abundance in oil field brines in Oklahoma, *Science*, Vol. 156, pp. 934–935.

Fuge, R. (1974). Iodine, in *Handbook of Geochemistry*, Vol. II, Part 4, ed., K. H. Wedepohl, Springer-Verlag, New York, Chap. 53.

Gallagher, A. V. (1984). Iodine: A pathfinder for petroleum deposits, in *Unconventional Methods in Exploration III*, Southern Methodist University, Dallas, TX, pp. 148–159.

Gallego, R. and S. Oliver (1959). Estudio sobre yodo en suelos, *Ann. Edafol. Fisiol. Veg.*, Vol. 18, pp. 207–238 (in Spanish).

Glushenko, A. V., N. G. Zyrin, and T. K. Imadi (1964). Iodine content of light soils of terraces of the ancient valley of the Moskva, *Vestn. Moskov. Univ. Ser. Biol. Pochvoved.*, Vol. 6, pp. 74–80 (in Russian).

Goldsmidt, V. M. (1954). *Geochemistry*, Clarendon Press, Oxford, England.

Gordon, T. L. and M. Ikramuddin (1988). The use of iodine and selected trace metals in petroleum and gas exploration. *Geologic Society of America Abstracts with Programs*, Vol. 20, No. 7, p. 228.

Hagni, R. D., M. R. Bradley, R. G. Dunn, P. E. Gerdemann, J.

M. Gregg, T. D. Masters, C. G. Stone, and H. M. Wharton (1986). Sediment Pb-Zn-Ba deposits of the Midcontinent, Premeeting field trip No. 1 Nov. 3–8, San Antonio, TX, Geological Society of America, Boulder, Colorado, p. 148.

Hitchon, B., G. K. Billings, and J. E. Klovan (1971). Geochemistry of formation waters in the western Canada sedimentary basins. III. Factors controlling chemical composition. *Geochimica, Cosmochim. Acta*, Vol. 35, pp. 567–598.

Kabata-Pendias, A. and H. Pendias (1992). *Trace Elements in Soils and Plants*, 2nd ed., CRC Press, Boca Raton, FL, pp. 254–258.

Karelina, L. (1961). Iodine in the soils of the Latvian SSR and occurrence of endemic goitre. *Mikroelem. Urozh*, Vol. 3, pp. 233–255 (in Russian).

Karelina, L. (1965). Total iodine in soils of the Latvian SSR., *Mikroelem. Prod. Rast.*, pp. 249–270 (in Russian).

Katalymov, M. V. (1964). Iodine problem in geochemistry, *Agrarchemie*, pp. 69–80 (in Russian).

Koch, J. T. and B. D. Kay (1987). Transportability of iodine in some organic materials from the Precambrian Shield of Ontario, *Canadian Journal of Soil Science*, Vol. 67, p. 353.

Kovda, V. A., and V. D. Vasil'eyvskaya (1958). A study of minor element contents in soils of the Amur River area, *Soviet Soil Science*, No. 12, pp. 1369–1377.

Kudel'sky, A. V. (1977). Prediction of oil and gas properties on a basis of iodine content of subsurface waters, *Geologiya Nefti i Gaza*, No. 4, pp. 45–49.

Levinson, A. A. (1980). *Introduction to Exploration Geochemistry*, Applied Publishing, IL, p. 924.

Maybe, W. and T. Mill (1978). Critical review of hydrolysis of organic compounds in water under environmental conditions, *Journal of Physical and Chemical Reference Data*, Vol. 7, pp. 383–415.

Means, J. L. and N. J. Hubbard (1985). The organic geochemistry of deep ground waters from the Palo Duro Basin, Texas: Implications for radionuclide complexation ground water origin, and petroleum exploration, Batelle Memorial Institute, Columbus, OH. Technical Report.

Mello, M. R., F. T. T. Gonclaves, and N. A. Babinski (1992). Hydrocarbon Exploration in the Amazon Rain Forest: A nonconventional approach using prospecting geochemistry, microbiology and remote sensing methodologies. American Association of Petroleum Geologists, National Meeting, Calgary, Alberta.

Moore, J. W. and S. Ramamoorthy (1984). *Organic Chemicals in Natural Waters, Applied Monitoring and Impact Assessment*, Springer-Verlag, New York, p. 289.

Prince, N. B. and S. E. Calvert (1973). The geochemistry of iodine in oxidized and reduced marine sediments, *Geochimica, Cosmochim. Acta*, Vol. 37, p. 2149.

Rao, S. S., S. K. De, C. M. Tripathi, and C. Rai (1971). Retention of iodide in soil clays, *Indian Journal of Agricultural Chemistry*, Vol. 4, pp. 44–49.

Runyon, H. E. and R. Rankin (1936). The bromine and iodine content of the subsurface waters of Russell, Ellis and Trego Counties, Kansas, *Transactions of the Kansas Academy of Science*, 39, pp. 127–128.

Shacklette, H. T. and J. G. Boergnegen (1984). *Element Concentration in Soils and Other Surficial Materials of the Counterminous United States*, U.S. Geologic Survey Professional Publication 1270, p. 105.

Singh, R. R., J. G. Saxena, S. K. Sahota, and K. Chandra (1987). On the use of iodine as an indicator of petroleum in Indian basins, *1st India Oil and Natural Gas Comm. Petroleum Geochemistry and Exploration in the Afro-Asian Region International Conference Proceedings*, pp. 105–107.

Tainter, P. A. (1984). Stratigraphic and paleostructural controls on hydrocarbon migration in Cretaceous D and J sandstone of the Denver Basin, in *Hydrocarbon Source Rocks of the Greater Rocky Mountain Region*, eds., J. F. Woodward, F. Meissner and J. Clayton, Rocky Mountain Association of Geologists, Denver, CO, pp. 339–354.

Tedesco, S. and C. Goudge (1989). Application of iodine surface geochemistry in the Denver-Julesburg Basin, *Association of Petroleum Geochemical Explorationists Bulletin*, Vol. 5, No. 1, pp. 49–72.

Tikomirov, R. A., S. V. Kaspatov, B. S. Prister, and V. G.

Salinikov (1980). Role of organic matter in iodine fixation in soils, *Soviet Soil Science*, Vol. 12, pp. 64–72.

Tinsley, I. J. (1979). *Chemical Concepts in Pollutant Behavior*, John Wiley & Sons, New York, p. 265.

Vil'gusevich, I. P. and N. P. Bulgakov (1960). Microelement content in the soils of Belorussia, *Soviet Soil Science*, Vol. 3, pp. 319–326.

Vinogradov, A. P. (1959). The geochemistry of rare and dispersed chemical elements in soils, Consultants Bureau, New York, p. 209 (translated from Russian).

Whitehead, D. C. (1978). Iodine in soil profiles in relation to iron and aluminum oxides and organic matter, *Journal of Soil Science*, Vol. 29, pp. 88–94.

Zimovets, B. A. and A. I. Zelenova (1963). Iodine content in the soils of the Amur Basin, *Soviet Soil Science*, Vol. 11, pp. 1031–1039.

Zyrin, N. G. and L. N. Bykova (1960). Iodine in some soils of Moscow Region, *Vest. Moskv. Univ. Ser. Biol. Poch.*, Vol. 6, pp. 55–66 (in Russian).

8

Major and Minor Elements

Introduction

Trace and major elements, certain isotopes, and the compounds in which they are incorporated have been used to identify petroleum microseepage. Applying these forms is a departure from typical surface geochemical methods and represents an attempt to find more reliable techniques, other than soil gas, to target petroleum accumulations. The results have been encouraging and have added information to the understanding of microseepage and the changes it causes in the near-surface. However, the results have not been sufficiently reliable, repeatable, or cost-effective to warrant replacing soil gas, radiometrics, or iodine.

The majority of recent work has focused on magnetic and nonmagnetic iron minerals. The literature also suggests that anomalous amounts of carbonates, delta C (carbon), vanadium, chromium, nickel, cobalt, manganese, mercury, copper, molybdenum, uranium, zinc, lead, and zirconium are positive indicators of petroleum deposits (Alekseev et al., 1961; Miodrag, 1975; and Duchscherer, 1984). The presence of barium, strontium, boron, sodium, or potassium is considered a negative indicator of hydrocarbons in the soil although soils may naturally contain many of these elements in relative abundance. Normal soil processes can cause their accumulation irrespective of the presence of migrating hydrocarbons. Therefore, the main problem with using trace or major elements in the search for petroleum is determining if their concentrations result from hydrocarbons seeping into the soils or from some other cause. Consequently, soil chemistry, volume of organic matter, and clay composition must be quantified in order to determine effectively if element and compound accumulations are products of migrating hydrocarbons or some other agent.

Relationship to Mineral Deposits

Matveeva (1963) determined that migrating trace concentrations of heavy metals precipitate under favorable conditions. This results in aqueous dispersion halos of these metals. Precipitation can occur along a geochemical barrier, such as an oxidation/reduction interface, or a pH-Eh barrier (Garrels and Christ, 1965). This concept is based on dispersion halos associated with mineral deposits, but no evidence suggests that it is not applicable to petroleum microseepage. The source of the metals is the soil itself, and a reducing environment increases the elemental solubility. Their accumulation is a function of ease of transport under oxidizing/reducing conditions, and of complexing with organic matter, moisture, and clay. Reducing conditions normally are formed in environments containing organic materials, and these conditions are complemented by an increase in carbon dioxide and hydrogen sulfide and by a negative Eh value.

Geochemical patterns can occur over concealed mineral deposits. The patterns range in shape from halos to fans, depending on lateral groundwater flow or upward movement of soil moisture (Hawkes and Webb, 1962). Anomalous subsurface occurrences of gold, silver, molybdenum, mercury, copper, lead, and zinc have been individually observed in drill hole cores taken from 100 to 1200 ft below the surface to assess local mineral potentials (Ketner et al., 1968). Sulfide ore bodies are postulated to be natural galvanic cells that cause extensive local solution and transportation of metals and may be responsible for heavy-metal dispersion halos at the surface over ore bodies 100 to 150 m deep (Sveshnikov and Ryss, 1964). Even surface vegetation in areas of oxidized ores is markedly affected by the presence of uranium-vanadium deposits (Cannon, 1952).

Groundwater Analysis

The chemical characteristics of subsurface waters in the vicinity of oil and gas fields have been investigated because it was thought that the analyses could be useful in petroleum exploration. Ammonium contents in groundwater are said to increase 100% in the vicinity of oil and gas formations. The trace-element data from the Saratov-Volograd Field of the former Soviet Union show concentrations of manganese, strontium, barium and, to a lesser extent, lithium in groundwater inside the reservoir. Strontium was found to be three to five times more concentrated around the oil reservoir than in barren country rock.

Relationship to Deposits/Microseepage

Many surface and subsurface geochemical anomalies have been observed to have direct relationships to hydrocarbon deposits. Manganese, vanadium, nickel, and copper trace elements, as well as the radioactive elements uranium and radium, were found to be absorbed to a greater extent around the periphery of the Kyurov-Dag Oil Field in the former Soviet Republic of Azerbaijan than actually over the field (Alekseev et al., 1961). In some basins, copper and manganese were shown to increase in concentration near oil and gas structures, probably as the result of a change in the local groundwater redox potential (Eh). Sikka (1964) listed 25 oil and gas deposits found by trace-metal analysis of the soil. Preliminary results by Zak (1964) failed to show any trace-element anomaly resulting from migration of nickel, vanadium, zinc, chromium, copper, or uranium from the underlying oil reservoir to the surface in the soil samples over the Heletz Oil Field, Israel.

Fluorometric analysis has been applied to petroleum exploration in the examination of sulfur compounds extracted from soil bitumen (Johnson, 1970). A 35% success rate was claimed during a six-year period of field-testing the surface geochemical method involving heavy-metal detection. According to theory, heavy metals of low mobility can be moved vertically by expelled fluids as a result of compaction and can be deflected by the prospective hydrocarbon barrier. These heavy metals would form anomalous halos in mineralized chimneys surrounding the deposit. Duchscherer (1984, 1985) also proposed that the development of the halo type of anomaly was due to a diversion of the vertical migrating hydrocarbons around the field because cementation developed above the field in a chimney-type structure. This contradicts other authors who believe that the most intense oxidation of hydrocarbons by bacteria occurs directly over the field and that, therefore, reduction of other elements occurs. The center of the field should be the area of greatest anomalous carbonate production, but this is not the case.

The accumulation of anomalous trace and major metals, carbonates, magnetic minerals, and carbon isotopes in relation to microseepage has to be consistent with the known conditions of the geochemical environment and other facts supporting the theory. Deposits of carbonate cements occurring over petroleum accumulations have been cited. But, since hydrocarbons are still leaking to the surface subject to the density of fracture pathways present, cementation has probably not occurred consistently in relation to microseepage. The hydrocarbons create a reducing environment in the soil and subsurface, which increases the solubility of many trace and major elements (Fig. 8-1; see color plate).

As local groundwater moves these elements out to the edges of the reducing zone, the aqueous system becomes saturated, and the elements drop out in the form of hydroxides, oxides, sulfides, and so forth. Therefore, a halo anomaly is not due to deflection but to movement of these elements from a reducing to an oxidizing environment, where they will then accumulate. Carbonate and trace-metal methods will not work if the soils are already acidic or are in a reducing state because adding hydrocarbons may not significantly change the pH or oxidation state of the aqueous solution to cause precipitation of metals.

Carbon isotopes are more directly related to the hydrocarbons migrating through the soil than are other isotopes or elements. Carbon isotope surveys attempt to identify what percentage of the methane present is thermogenic or biogenic. Neither carbon nor oxygen isotopes have been reported successful in identifying a petroleum accumulation prior to discovery.

There are limited data available that link trace- and major-element concentrations directly to petroleum seepage and record the responses of the elements to the reduction processes. Many elements have still not been researched adequately with respect to soil chemistry. If elements are restricted to those used to type petroleum, the scope of the investigation would be limited to nickel (Ni) and vanadium (V) (Waples, 1981; 1985) and copper (Cu). The literature has focused on iron (Fe), stable isotopes of oxygen and carbon, and delta carbon. Consequently, this chapter will concentrate on these elements and isotopes.

Influence of Humic Complexes/Bacteria

Humic substances are the product of microbial activity and the reduction of organic matter, which is dependent on the continuing presence of water in the soil. Depending on the climate, the humic complexes will either migrate directly into the drainage system, or their progress will be impeded. In wet climates, humic substances enhance the solubility of trace and major elements from the soil-rock boundary and continue to affect their migration in groundwater through the pedologic section. In dry climates, these processes are greatly reduced. Predicting solubilities of inorganic compounds may be feasible if the elements are in water alone but, if dry conditions occur, the humic substances are nearly inactive. Prediction becomes irrelevant when wet conditions exist and involve humic material, whose effect on solubility is profound. An example of this is vanadium (V), which was found enriched in peat 50,000 times over its concentrations in associated water (Szalay and Szilagyi, 1967). Vanadium is transported as colloidals and is modified by the types of humic substances present. The humic complexes of Ca and Ni display mobility in distilled water but have little mobility when dry (Baker, 1966). Also impacting mobility are pH, humic particle size, and concentrations of the elements present. Therefore, the best use of trace and major elements in petroleum exploration is in environments that have a minimum of humic production and accumulation so that the presence of chelated elements can be related to microseepage and not to humic acid.

In environmental geology, the methyl derivatives $(CH_3)_nM$ and related salts $(CH_3)_nMX_2$ are a concern because they are toxic. Methylization is the transfer of a methyl group from one compound to another. It occurs because of both biological (biomethylization) and chemical processes. Methylization is important in environmental studies, especially with respect to Pb, Zn, Ti, Co, Cd, Hg, Sb, As, Se, and Bi. The majority of these elements are not currently pursued in the search for petroleum seepage, with the rare exception of Hg. If methane is seeping into the soil, methylization of Hg occurs as follows:

Chemical: $2CH_3Hg^+S^{2-} \rightarrow (CH_3Hg)_2S \rightarrow (CH_3)_2Hg + HgS$
Bacterial: $CH_3Hg+ \; bacteria \rightarrow CH_4 + Hg(O)$

Mercury concentrations due to methylization increase in conjunction with increase of sulfides but, once sulfide concentrations reach a specific level, the process seems to stop. It can be assumed that methylization of trace and major elements stems from the presence of seeping hydrocarbons. However, the research has not yet determined either general or specific relationships.

Iron and Related Minerals

Iron (Fe), specifically in magnetic assemblages such as monoclinic pyrrhotite (Fe_7S_8), hematite (Fe_2O_3), magnetite (Fe_3O_4), and greigite (Fe_3S_4), has been the main focus of most geochemical surveys. Donovan et al. (1979, 1984), Foote (1984), and Saunders and Terry (1985) have used soil magnetic susceptibility surveys as a supplement to other geochemical methods of finding oil and gas production. The assumption is that the hydrocarbons create a reducing environment in which some Fe compounds are altered into a magnetic form. These minerals can be formed by bacterial processes under naturally occurring anaerobic and reducing conditions (Machel and Burton, 1991). Thermodynamic equilibrium must be present to form any particular mineral but may not be possible because of chemical imbalance and microbial activity. Increases in magnetic minerals have been reported in drill cuttings from various strata above a petroleum accumulation (Donovan, 1974; Donovan et al., 1979, 1986). In dry wells, no significant increases in magnetic or iron-rich minerals have been reported. Reynolds et al. (1991) found that three different areas (Cement Oil Field, Oklahoma; Simpson Oil Field, Alaska; and the Wyoming-Idaho-Utah thrust belt area) had three different magnetic assemblages. The magnetic anomalies in two of the areas were possibly caused by authigenic minerals related to microsee-

page. Some geologic environments may enhance magnetic mineral formation.

Iron constitutes approximately 5% of the lithosphere and is concentrated mainly in mafic rocks. Iron chemistry is relatively complex, and the iron changes valence states with relative ease (Fig. 8-2). Iron is closely associated with carbon, oxygen, and sulfur. In the weathering cycle, reactions involving Fe are closely associated with Eh-pH and oxidation conditions. Oxidation or alkaline conditions promote the precipitation of Fe. Reducing/acidic conditions increase solubility, and the iron remains chelated. Iron compounds typically precipitate as oxides and hydroxides. Iron can be incorporated in carbonate minerals, can replace manganese and alumina, and can complex with organics. Because seeping hydrocarbons create a reducing environment, we would expect Fe to be present in the chelated form. Table 8-1 is a summary of the major iron compounds and the soils and climates in which they are typically found.

The mobility of Fe in the soil is controlled by the solubility of Fe^{+3} and Fe^{+2} amorphous hydrous oxides. Solubilities are modified by the presence of Fe phosphates, sulfide, and carbonates. Figure 8-3 indicates the variance in the type of minerals deposited or impeded under different concentrations of HS, HCO_3, and Fe^{+2}. Iron compounds that are

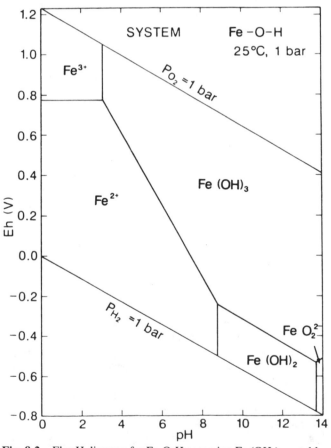

Fig. 8-2 Eh-pH diagram for Fe-O-H assuming Fe $(OH)_3$ as stable Fe(III) phase. Assumed activity of dissolved Fe = 10^{-6}. (after Brookins, 1987, reprinted with permission from Springer-Verlag).

Table 8-1 Major iron compounds and their climates

Mineral	Chemical	Climate
Hematite	Fe_2O_3	Arid, semiarid, tropical
Maghemite	Fe_2O_3	Tropical: highly weathered soils
Magnetite	Fe_3O_4	Inherited from the parent material
Ferrihydrite	Fe_2O_3 nH$_2$O	Goes to hematite in warm climates, goethite in humid climates
Lepidocricite	FeOOH	Humid, temperate, low pH and low temperature; poorly drained soils
Illmenite	$FeTiO_2$	Resistant to weathering, related to parent material
Pyrite	FeS_2	Submerged soils

present in a soil are responsible for its color. Iron is closely involved in the behavior of other trace elements and micronutrients. The distribution of Fe in the soil profile represents different chemical processes, and its accumulation in various layers and horizons will modify the geochemical environment.

Iron in the soil is related to the parent material and the soil processes. Some researchers have evaluated Fe in relation to microseepage as the total amount of the element present, but

only a small percentage of it is actually related to petroleum microseepage.

Magnetic forms of iron minerals have been aggressively studied through airborne and ground magnetic methods and by magnetic susceptibility measurements. Under reducing conditions, magnetite and other associated magnetic minerals are formed and are deposited in the soil substrate and at the surface. The theory of magnetism and its history and use in natural resources exploration can be found in other texts. The airborne and ground magnetic methods measure minerals in the soil resulting from petroleum seepage or the presence of metal deposits. Magnetic minerals can be formed from the cap rock to the surface depending on the presence of reducing conditions and processes favorable to their formation and preservation in the strata and soil through time.

Factors affecting the use of this method are the presence of microbes that either digest or produce magnetite or other iron minerals. Also, normal chemical processes can create these minerals. There are formations that are naturally high in magnetite, specifically, coarse lag deposits in buried channel systems or phosphates/iron-rich carbonates. Magnetic measurements must be confirmed by sampling at 50 to 150 m (150–500 ft) from the surface. When existing oil fields are sampled, the anomalies present probably relate to the production pipe from the producing wells.

Soils sampled for magnetite susceptibility measurements are typically taken at very shallow depths from zero to

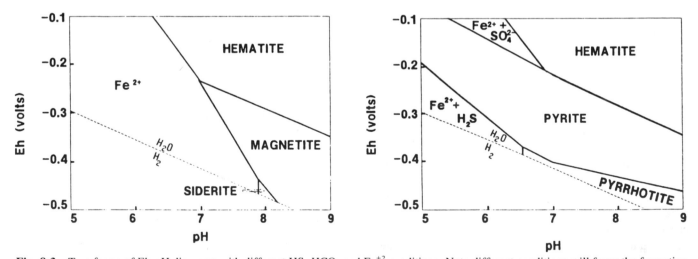

Fig. 8-3 Two forms of Eh-pH diagrams with different HS, HCO$_3$ and Fe^{+2} conditions. Note different conditions will favor the formation of certain minerals and impede the development of others. (A) Dependence of thermodynamic stabilities of iron minerals on pH and Eh in an aqueous environment with $a_{HS^-} = 10^{-12}$; $a_{HCO3-} = 10^{-2}$; $a_{Fe2+} = 10^{-6}$; 25°C; 1-atm total pressure. Dashed line is the limit for stability of liquid water relative to hydrogen gas. Over most of the Eh and pH range shown, Fe^{2+} ions in aqueous solution are favored. In the upper part of the Eh range, at pH greater than 6.5, hematite precipitates but, as conditions become more reducing, magnetite and then siderite are relatively more stable. The field for siderite is very small and limited to pH > 7.8. At pH < 7.8, the fluids become undersaturated with respect to siderite. (B) Dependence of thermodynamic stabilites of iron minerals on pH and Eh in an aqueous environment with $a_{HS-} = 10^{-6}$; $a_{HCO3-} = 10^{-2}$; $a_{Fe2+} = 10^{-6}$; 25°C; 1-atm total pressure. Dashed line is the limit for stability of liquid water relative to hydrogen gas. In a comparison to Fig. 10-3A, the stability fields for pyrite and pyrrhotite replace those of magnetite, siderite, and much of the Fe^{+2} field. The upper limit for pyrite stability is defined by the oxidation of H$_2$S and HS$^-$ to SO$_4^{2-}$. Above this limit, hematite precipitates, except at pH < 7, where hematite dissolves and fluids contain Fe^{2+} and SO$_4^{2-}$ ions. At low pH and Eh, pyrite dissolves to form Fe^{2+} ions and H$_2$S. Pyrrhotite requires more reducing conditions than pyrite, and also needs pH values greater than about 6.6 to be stable (after Machel and Burton, 1991; reprinted with permission from the Society of Exploration Geophysicists).

0.3 m (0–1 ft). The sample is sieved. Heavy minerals are segregated out and are analyzed using a magnetic susceptibility unit. This is an inexpensive way to collect magnetic data, but numerous factors affect its use. Samples taken at the surface are biased as a result of water runoff concentrating heavy minerals in streamlet areas. Sampling has to avoid these concentration localities. Soils must be analyzed to determine the normal range of magnetic minerals present and the composition of the parent material undergoing weathering. The parent material may create local anomalies because of lithologic variations.

Well cuttings have been used for magnetic susceptibility measurements (Donovan and Roberts, 1980). The Cement Field, Oklahoma, is the often-used example that shows drill cuttings from producing wells that exhibit magnetic anomalies. Adjacent dry holes indicate only background magnetic mineral assemblages. Subsequent work found that the majority of the magnetic minerals present were a product of drilling. In the Las Animas Arch, southeastern Colorado, only 60% of the fields producing from the Morrow sandstone have a magnetic signature as determined by either micromagnetic or magnetic susceptibility measurements. It has been concluded that the presence of iron compounds depends on water chemistry, reduction/oxidation conditions, and Eh-pH in the strata and soils above the reservoir.

Manganese

Manganese is a common metal that has been noted in halolike anomalous amounts around petroleum accumulations. The metal has numerous oxidation states. Specifically, manganese dioxide (MnO_2) is found in soils in either a crystalline or amorphous phase (Clark, 1992). The amorphous phase contains a tremendous surface area because of its irregular shape and can incorporate numerous metal and nonmetal ions. Clark found that the MnO_2 amorphous form contained anomalous amounts of iodine, Br, Cl, Ur, and thorium, in an apical form across the Sleeper Mine, Nevada, where arsenic, selenium, and molybdenum formed apical anomalies. Additional data were presented to suggest that a similar phenomenon was occurring in and around petroleum accumulations as well. The amounts of ions, such as iodine and uranium, were considerably less than those detected and described in the previous chapter but represented anomalous accumulations.

The method of detection is by enzyme leach, whereby only the MO_2 oxide coatings are dissolved. It was found that these anomalous accumulations did not exist in the A and C horizons. Thus, detail soil profiling to identify the B horizon is required.

Carbonate Minerals and Delta Carbon

The oxidation caused by microseeping hydrocarbons produces carbon dioxide, which dissolves into the groundwater and into the water present in the pores of the soil. The dissolved carbon dioxide reacts with calcium to produce calcium carbonate or calcite. In the process, other minor elements such as manganese or reduced iron will be incorporated into the carbonate lattice. Visible carbonate deposits have been used in at least one case to direct exploration at a major oil field. The Rangely Field, located in Townships 2 and 3 North, Range 101 to 103 West, Rio Blanco County, Piceance Basin, Colorado, is a large anticlinal structure that has produced over 600 MMbbls of oil and 700 Bcf of gas. During early development, one of the techniques for determining the viability of a site was to ascertain the presence or absence of calcite veins in the soil. The presence of numerous calcite veins typically implied a productive site.

The oxidation of the microseeping hydrocarbons creates reducing conditions in the soil. Iron is more soluble in the reduced (ferrous) state than in the oxidized (ferric) state. Reduced Fe in the soil can become incorporated into the carbonate lattice to produce ferroan calcite $(Ca, -x, Fex)CO_3$. Ferroan calcite (siderite) is not common in the near-surface, and its presence is the basis of the delta C or delta carbon geochemical method.

Duchscherer (1984, 1985) and Duchscherer and Mashburn (1987) published several papers on the delta C method, which was developed as an exploration tool in the 1940s. A soil sample is collected at depths of 0.3 m (1 ft) to 3 m (10 ft) with a 1.5 to 2 m (5–7 ft) sample preferred. The sample is dried and sieved, and the clay size fraction is retained. The analytical approach is based on thermal dissociation of carbonates in the soil sample. The sample is heated to 600°C, which will decompose siderite ($FeCO_3$). Other carbonate minerals, such as calcite or dolomite, do not decompose at this low temperature.

The siderite decomposition can be measured in one of two ways. The first is differential thermal analysis, or DTA. A small sample is heated at a uniform rate, and temperatures are measured over time. Reactions are either exothermic or endothermic. An exothermic reaction releases heat and is oxidizing, which causes decomposition, combustion, and reconstruction of the crystal structure. An endothermic reaction absorbs heat and is reducing, which causes loss of water (dehydration), structural decomposition, and transformation. Changes in the shape, number, and position of the peaks through time with respect to furnace temperatures allow for mineral identification. Figure 8-4 is a typical thermal-dissociation curve for the delta C method. It shows the decomposition of siderite ($FeCa(CO_3)_2$), which is the carbonate that is the target of this method. The increase in siderite reflects excess Fe being incorporated with the carbonate minerals that are being deposited. The decomposition of siderite occurs from 450°C to 650°C, with a maximum endothermic effect at 580°C. During the endothermic reaction, the clays dehydrate and decompose, and the lattice structure of the ferroan calcite is broken down, which releases CO_2. Decomposition rates may vary in response to the rate of increases in furnace temperature, atmospheric

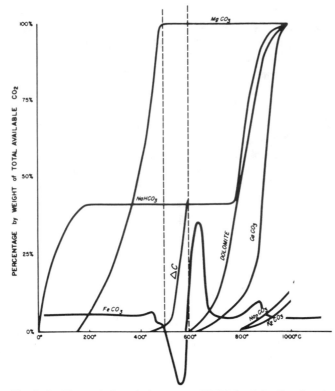

Fig. 8-4 Thermal-dissociation curves (DTA) for delta C and various carbonate minerals (Duchscherer, 1984; reprinted with permission of Pennwell Publications).

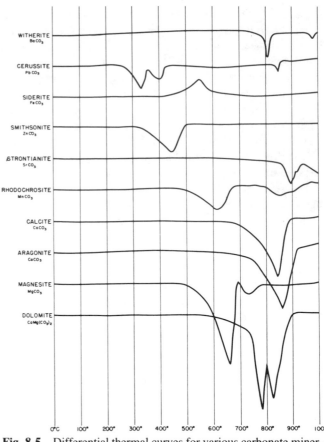

Fig. 8-5 Differential thermal curves for various carbonate minerals (Duchscherer, 1984; reprinted with permission of Pennwell Publications).

content in the furnace, sample amounts, porosity and permeability of the sample, variations in grain size, and the presence of inert material covering the sample. Typically, the temperature window for generating CO_2 from a soil sample is 500°C to 600°C. Figure 8-5 is an actual delta C curves used in analysis.

The second method for decomposing siderite uses the same principle but detects the decomposition in a different way. The decomposition of the carbonate results in the generation of carbon dioxide. The carbon dioxide can be detected by using a thermal-conductivity detector (TCD) gas chromatograph. In this case, the column for separation of gases is not used; only the detector is used. If the evolved gas, such as carbon dioxide, flows through the detector, the conductivity changes are a measure of the amount of carbon dioxide and the amount of siderite (ferroan calcite) present. Duchscherer called this anomalous carbonate *delta C*.

Duchscherer (1985) published the results of a survey over the Bell Creek Oil Field in the Powder River Basin in southeastern Montana (Fig. 8-6). Production is from the Cretaceous muddy sandstone, which is a marine deltaic and barrier bar reservoir. Depth of production is approximately 1360 m (4500 ft). Soil samples were collected as part of a U.S. Geologic Survey study and are from a depth of 0.3 m (1 ft) instead of the standard 1.5 to 2 m (5–7 ft). Figure 8-7 shows the contoured delta C values. The units are not given but probably represent changes in the electrical current as mea-

sured by the detector in millivolts of the carbon dioxide evolved. Selection of the contour interval is based on means, standard deviations, and histograms used in the study. The contour selected is equal to the mode plus 1.5 times the standard deviation. Figure 8-7 indicates a halo around the Bell Creek Field. There is another field to the southwest, and a partial halo occurs in the small corner of the field, which is covered by the survey.

Saeed (1988) indicated that the carbonate deposition occurs over the petroleum seepage rather than on the flanks. Several authors have cited carbonate deposition in the strata, based on the rate of drill-bit penetration, porosity logs, and samples. Others, however, have noted no changes in penetration rate, porosity, or changes in cements in the strata overlying the field. Therefore, the presence and the location of carbonates may actually be related to chemical processes in the subsurface and show only an occasional relationship to petroleum seepage.

The consistent geochemical response through all Duchscherer's studies is the halo anomaly. Many of the reported anomalies would be difficult to interpret without the presence of existing fields. With the survey methods employed, it may not be possible to discern if the surface geochemical expressions of the fields are anomalous. Many

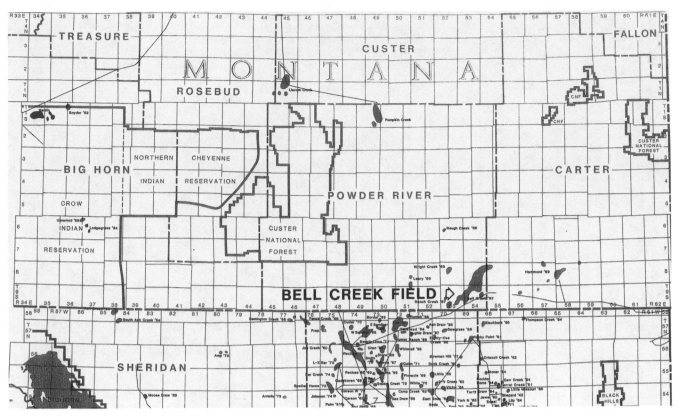

Fig. 8-6 Location of Bell Creek Field, Powder River Basin, Montana (reprinted with permission of Promap Corporation).

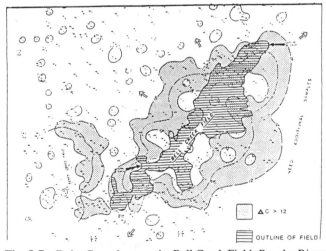

Fig. 8-7 Delta C results over the Bell Creek Field, Powder River Basin, Montana. The values are in millivolts, and >12 mV is anomalous. The anomaly implies a halo. The grid density is widely spaced (Duchscherer, 1981; reprinted with permission of the Association of Petroleum Geochemical Explorationists).

of the examples used widely spaced grid densities, which preclude many of the interpretations that were presented. Recently, the author learned that an exploration company attempted to confirm several delta C anomalies in Nebraska. The company was not able to verify them with soil-gas or iodine surveys. This suggests that caution must be taken when using the delta C method.

Trace Metals

A number of metals are known to be absorbed by petroleum. The ratios and occasionally the amounts of these metals have been used to type a specific oil or family of oils. Nickel, vanadium and, to a lesser extent, copper are the three most common metals identified as useful. Nickel and vanadium have been associated with seeping hydrocarbons in the soil.

Nickel

Nickel (Ni) ranges from 5 to 15 ppm in granite, 5 to 90 ppm in sedimentary rocks, and 1400 to 2000 ppm in ultramafic rocks. It is closely associated with Fe and S and typically is organically bound in soils. Nickel is easily weathered. It is relatively stable in aqueous solutions and is capable of migration over long distances. Nickel has a strong association with Mn and Fe oxides but is not as strongly fixed as some other elements. The soil profile of Ni is typically related to changes in organic matter, clay content, and amorphous oxides.

The Ni content of soils is highly dependent on the parent material. Organic acids and the pH, which affect the ability of Ni to be absorbed by clays, increase the release of Ni from carbonates and oxides (Fig. 8-8). Under reducing con-

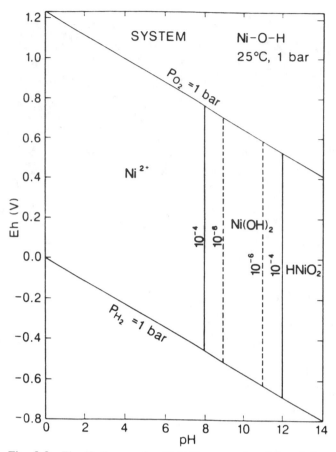

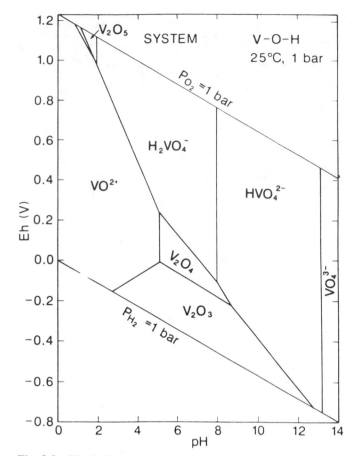

Fig. 8-8 Eh-pH diagram for Ni-O-H, assumed activity of dissolved Ni $= 10^{-4, -6}$ (after Brookins, 1987; reprinted with permission from Springer-Verlag).

Fig. 8-9 Eh-pH diagram for V-O-H system, assumed activity for dissolved V $= 10^{-6}$ (after Brookins, 1987; reprinted with permission from Springer-Verlag).

ditions, Ni is in a +2 state and, above neutral pH, it forms $Ni(OH)_2$. Various soils have a wide range of Ni content, <5 to 700 ppm, with an average of 19 ppm total (Shacklette and Boerngen, 1984). Nickel-rich soils are typically clay and loamy. Elevated Ni contents can be found in organic-rich soils in semiarid and arid regions.

Vanadium

Vanadium (V) is concentrated in mafic rocks and shales. Its ratio with Ni has been used to type oils. Vanadium generally replaces other metals (Fe, Ti, and Al) in soils, and its behavior is dependent on its oxidation state, which can be +2, +3, +4, or +5. In soils, V shows a strong relationship with Mn and K. Vanadium can be absorbed by organics, especially coal and petroleum, and by Fe oxides. Humic acids form complexes with V, which makes it highly mobile. Figure 8-9 is the Eh-pH diagram for V. Vanadium is found rather uniformly in most soil profiles. Variations are specifically related to the parent material. Typically, V ranges from 7 to 500 ppm, with an average of 80 ppm total (Shacklette and Boerngen, 1984). Loamy and silty soils

contain amounts of V (150–460 ppm) that exceed the concentrations in the parent material. Peat soils seem to have the lowest concentration (5–25 ppm).

Copper

Copper (Cu) is abundant in mafic and intermediate rocks and is very low in carbonates. Copper is a highly mobile element. It complexes with many other elements and is easily solubilized during the weathering process, especially in acid environments. Copper precipitates out in carbonates, hydroxides, and sulfides. Figure 8-10 is the Eh-pH diagram for Cu. Across soil profiles, Cu shows relatively little variation. However, concentration typically occurs in the upper layers, and migration downward is minimal. Copper ranges from 1 to 700 ppm, with an average of 25 ppm total for soils (Shacklette and Boerngen, 1984). The Cu content of any particular soil is controlled by the parent material and ongoing processes. Organic acids will form stable complexes with Cu. Microbial fixation of Cu, which is affected by the growing season, also plays an important role in its accumulation.

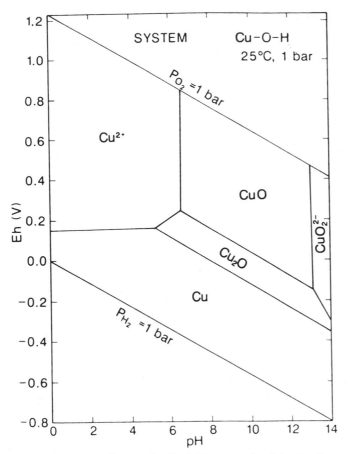

Fig. 8-10 Eh-pH diagram for Cu-O-H, assumed activity for dissolved Cu $= 10^{-6}$ (after Brookins, 1987; reprinted with permission from Springer-Verlag).

Collection

Samples for trace metals are typically taken from 0.3 to 1 m (1–3 ft) in depth. The measurements of Ni, V, and Cu usually occur as a secondary analysis in association with delta C. The clay fraction is separated and analyzed on a mass or plasma spectrometer. All the elements are reported as total amounts present, which are not a reflection of the amounts caused to accumulate and associate with microseepage. DPTA and EPTA methods identify only those that are in the soluble form. The reducing environment caused by seeping hydrocarbons should show increased amounts of soluble Ni, V, and Cu.

Discussion

Accumulations of Fe, delta C, Ni, V, and Cu have been observed but not described in terms of the conditions that cause their increase. The failure to investigate the processes under which these elements are anomalous has led to many misconceptions about their importance to microseepage. All these elements have been found in anomalous amounts in

association with uranium roll-front deposits formed under redox conditions. Studying this analogy can further the understanding of the conditions that affect this method. The close association of trace elements with radioactive minerals further suggests that they represent an important clue to understanding radiometrics.

The uranium roll-front deposits occur in the Shirley Basin, Township 27 and 28 North, Range 78 West, Carbon County, Wyoming. The roll-fronts have undergone considerable investigation (Harshman, 1972), especially concerning their associated major and minor elements. Figure 8-11 is the location of uranium roll-front deposits found at the Petrotomics Co. mine in Section 9, Township 27 North, Range 78 West. In Fig. 8-12, note the zone of mineral deposition indicating increases of FeS_2, UO_2, $CaCO_3$, and Fe_2O_3 across the ore body. These types of roll-front deposits are formed under low pH (less than 3), Eh (less than +3 V), temperatures and pressures, and in the presence of CO_2 (Hostetler and Garrels, 1962). All these conditions control solubility transport. The ore-rich fluids in the Shirley Basin roll-front deposits were alkaline and moved through permeable beds in the hydrologic system until they entered the reducing zone, where deposition of the minerals occurred. The reducing zone is a product of mixing meteoric waters and groundwater. Uranium is transported in the chelated or colloidal form as part of organometallic and carbonate complexes. Precipitation is dependent on Eh and pH changes, clay ab-

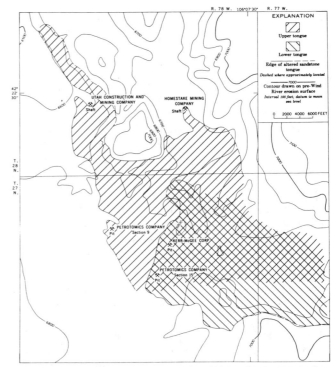

Fig. 8-11 Uranium roll-front deposits, Shirley Basin, Wyoming. The map indicates the relation of altered sandstones to paleotopography. Location of principal mines are at the edges of altered sandstone.

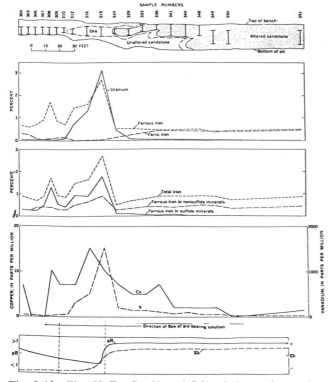

Fig. 8-12 Eh, pH, Fe, Cu, V, and C in relation to the uranium roll-front deposit, Shirley Basin, Wyoming. There is a strong correlation with the ore deposit and several geochemical elements and parameters.

sorption capacity, bacterial reduction, and the presence of humic materials and volatile hydrocarbons (Boyle, 1982). Figure 8-12 shows the Eh, pH, Fe, Cu, V, and carbon relationships with the roll-front deposit. Nickel is reported to accompany U in solution. Eh and pH changes control the deposition in the updip area of the roll-front. Where Fe, Cu, and V are present, they act as binding agents, and uranium is not leached from the surface. However, uranium will be leached from the near-surface over the ore deposits if there is high rainfall or if the binding elements are absent. The roll-front typically has goethite and ilmenite on the redox side of the deposits.

The uranium roll-front model indicates that certain conditions must exist for these deposits to occur. The major factor affecting transport is oxidation with several elements, Fe, V, Ni, Cu, and Ca, carried in conjunction with the uranium. Petroleum and humic and carbonaceous materials help transport, and are accumulated with, these elements, especially in the areas of deposition. Deposition occurs where meteoric recharge interfaces with hydrodynamically migrating groundwater. In the case of petroleum microseepage, which causes a reducing environment, solubility increases and transport occurs. There are distinct differences, but the end result implies that the boundaries between the two environments create an area in which deposition of carbonate, oxide, and sulfide minerals occur. In numerous papers on roll-

front deposits, organics are cited as either the primary or secondary agent of transport and accumulation. The relationship has not been clearly determined but is fundamental for the mechanism's function.

Petroleum seeps into the soil substrate, which alters the existing Eh and pH conditions. In cases of high-alkaline or acidic soils, the petroleum may not alter pH, Eh, and the solubilities significantly. The amount of petroleum seepage that is occurring will also determine the amount of variation in Eh and pH. Other factors to take into account are the amounts of humic (and fulvic) acids and organic matter, clay types and their absorption capacity, availability of trace and major elements, microbial activity, and water content. The reducing environment created by the petroleum oxidizes carbonates and mobilizes various elements. The hydrodynamic flow in the soil moves the solubilized elements around although they do not necessarily leave the reduced area. Concentration into the "anomalous" areas occurs where reduced groundwater is low in pH and Eh and intermingles with groundwater of higher Eh and pH. The result is deposition of carbonate, sulfide, phosphate, and oxide minerals, which incorporate the trace elements into the compounds. However, if the pH of the reducing zone is only slightly different from the pH of the oxidizing zone (e.g., 6.5 vs. 6.8 pH), the differences in the concentrations of these elements might not be statistically significant enough to show the presence of two populations and exceed the range of statistical error.

Stable Isotopes

The gases trapped in secondary calcite cements have been studied in a completely different way. Analyzing isotopes has been identified as a method of distinguishing between biogenic and thermogenic methane. The methane present in a head-space gas sample can be analyzed to determine the percentages derived from either biogenic or thermogenic sources. This indicates whether methane anomalies are due to biological activity or seepage. The presence of oxygen has also been cited as signifying that secondary calcite is a product of microseepage.

Isotopes are different forms of the same chemical element. Physically, two isotopes of the same element differ, but their chemical properties are almost identical. For example, carbon has three isotopes: C-12, C-13, and C-14. Isotopes can be stable or unstable (radioactive). Only 21 elements are pure; all others are made up of two or more isotopes. In most cases, one isotope is predominant, and the others are in trace amounts. Carbon 14 is radioactive, decaying to nitrogen 14, which is the basis for carbon 14 dating. Carbon 12 and carbon 13 are stable, and both are present in nature at a ratio of approximately 100:1. Carbon 12 is made up of six protons (atomic number) and six neutrons for an atomic weight of 12. Carbon 13 has six protons and seven neutrons for an atomic weight of 13. (Hereafter these isotopes will be designated ^{12}C and ^{13}C, and so forth, according to preferred international usage.)

Different physical and chemical processes alter the average ratio of ^{12}C to ^{13}C through an equilibrium exchange process. The changes in the ratio of isotopes present can cause changes in the rates of chemical reactions. A given process may have a slight preference for one isotope of carbon over the other isotope. This preference can be measured as a fractionation factor, usually labeled *alpha*. Alpha is the ratio of two isotopes in one material to the ratio of the same isotopes in another material, according to the reaction:

$$A - B = \frac{R_A}{R_B}$$

R_A is the ratio in one chemical compound, and R_B is the ratio in the other.

The standard usually used for carbon isotopes is the University of Chicago's reference calcite. Isotopic analyses are reported in the form:

$$^{13}C = 1000 \times \frac{R_A - 1}{R_B}$$

If the value of the ^{13}C is small or negative, the sample is said to be isotopically light; this means that it has relatively more ^{12}C. If the ^{13}C value is larger, the sample is said to be isotopically heavy; that is, it has relatively more ^{13}C.

The isotope ^{13}C has been used to identify the origin (thermal vs. biogenic), generation, migration, and accumulation of hydrocarbons (Fuex, 1977). Isotopic composition is measured by mass spectrometry. The carbonate mineral is converted to carbon dioxide by the use of an acid. The carbon dioxide containing ^{12}C is lighter than the carbon dioxide containing ^{13}C and is measured by mass. The ratio of the amounts of the two are submitted into the equation.

Carbon is a trace element in the earth but an abundant element in the universe. Carbon's two isotopes of importance are: ^{12}C (98.89% of all carbon) and ^{13}C (1.11% of all carbon). Ratios of ^{13}C to ^{12}C are indicated in Table 8-2. Samples with strongly negative ^{13}C values are associated with organic carbon. Water washing and biodegradation have little effects on the isotopic composition in relation to petroleum. The use of C ratios along with isotope data is a powerful tool for source-rock evaluation.

Organisms have the ability to fractionate isotopes to a slight degree. In the case of carbon, there is a preference for the lighter isotope, ^{12}C. If a carbonate cement is produced from carbon dioxide that was derived from decay of vegetation at the surface, the cement will have more ^{12}C or be isotopically light or have more negative ^{13}C values. If the carbon dioxide is derived from the oxidation of hydrocarbons originally produced by thermogenic processes, the cement will be isotopically heavy. The cement derived from micro-

Table 8-2 Carbon Sources and $^{13}C/^{12}C$ Ratios

Source of Carbon	Range of ^{13}C Values
Organic carbon	79–62
Marine limestones	5–5
Atmospheric carbon	10–8
Living marine organisms	20–8
Land plants	30–7
Hydrocarbons from marine source	27–18
Hydrocarbons from nonmarine source	30–22
Hydrocarbons from coal	40–20
Methane (all sources)	90–25
Gas deposits	78–28
Bacterial methane	90–50
Dry gas (shallow)	75–60
Dry gas (deep)	40–25
Gas associated with oil	60–40
Geothermal methane	30–25
Ethane	45–25
Propane	38–20
Butane	36–23

Adapted from Fuex, 1977.

seepage will also be isotopically heavy. The carbon and oxygen isotopic composition of the calcite is used to infer its origin.

Donovan (1985) determined the isotopic composition of secondary calcite cements over an area suspected of having microseepage. The area is located in the Denver-Julesberg Basin, Boulder County, Colorado, and the production is from the Cretaceous Dakota sandstone at a depth of 7000 ft. Figure 8-13 is a map of del ^{13}C for the secondary calcite

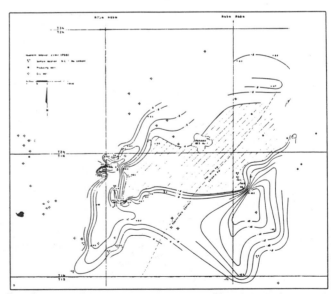

Fig. 8-13 Carbon 13 map along the outcrop of the Dakota sandstone, Gun Barrel Hill area, Colorado. The hatched area is considered prospective. The well control to date is also indicated (after Donovan, 1985; reprinted with permission of the Association of Petroleum Geochemical Explorationists).

cements in the outcropping sandstones. The research was done in 1975. Near the center of the map, the calcite cement is isotopically heavy relative to the surrounding area. This is taken as direct evidence of hydrocarbon microseepage. There are some producing wells in the eastern part of the area and numerous dry holes to the west. The central area is considered prospective.

There are stable isotopes of oxygen, including ^{16}O, ^{17}O, and ^{18}O. Oxygen is the most abundant element on earth, and it occurs in a variety of forms. Of all oxygen, ^{16}O makes up 99.763%, ^{17}O 0.0375%, and ^{18}O 0.1995%. A similar fractionation occurs with the ratio of ^{18}O to ^{16}O and is expressed as alpha ^{18}O. Organisms again have a preference for the lighter isotope, ^{16}O. If microorganisms are producing carbon dioxide from vegetative material on the surface, the carbon dioxide will be isotopically light with respect to oxygen. If the secondary calcite cement is isotopically heavy with respect to oxygen, it is possibly from hydrocarbon microseepage. Figure 8-14 is the oxygen isotope map for the carbonate cements in the same Dakota sandstone surface as Fig. 8-13. Again, a similar area of isotopically heavy cements occurs and is considered prospective for petroleum.

Subsequent drilling has proved the prospective area. Figure 8-14 shows the original anomalous area as defined by the isotopic composition of the cements and the drilling through 1984. A significant amount of new production has been found in and around the prospective area (Fig. 8-15).

Collection Methods for Isotopes

The forms of isotopic sampling are typical soil-gas collection techniques; either head-space or absorbed methods are

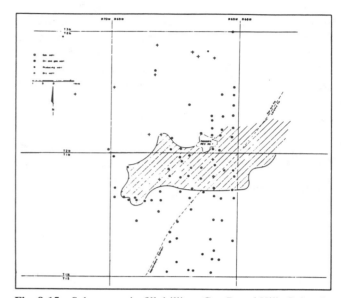

Fig. 8-15 Subsequent in-fill drilling, Gun Barrel Hill, Colorado. The new producing wells indicate no definitive correlation with the anomaly (after Donovan, 1985; reprinted with permission of the Association of Petroleum Geochemical Explorationists).

used. In some cases, samples can be taken from drill cuttings. Absorbed methane consistently produces heavier isotopic concentrations (Woltemate, 1982; Horvitz, 1985).

Discussion of Isotopes

Isotopes are considered a viable tool for separating biogenic from thermogenic methane gas, especially in marine sediments. Abrams (1989) analyzed 700 cores using a variety of analyses of marine sediments from offshore Alaska. The conclusion of the study was that isotopic data can be useful in identifying the source of the methane. However, different forms of sampling yield different results. Interpretation is dependent on careful consideration of the sampling technique and the location of the sample in the soil or bottom sediments.

Case Histories

Sixto Oil Field, Starr County, Texas

The Sixto Field was discovered in 1964 and is located in Starr County, Texas (Fig. 8-16). The discovery was based on subsurface geology and a delta C geochemical survey (Ransome, 1969). Figure 8-17 presents the structure on top of the productive sandstone at 660 m (2170 ft). Figure 8-18 represents the predrilling interpretation of the delta C survey. Subsequent drilling of the field resulted in a reinterpretation of the anomaly to extend the halo outline to the north (Fig. 8-19). The anomaly covering the southern part

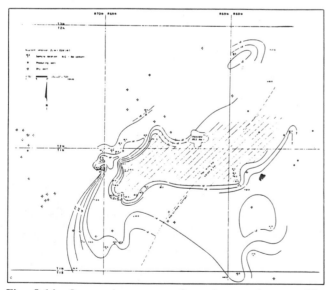

Fig. 8-14 Oxygen isotope map for carbonate elements. The hatched area indicates Gun Barrel Hill, Colorado, and the prospect recommended for drilling (after Donovan, 1985; reprinted with permission of the Association of Petroleum Geochemical Explorationists).

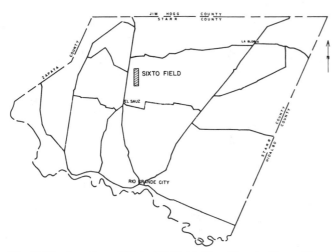

Fig. 8-16 Location of Sixto Oil Field, Starr County, Texas (Ransome, 1969; reprinted with permission from the Institute for the Study of Earth and Man, Southern Methodist University).

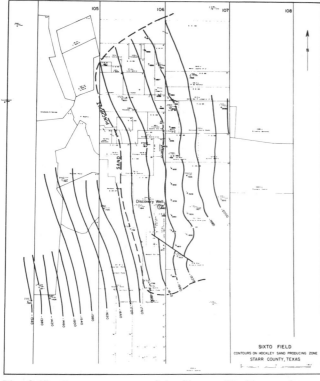

Fig. 8-17 Structure on top of the Sixto or Hockley sandstone. Contour interval is 6 m (25 ft) (after Ransome, 1969; reprinted with permission from the Institute for the Study of Earth and Man, Southern Methodist University).

of the field forms a halo, but the remaining part of the survey could not be interpreted until after drilling.

Caledonia and Walbanger Fields, Denver-Julesberg Basin, Colorado

Duchscherer and Mashburn (1987) described a delta C survey conducted over Caledonia and Walbanger fields in

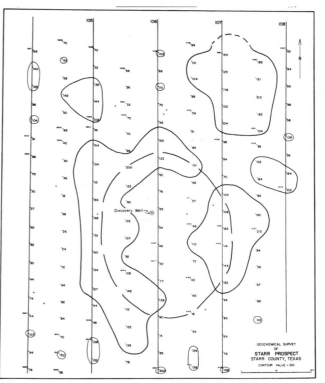

Fig. 8-18 Delta C survey, interpretation before drilling. Values are in millivolts, and a halo anomaly is indicated (after Ransome, 1969; reprinted with permission from the Institute for the Study of Earth and Man, Southern Methodist University).

the Denver-Julesberg Basin, Colorado (Fig. 5-57). The fields were discovered in 1981 and 1982, respectively. The delta C survey was done in 1986. Production is from the D and J sands of the Cretaceous Dakota Formation at a depth of 2480 to 2550 m (8200–8400 ft). Figure 8-20 is a structure contour map drawn on the top of the J sandstone. The fields are in a channel sand that is offset by a small displacement growth fault. Figure 8-21 is the delta C survey. A halo type of anomaly is clearly shown over the Walbanger Field. Selection of the contour interval was based on the method previously described for the Bell Creek Field. Other possible halo anomalies exist to the southeast. There is a dry well in one of the halos.

Duchscherer and Mashburn (1987) published trace-element data from the soils of the Caledonia and Walbanger fields. They used deeper soil samples than those collected for the delta C method. The sample digestion method was much stronger, approaching total digestion, although Duchscherer and Mashburn do not specify the details. Figure 8-22 is a map of iron in the soils over the Caledonia and Walbanger fields. The halo pattern is not readily apparent. The analytical method was emission spectroscopy, in which the concentration is commonly reported in class intervals as indicated in the map data. This makes interpretation more difficult. The normal concentrations for total iron in soil derived from sedimentary rocks are in the 10,000- to 30,000-

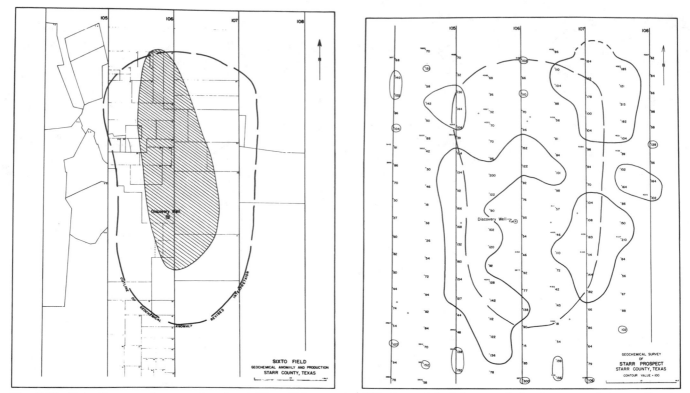

Fig. 8-19 Delta C survey, interpretation after drilling. Values are in millivolts, and the halo anomaly has been extended to outline the field as defined by drilling (after Ransome, 1969; reprinted with permission from the Institute for the Study of Earth and Man, Southern Methodist University).

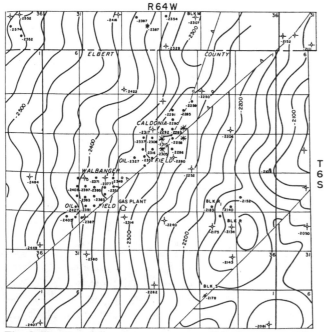

Fig. 8-20 Structure contour map on top of the J sandstone. Contour interval is 25 ft (7.6 meter) (after Duchscherer and Mashburn, 1987; reprinted with permission of the Association of Petroleum Geochemical Explorationists).

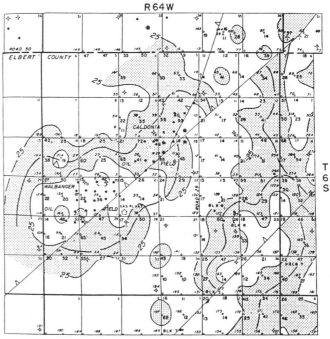

Fig. 8-21 Delta C map for the Caledonia and Walbanger area, Denver-Julesberg Basin, Colorado. Contour value is 25 mV. The anomalous area is identified by the stippled pattern. A halo anomaly is indicated (after Duchscherer and Mashburn, 1987; reprinted with permission of the Association of Petroleum Geochemical Explorationists).

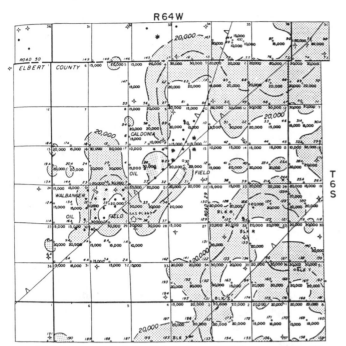

Fig. 8-22 Iron map of the Caledonia and Walbanger fields. The contour interval is 20,000 ppm, and the anomalous areas are indicated by the stippled pattern. A halo anomaly is indicated (after Duchscherer and Mashburn, 1987; reprinted with permission of the Association of Petroleum Geochemical Explorationists).

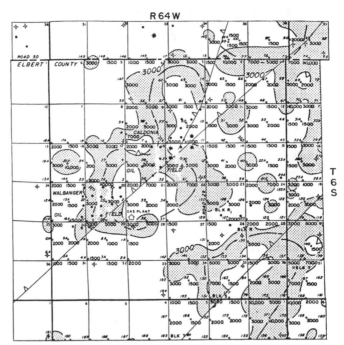

Fig. 8-23 Total titanium in the soils from Caledonia and Walbanger fields. The contour interval is 3000 ppm, and anomalous areas are indicated by a stippled pattern. A halo anomaly is indicated but does not clearly define the fields (after Duchscherer and Mashburn, 1987; reprinted with permission of the Association of Petroleum Geochemical Explorationists).

ppm range. Roeming and Donovan (1985) used a contour internal of 11 ppm at the Bell Creek Field. Nearly all the iron in the soil is still locked in silicate minerals and is not available. Only the iron in a mobile form is susceptible to reduction by hydrocarbon microseepage, by plants, and by DPTA extraction. It does not seem likely, from a geochemical point of view, that all, or even a significant fraction, of the iron in the soils will be mobilized by microseepage. Therefore, a determination of total iron in the soils sheds little light on this element's relationship with microseepage.

Figure 8-23 is the map for total titanium in soils from Duchscherer and Mashburn (1987). The data are reported in class intervals as they were for iron. Titanium is a very insoluble element at the pH expected for these soils in the climate of the Colorado plains. Large-scale mobilization and migration of a large proportion of the titanium to form a halo or an apical pattern is geochemically unrealistic.

It has not been clearly demonstrated in the literature that trace- and major-element concentrations indicate hydrocarbon microseepage. Depth of sampling, method of analysis, climate impact, soil compositions, moisture content, and so forth, have not been considered for their effects on element concentrations and locations.

Railroad Valley, Basin and Range Province, Nevada

Saeed (1988) and Klusman et al. (1992) used the DPTA extraction procedure in a study area at Eagle Springs Oil Field in Railroad Valley, Nevada. The DPTA extractable iron and manganese (Figs. 8-24 and 8-25) did not show a pattern that could be interpreted as anomalous and associated with the field. Saeed and Klusman et al. stated that the depth of the carbonate formation was too great to be reached by hand tools and that, therefore, this form of sampling was unreliable. However, analyzing plants whose root systems reached the depth of the carbonate formation was effective in delineating the area of microseepage. The researchers found that there were high concentrations of Fe, Mn, Ca, and Ti in a specific type of plant over the Eagle Springs Oil Field. This reflected enrichment in the chelate form, which increased uptake of these elements by the plants. Aluminum lows in these same areas suggested carbonate deposition. This is one of the few cases in which carbonate deposition is indicated in the area of microseepage.

The difficulty with this example is that there is no clear definition of the existing field or any evident overlap between the different elements that defines other potential target anomalies. Even though the anomalous areas are identified as associated with microseepage, it is uncertain if the causes are actually related to microseepage or to soil variations. It is known that the Basin and Range Province has a considerable amount of macro- and microleakage that has been, for the most part, related to fault and fracture trends. The few fields that have been found suggest that the presence of leakage is rarely indicative of accumulation.

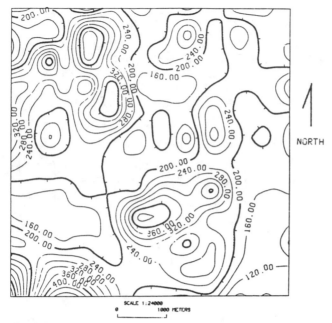

Fig. 8-24 Extractable iron in Railroad Valley, Nevada. The map is generated by computer, and no data are presented (after Saeed, 1988; reprinted with permission from the Colorado School of Mines).

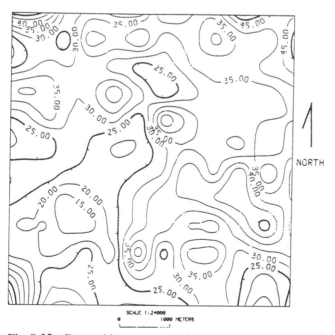

Fig. 8-25 Extractable manganese in Railroad Valley, Nevada. The map is generated by computer, and no data are presented (after Saeed, 1988; reprinted with permission of the Colorado School of Mines).

Dolphin Oil Field, Williston Basin, North Dakota

The Dolphin Field is located in Sections 29, 30, and 32 of Township 161 North, Range 95 West and Section 2 of Township 160 North, Range 96 West, Divide County,

Fig. 8-26 Location of Dolphin Field, Divide County, Williston Basin, North Dakota (Ma and Horvitz, 1989; reprinted with permission from the Association of Petroleum Geochemical Explorationists).

Williston Basin, North Dakota (Fig. 8-26). The field, which was discovered in 1986, produces from the Dawson Bay Formation (Middle Devonian) and has an estimated recovery of 3 MMbbls of oil and 7.5 BCF of gas.

Ma and Horvitz (1989) collected eight samples to determine the ratio of ^{13}C to ^{12}C and the thermogenic vs. biogenic gas. Samples were taken from 3 m. Approximately 100 to 250 g of soil were mixed in a blender with distilled water and stirred for approximately 3 min. The slurry was sieved through a 63-μ sieve. The sand or coarse fraction was dried, then ground, and used for analysis. The clay fraction was analyzed for light hydrocarbons using the acid extraction method. Methane gas was extracted from the coarse sample using liquid nitrogen. The methane was then combusted with a wire burner to form H_2O and CO_2. The CO_2 was separated and analyzed by mass spectrometry. The data were reported in ^{13}C values per milliliter for the absorbed methane in the sand and clay fractions.

Figure 8-27 illustrates the results for the clay fraction; Fig. 8-28 shows the results for the sand fraction. Both data sets indicate a low in association with the field outline. However, the data sets are too small to lead to a definitive evaluation.

Summary

Trace and major elements have been used occasionally in exploration for petroleum accumulations. The majority of the work has focused on magnetic minerals, isotopes, delta C, and total amounts of trace elements present. Trace and major elements and isotopes have been the subject of controversy and have had limited acceptance. The element concentrations have many sources of accumulation or causes of depletion and need to be confirmed by other geochemical exploration methods. The forms of analysis for major and trace elements range from simple to complex. The correct form of extraction to prove that the anomaly is related to petroleum seepage has not been clearly identified. Soil com-

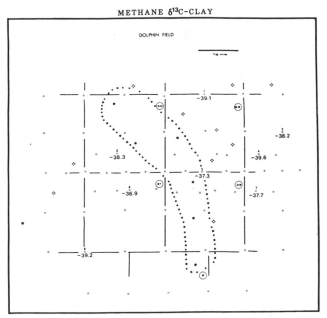

Fig. 8-27 Carbon 13 isotopic data in milliliters for the clay fraction. Data density is small, but there is a general decrease toward and around the Dolphin Field (after Ma and Horvitz, 1989; reprinted with permission from the Association of Petroleum Geochemical Explorationists).

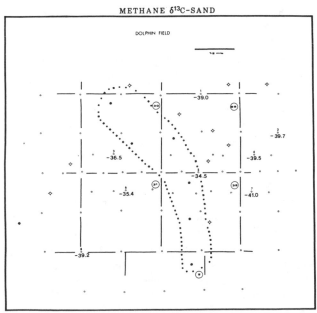

Fig. 8-28 Carbon 13 isotopic data in milliliters for the sand fraction. Data density is small, but there is a decrease toward the Dolphin Field (after Ma and Horvitz, 1989; reprinted with permission from the Association of Petroleum Geochemical Explorationists).

position, moisture, organic matter, clay type, and pH and Eh must be studied to understand element variations across a survey. Otherwise, the analytical results will not have a high level of confidence.

References

Abrams, M. A. (1989). Interpretation of methane carbon isotopes extracted from surficial marine sediments for detection of subsurface hydrocarbons, *Association of Petroleum Geochemical Explorationists Bulletin*, Vol. 5, pp. 139–166.

Alekseev, F. A. (1959). Radiometric method of oil and gas exploration, nature of radiometric anomalies and radiogeochemical anomalies in the region of oil and gas fields, in *Yadernaya Geofizika*, Gostoptekhizdat, Moscow, pp. 3–26; also an abstract in *Geophysical Abstracts*, No. 183–525, Oct.–Dec. 1960, p. 607.

Alekseev, F. A., R. P. Gottikh, and G. D. Sundukove (1961). Results of radiogeochemical investigations of the Kyurov-Dag Oil Field., in *Yadernaya Geofiziika*, Gostoptekhizdat, Moscow, pp. 160–176; also as an abstract in *Geophysical Abstracts*, (1963). No. 199–286, Aug., pp. 728–729.

Baker, W. E. (1966). Humic substances and their role in the solubization and transport of metals, in D. Carlisle, ed., *Mineral Exploration, Biological Systems and Organic Matter*, Prentice-Hall, Englewood Cliffs, NJ, pp. 377–407.

Barton, R. H., (1990). Relationship of surface magnetic susceptibility variations in hydrocarbons and subsurface structures, *Association of Petroleum Geochemical Explorationists Bulletin*, Vol. 6, No. 1, pp. 1–11.

Bolt, G. H. and M. G. M. Bruggenwart (1976). *Soil Chemistry, A. Basic Elements*, Elsevier Scientific Publishing Co., New York.

Boyle, R. W. (1982). *Geochemical Prospecting for Thorium and Uranium Deposits*, Elsevier Scientific Publishing Co., New York.

Brookins, D. G. (1987). *Eh-pH Diagrams for Geochemistry*, Springer-Verlag, New York.

Cannon, H. L. (1952). The effects of uranium-vanadium deposits on the vegetation of the Colorado Plateau, *American Journal of Science*, Vol. 250, pp. 737–770.

Dalzeil, M. C. and T. J. Donovan (1984). Correlations of suspected petroleum-generated biogeochemical and aeromagnetic anomalies, Bell Creek Oil Field, Montana, in *Unconventional Methods in Exploration for Petroleum and Natural Gas Symposium III*, Southern Methodist University, Dallas, TX, pp. 59–69.

Donovan, T. J. (1974). Petroleum microseepage at Cement, Oklahoma: Evidence and mechanism, *American Association of Petroleum Geologists Bulletin*, Vol. 58, pp. 429–446.

Donovan, T. J. (1985). Stable isotopes in petroleum exploration—oxygen isotopes: Special Publication No. 1, in *Surface and Near-surface Geochemical Methods in Petroleum Exploration*, Association of Petroleum Geochemical Explorationists, pp. F1–F9.

Donovan, T. J., R. L. Noble, R. L., I. Friedman, and J. D. Gleason (1975). A possible petroleum-related geochemical anomaly in surface rocks, Boulder and Weld Counties, U.S. Geologic Survey Open-File Report No. 75-47.

Donovan, T. J., R. L. Forgey, and A. A. Roberts (1979). Aeromagnetic detection of digenetic magnetite over oil fields, *American Association of Petroleum Geologists Bulletin*, Vol. 63, pp. 245–248.

Donovan, T. J. and A. A. Roberts (1980). Stable isotope anomalies in surface rocks and helium anomalies in soil-gas over fields—Causes and correlations (Abstract), American Chemical Society Meeting, Houston, TX, March.

Donovan, T. J., J. D. Hendricks, A. A. Roberts, and P. T. Eliason (1984). Low altitude aeromagnetic reconnaissance for petroleum in the Artic National Wildlife Refuge, *Geophysics*, Vol. 49, No. 8, pp. 1338–1353.

Donovan, T. J., D. P. O'Brien, J. G. Bryan, and K. I. Cunningham (1986). Near-surface magnetic indicators of buried hydrocarbons: Aeromagnetic detection and separation of spurious signals, *Association of Petroleum Geochemical Explorationists Bulletin*, Vol. 2, No. 1, pp. 1–20.

Duchscherer, W. (1984). *Geochemical Hydrocarbon Prospecting*, Pennwell Books, Tulsa, OK.

Duchscherer, W. (1985). Bell Creek Oil Field, a further geochemical confirmation, *Association of Petroleum Geochemical Explorationists Bulletin*, Vol. 1, pp. 57–84.

Duchscherer, W. (1988). Organic carbon contamination and the delta C method of geochemical hydrocarbon prospecting, *Association of Petroleum Geochemical Explorationists Bulletin*, Vol. 4, No. 1, pp. 1–29.

Duchscherer, W. and L. Mashburn (1987). Application of the Delta C method to Caledonia and Walbanger fields, Elbert County, Colorado, *Association of Petroleum Geochemical Explorationists Bulletin*, Vol. 3, pp. 15–39.

Elmore, R. D., R. McCollum, and M. H. Engel (1989). Evidence for a relationship between hydrocarbon migration and digenetic magnetic minerals: Implication for petroleum exploration, *Association of Petroleum Geochemical Explorationists Bulletin*, Vol. 5, No. 1, pp. 1–17.

Foote, R. S. (1984). Significance of near-surface magnetic anomalies, Unconventional Methods in Petroleum and Natural Gas Exploration, Symposium III, Dallas, Texas, pp. 12–24.

Foote, R. S. and G. J. Long (1988). Correlations of oil and gas producing areas with magnetic properties of the upper rock column, Eastern Colorado, *American Association of Petroleum Geochemical Explorationists Bulletin*, Vol. 4, No. 1, pp. 47–61.

Fuex, A. N. (1977). The use of stable carbon isotopes in hydrocarbon exploration, *Journal of Geochemical Exploration*, Vol. 7, pp. 155–188.

Garrels, R. M. and C. L. Christ (1965). *Solutions, Minerals, and Equilibria*, Harper and Row, New York, pp. 136–139.

Harshman, E. N. (1972). Geology and Uranium Deposits, Shirley Basin, Wyoming, U.S. Geologic Survey, Professional Paper 754.

Hawkes, H. E. and J. S. Webb (1962). *Geochemistry in Mineral Exploration*, Harper and Row, New York, pp. 70–238.

Henry, W. E. (1989). Magnetic detection of hydrocarbon seepage in a frontier exploration region, *Association of Petroleum Geochemical Explorationists Bulletin*, Vol. 5, No. 1, pp. 18–29.

Hoefs, J. (1987). *Stable Isotope Geochemistry*, 3rd Ed., Springer-Verlag, New York, p. 223.

Horvitz, L. (1985). Stable carbon isotopes and exploration for petroleum: presented at the Association of Petroleum Geochemical Exploration, Short Course, Rocky Mountain Section AAPG SEPM-EMD, Denver, CO, June 1985.

Hostetler, P. B. and R. M. Garrels (1962). Transportation and precipitation of uranium and vanadium at low temperatures with special reference to sandstone type uranium deposits, *Economic Geology*, Vol. 57, pp. 137–167.

Johnson, A. C. (1970). How to hunt oil and gas using the inorganic surface geochemical method, *Oil and Gas Journal*, Tulsa, OK, Vol. 68, No. 49, Dec. 7, pp. 110–112.

Ketner, K. B., J. G. Evans, and T. D. Hessin (1968). Geochemical anomalies in the Swales Mountain area, Elko County, Nevada, U.S. Geologic Survey Circular 588.

Klusman, R. W., M. A. Saeed, and M. A. Abu Ali (1992). The potential use of biogeochemistry in the detection of petroleum microseepage, *American Association of Petroleum Geochemical Bulletin*, Vol. 76, pp. 851–863.

Ma, M. and E. P. Horvitz (1989). $^{13}C/^{12}C$ isotope ratios of methane in near surface sand: A study of the Dolphin Field in North Dakota, *Association of Petroleum Geochemical Explorations Bulletin*, Vol. 5, pp. 167–176.

Machel, H. G. and E. A. Burton (1991). Chemical and microbial processes causing anomalous magnetization in environments affected by hydrocarbon seepage, *Geophysics*, Vol. 50, No. 5, pp. 598–605.

Matveeva, L. A. (1963). Hydrolytic precipitation of heavy metals, in G. V. Bogomolov and L. S. Balashov, eds., *Hydrogeochemistry, Akademiya Nauk SSSR*, Moscow, pp. 78–85.

Miodrag S. (1975). Should we consider geochemistry an important exploratory technique? *Oil and Gas Journal*, Aug. 4, pp. 106–110.

Ransome, W. R. (1969). Case history of the Sixto Oil Field, Starr County, Texas; A geochemical discovery, in W. B. Heroy, ed., *Unconventional Methods in Exploration for Petroleum and Natural Gas*, Southern Methodist University, Dallas, TX, pp. 187–196.

Reynolds, R. L., N. S. Fishman, and M. R. Hudson (1991). Sources of aeromagnetic anomalies over Cement Oil Field (Oklahoma), Simpson Oil Field (Alaska), and the Wyoming-Idaho-Utah thrust belt, *Geophysics*, Vol. 56, No. 5, pp. 606–617.

Roeming, S. S. and T. J. Donovan (1985). Correlations among hydrocarbon microseepage, soil chemistry, and uptake of micronutrients by plants, Bell Creek Oil Field, Montana, *Journal of Geochemical Explorationists*, Vol. 23, pp. 139–162.

Saeed, M. A. (1988). Relationship of trace elements in soils and plants to petroleum production in Eagle Springs Oil Field, Railroad Valley, Nevada, unpublished Master's thesis, Colorado School of Mines, Golden, CO.

Saunders, D. F. (1989). Simplified evaluation of soil magnetic susceptibility and soil gas hydrocarbon anomalies, *Association of Petroleum Geochemical Explorationists Bulletin,* Vol. 5, pp. 30–44.

Saunders, D. F. and S. A. Terry (1985). Onshore exploration using the new geochemistry and geomorphology, *Oil and Gas Journal,* Sept. 16, pp. 126–130.

Shacklette, H. T. and J. R. Boerngen (1984). Element concentrations in soils and other surficial material of the conterminous United States, U.S. Geologic Survey Professional Paper 1270.

Sikka, D. B. (1964). Possible modes of formation of radiometric anomalies, *Geophysical Abstracts,* No. 206–265, p. 227.

Sveshnikov, G. B. and Y. S. Ryss (1964). Electrochemical processes in sulfide deposits and their geochemical significance,

Geokhimiya, No. 3, pp. 208–218; translation available in *Geochemistry International,* Vol. 1, 1964, pp. 198–204.

Szalay, A. and M. Szilagyi (1967). The association of vanadium with humic acids, *Geochem Cosomochim Acta,* Vol. 31, pp. 1–6.

Waples, D. (1981). *Organic Geochemistry in Exploration,* Burgess Publishing Co., Minneapolis, MN, p. 157.

Waples, D. (1985). *Geochemistry in Petroleum Exploration,* International Human Resources Development Corp., Boston.

Woltemate, I. (1982). Isotopische Untersuchungen zur bakteriellen Gasbildung in cinem Subwasseree: Diplomarbeit, Universitas Clausthal, Clausthal Zellerfeld, Germany.

Zak, I. (1964). Geochemical study of soils in the Heletz Oil Field, *Israel Journal of Earth Sciences,* Vol. 13, pp. 183–184.

9

Microbiological Methods

Introduction

Bacteria that digest hydrocarbons have been used by several investigators to find petroleum accumulations. Early work was done by European and Russian scientists, but it was not until the 1940s that the U.S. scientific community recognized the existence of bacteria that thrive by using petroleum as their sole source of carbon. Microbes are also used effectively in the search for other natural resources such as gold (Parduhn, 1991). Many microbial families, genera, and species have a specific, critical, or unique element or compound that they have adapted to their metabolic requirements. The volume of certain bacteria populations increases in conjunction with the rise in concentration of methane or other hydrocarbons that enter the soil substrate. Three methods were developed by the Soviets for microbial detection. They are: (1) microbial soil surveys, (2) microbial surveys of drill cuttings or cores from nondevelopment wells, and (3) groundwater microbial surveys. The third has become the most common microbial method used in the former Soviet Union, but only the first method will be discussed here because it is the dominant one used in the United States.

The basis for microbial detection of petroleum lies in the bacteria's ability to metabolically oxidize some portion of the hydrocarbons migrating through and out of the soil substrate. The number of bacterial species able to break down heavier hydrocarbons has been assumed to diminish with chain length. At present, several types of bacterial species have been mutated in order to help in the cleanup of the petroleum spills that damage the environment. However, the number of species that have the ability to utilize heavy hydrocarbons in the soil is thought to be low to nonexistent. The heavier hydrocarbons are also more likely to convert from a gaseous state to a liquid or solid one under the rapid changes in pressure, temperature, and chemical conditions that are typically found in the shallow soil or rock substrate. Recent evidence indicates, however, that bacteria can adapt to new food sources by undergoing high rates of mutation. These mutations allow a species, group, or family of bacteria

to become opportunistic and, in this case, to use petroleum as a source of food. A variety of other bacteria are present that digest the by-products of petroleum reactions in the soil. These by-products are sulfates, iron compounds, trace elements, and heavy-metal compounds. This chapter is specifically concerned only with those bacteria that ingest petroleum directly or that use a reaction by-product in the near-surface soil environment. The bacteria living in the rock column can also be affected by petroleum migration, but microbial survey cores have generally not been considered a viable or economic exploration tool.

The two types of microbial population in the soil are (1) aerobic bacteria, which require oxygen to grow, and (2) anaerobic microorganisms, which need no oxygen when supplied with enough nutrients.

Microbial Activity

Microbial activity in the soil is dependent on the composition and changes of that medium on a daily or seasonal temperature basis (Frobisher, 1969). It is important to understand that not all species respond to the same set of environmental conditions. Certain conditions may be toxic to one species and not to another and thus encourage or discourage microbial growth. Microbes are very important in the development of soils because they influence many soil conditions. Several important factors that can eliminate, inhibit, or encourage microbial activity are listed below. These are constraints under which microbial activity will either oxidize or be unable to oxidize hydrocarbons.

1. The exposed surface area of the petroleum molecule mass controls the rate of its destruction by microbes. The greater the surface area, the faster the oxidation rate proceeds.

2. Moisture is vital to microbial metabolic activity for it must be present to transport the oil into the microbe and to remove waste products (and water). There is a strong interrelationship between microbial activity and moisture

content in the soil. Variations in moisture content can cause microbial increases or decreases that can be incorrectly perceived as anomalous.

3. The temperature must be greater than 0°C to allow population growth (Samtsevich, 1955). All microbes have a temperature envelope that has a minimum and maximum within which they can exist. In the envelope is the optimum temperature at which microbes will grow at a maximum rate.

4. The rate of oxidation is controlled by the amount of oxygen present. Reactions will occur under anaerobic conditions but at much slower rates for many bacteria.

5. Other mineral nutrients must be present for the bacteria to grow. These are usually satisfied by the composition of the soil itself, which provides phosphate, sulfate, iron, potassium, and other trace elements. Depletion or lack of any particular mineral nutrient may inhibit bacterial growth and thus inhibit the oxidation of hydrocarbons. If certain minerals and salts are missing from the soil, there may be no bacteria, regardless of the amounts of hydrocarbons passing through at that locale.

6. The type of hydrocarbons introduced into the soil can inhibit or encourage oxidation. Typically, aromatic hydrocarbons are more difficult to oxidize than paraffinic hydrocarbons.

7. Eh and pH conditions cause either a reducing, neutral, or oxidizing environment. The oxidation/reduction conditions will determine whether a microbe can grow in the soil. The Eh/pH conditions may fluctuate rapidly or remain unchanged for long periods of time, depending on climatic conditions.

8. Microorganisms can be in a sorbed state on soil (clay) particles, and this can have either a positive or negative effect, depending on the type of bacteria. Different clays can inhibit respiration, encourage or inhibit carbon dioxide evolution, suppress adenine decomposition, stimulate non-symbotic nitrogen fixation (bacteria-specific), and suppress ammonium oxidation. Soils with a trivalent cation state suppress metabolic and biological activity. Soils with nonvalent cations usually do not inhibit carbon dioxide evolution; they stimulate uric acid decomposition, nitrogen fixation (bacteria fixation), and sulfate reduction (McLaren and Skujins, 1971, A, B). Specifically, the presence of montmorillonite and kaolin clays causes the death of many bacteria when they enter a sorbed state. Other types of clays encourage growth. The size of the clay particles is also important because the finer particles encourage the growth of bacteria.

9. Metabolic activity and growth increase with increasing amounts of organic nutrients (in the form of humus). Initially, as stated earlier, the surface area of the molecule affects oxidation capabilities. However, the surface area is no longer a constraining or vital factor if the percentage of organics in the soil reaches a certain volume. This critical organic volume varies with each type of microorganism.

Source of Food

It is generally thought that most hydrocarbons in the soil have been derived almost entirely from petroleum migrated from depth. However, methane can be derived from oxidation of the organic matter in the soil itself. During the humification process, microbes convert plant material to humus and generate methane at the same time. Humus is a series of complex organic compounds that are important constituents of surface soils in agriculture.

Oxidation is the most common form of organic degradation, and it results in the end products CO_2, H_2O, and acetic acid. Reduction of organic matter is achieved only with difficulty. It usually involves cleavage of ether linkages to form hydrocarbon and alcohol units, hydrogenolysis of hydroxyl units, saturation of double bonds, reduction of carbonyl to alcohol, carbon-carbon single bond cleavage, or repolymerization (McLaren and Skujins, 1971B).

Generation of ethane and methane, and in some cases butane and propane, by bacterial or fermentation processes may in itself account for the presence of those microbes. Veber and Turkel'taub (1958) reported ethane, propane, and pentane in estuary sediments, and Juranek (1958) observed ethane, propane, and butane in amounts greater than 100 ppm from the fermentation of cellulose. Davis and Squires (1954) observed the generation of ethane, with minor amounts of other heavier hydrocarbons, from the methane fermentation of cellulose. The generated amounts of ethane and heavier hydrocarbons are small (Hunt et al., 1980), but the possible effects of these processes have to be assessed when the results of a survey are interpreted.

Anomalous bacterial accumulations can also be caused by other gas sources. Organic-rich soils and rocks that are augered for samples may release other gases used by microbes. Hot groundwater or hydrothermal fluids that come into contact with organic-rich rocks may also cause the release of gases that migrate to the surface and increase bacteria activity.

Microbial Adaptation

Specific microorganisms typically are associated with certain soil conditions and food sources that represent their position or niche in the food chain. A recent study by Riese and Michaels (1991) suggests that some organisms may adapt through genetic mutation to the presence of toxins, such as petroleum, in the environment. Thus, microbes could possibly adjust also to changes in food sources and supply through genetic mutation. One method for detecting petroleum-consuming microbes was developed using this concept; a variety of bacteria respond and adapt to using hydrocarbons as food under certain conditions. It was assumed that the adaptation by non-hydrocarbon-consuming microbes would occur over a long period of time. Therefore, a method was designed to address this assumption, measuring the targeted bacteria over a short period of time. The amount of petroleum in the soil may have an impact on the permanent, partial, or temporary adaptation of some microbes to the new food source.

Paraffin Dirt

The early literature describes a waxy-looking soil material known as *paraffin dirt*. Kartsev et al. (1959) and Davis (1969) describe it as the remains of oxidizing bacteria. This material is not composed of paraffins and has thus been misnamed. It is actually composed of carbohydrates left from the cell walls. Carbon 14 analysis of the material concluded that it is not modern, which is to be expected for recycled plant material. Its ancient age suggests that the carbon in this material came from microseepage (Burke and Meinschein, 1955). Paraffin dirt has overall low total microbial counts, but the percentage of methane-, ethane-, propane-, and butane-consuming bacteria to the total microbes present is exceptionally high. Bacteria oxidize the hydrocarbons as they seep up from the subsurface. Barton (1925) determined that the presence of methane and water alone could cause some of the bacteria to produce paraffin dirt without petroleum being present in the soil. Davis (1952) confirmed this evaluation of paraffin dirt in the laboratory.

Several examples of the use of paraffin dirt in petroleum exploration have been cited in the literature. Specifically, it was always found in association with petroleum production from salt domes of the U.S. Gulf Coast. However, the material was an unreliable exploration tool in the Mississippi Delta region because it did not always occur in association with petroleum reservoirs (Barton, 1925).

The early observations of paraffin dirt resulted in research to detect microorganism concentration levels that would indicate the presence of microseepage. There was much research in the 1940s and 1950s but there has been relatively little recently. The present level of microbial techniques is in the literature, published by companies offering commercial services.

Methane-Oxidizing Bacteria

Methane oxidizers were the first types of bacteria studied to identify the location of petroleum accumulations. The presence of methane-oxidizing bacteria in the soil, in the absence of cellulose-oxidizing bacteria, has been interpreted as indicating the presence of methane exhaled from the subsurface. Few bacteria are capable of using a one-carbon

compound on a regular basis. The bacteria that partly or exclusively use methane as a food appear to be pseudomonades. Characteristics of methane oxidizers are: (1) They are associated with cellulose-oxidizing bacteria, but they have rarely been found below 1.5 m (Mogilevski, 1938, 1940). (2) They have been found as deep as 10 m but tend to remain concentrated above 1.5 m. (3) They increase in concentration with increase in soil permeability. (4) Population growth is dependent on soil moisture content. When the moisture content decreases, the populations of methane oxidizers decreases also.

Subbota (1947A) found that seasonal variations in amounts of soil gases over an oil accumulation were rather small. He also determined that the population of methane-oxidizing bacteria did not fluctuate either (Subbota, 1947B). These findings are in conflict with the experiences of the author and of other investigators in studying soil-gas variations, especially methane, and in measuring bacteria from surveys taken at different times across the same survey area.

The problem with using methane oxidizers for petroleum exploration is that methane can be generated by microbes in the soil rather than by migrating hydrocarbons. Consequently, their value to explorationists for specifically targeting petroleum accumulations has generally been dismissed. Coleman et al. (1988) indicated that microbial methane can be distinguished from petroleum-generated methane by carbon and hydrogen isotope analysis. Thus, the type and source of the methane can be determined by taking a soil-gas sample from the same site as the microbe sample and analyzing the gas isotopically. A method was developed that placed a soil sample in an environment with methane and ^{14}C. It was assumed that only the petroleum bacteria would consume the methane and ^{14}C effectively. Microbial methane is generally the product of the oxidation of buried organic matter. Therefore, methane oxidizers should not be present in significant quantities at sample sites where organic matter is present in only small amounts in the soil, unless the source of the food is not indigenous.

There are several occurrences of economic biogenic gas deposits (Niobrara gas production in Yuma County, Colorado) to which methane oxidizers would have application rather than ethane, propane, or butane. Oxidizing bacteria should not be abundant in and around these deposits.

Ethane- and Propane-Oxidizing Bacteria

When the Soviets first used microbial prospecting, they established that certain bacteria specifically consume either ethane, propane, or butane, but not methane. These hydrocarbons are assumed to be from petroleum migrating from depth and are not associated with generation in the soil. Ethane and propane, as discussed previously, have been found in small amounts as by-products of the fermentation of organic matter. The process by which these compounds are formed from the degradation of organic matter is not clear. The Soviets have shown a preference for using propane-digesting bacteria because they determined that these were more abundant in their soils than the ethane-consuming bacteria. Davis (1969) advocated the use of ethane-digesting bacteria because he found that they dominated the soils in the United States. Price (1986) has indicated that it is unlikely that ethane- or propane-consuming bacteria will dominate or be absent either regionally or on a site-specific basis. The use of butane-digesting bacteria for microbial surveys has become more common in the United States. As discussed in Chapter 5, the successful comparison of soil-gas ratios to ratios of petroleum found at depth suggests that adequate food exists for all the different petroleum oxidizers. Therefore, the absence of any particular oxidizer is probably a result of unsuitable soil conditions or of poor sampling and analytical methods.

Iron, Sulfide, and Other Mineral-Consuming Bacteria

Microbes consume or generate iron and sulfur compounds either directly or indirectly as part of the reducing process. This microbial activity is a major problem in terms of corrosion of petroleum equipment. The surface geochemical explorationist is concerned with this same metabolic activity generating or removing nonmagnetic and magnetic iron minerals resulting from organisms dying in the soil. As discussed in Chapter 8, one of the surface geochemical methods currently being used in petroleum exploration is the detection of various iron compounds, either magnetic or nonmagnetic, in the shallow subsurface. It is not clear if these microbes have a significant impact on the use of magnetic methods. Roeming and Donovan (1985) indicated that extractable chelated iron (Fe^{+3}) and manganese were found in anomalous amounts in plants directly over, but more commonly on the edges of, the Bell Creek Oil Field, Montana. The non-petroleum-producing areas exhibited very low amounts of chelated iron and manganese. This suggests that chemical conditions had been changed by petroleum migrating to the surface. The amounts of chelated iron and manganese were too large for the indigenous biological activity to absorb.

The overall number of sulfur-reducing bacteria species is small, but they are widely distributed. Sulfur is generally derived from minerals and is released and transformed in the process of soil formation, but sulfur is also an integral part of petroleum. Some of the sulfur may actually have migrated to the surface with the hydrocarbons. The petroleum will also bond with sulfur already in the soil. The sulfur is utilized by special bacteria that create organic sulfur, which is used by other microbes and many higher organisms. Most sulfur in the soil is in the organic form, suggesting a close association with the carbon cycle. Bacteria do play a vital role in the formation of some economic sulfur

deposits (Thode et al., 1954). Figure 9-1 is a general outline of the sulfur cycle in the soil.

Microbial corrosion or weathering of iron compounds occurs either directly by metabolic reactions or indirectly by promoting depolarization of the anodic and cathodic surfaces, which drives the reactions. The main cause of the corrosion or weathering of iron compounds is sulfur-reducing bacteria, especially in the presence of water with dissolved oxygen under anaerobic conditions. Reactions are typically anaerobic, but there are some minor reactions that are aerobic. The iron compounds then move downward through the soil and accumulate with organic material in the lower horizons.

Several microbes can digest hematite via the production of substances that complex Fe^{+3} (iron chelate) during the bacterial metabolic process. Other bacteria deposit the iron in their cell wall and create a magnetic fossil at the time of the microbe's death. Fassbinder et al. (1990) described a living magnetic bacterium in the A horizon of soils in Bavaria that is similar to those found in fresh and salt waters. They concluded that this bacterium contributes to the magnetic properties of the soil either while alive or eventually as fossils. The contribution is dependent on moisture, temperature, and overall soil conditions. Ancient iron deposits created by microbes concentrating iron through everyday metabolic processes are an extreme example of this process. These bacteria have been found in a wide variety of environments where they precipitate and accumulate iron hydroxides and oxides. Machel and Burton (1991) contend that these bacteria are important in a number of geologic settings but are not important in environments found in association with petroleum accumulations. They indicate that Fe^{+3} is

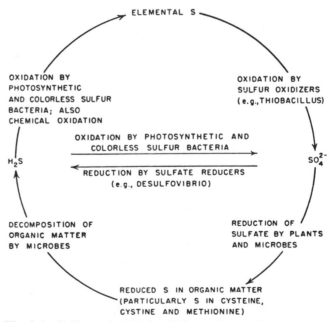

Fig. 9-1 Sulfur cycle (Davis, 1969; reprinted with permission from Elsevier Scientific Publication).

chemically formed and then dissolved. This process occurs in many geologic environments along with petroleum accumulations, and thus there is the likelihood that the first does relate to the second.

The formation and destruction of magnetic and nonmagnetic minerals in the soil are ongoing processes that are independent of petroleum-related reactions. Even if iron compounds or ions are actively being created by a petroleum-caused reducing environment, microbial activity could be effective in removing them and eliminating their effects. On the other hand, a significant number of iron compounds may be the result of microbial activity and have no relationship to hydrocarbons. Therefore, detection of iron compounds in the shallow soil has limited value, but their detection could be more useful well below the soil environment in the rock strata. If there are microbes present at depth, they are more likely to be related to petroleum seepage than to other biological activity.

Iron bacteria are of great importance in the geologic cycle because they are responsible for soil formation and the development of several iron-rich mineral deposits. The activities of the microorganisms determine their ability to produce various compounds of iron. Microorganisms can be separated into two groups: those that are present only in certain stages of the iron cycle in the soil and those that are iron-consuming bacteria.

The formation of iron minerals in the soil is controlled by Eh/pH. The minerals typically are insoluble compounds in the form of oxides, sulfides, and carbonates. Ionic iron is generally found only in acidic environments. The iron bacteria form iron products at the expense of the energy of the ferrous iron oxidation. The reaction is difficult to continue outside this area, and by-products are not accumulated because of the lack of metabolic activity.

There are four groups of microorganisms that utilize iron under different conditions. (1) The first takes advantage of saturated oxygen environments that are in the stability field of ionic iron. (2) The second group develops under low oxygen pressure, neutral to slightly acidic environments, and decreased Eh. Concentrations of soluble iron are extremely low, and there must be a continuous supply of substrate. (3) The third group is oxidizers of organic complexes of iron rather than simple mineral substances. These are usually in the form of chelated iron, are stable, and are not subject to spontaneous oxidation. (4) The fourth group is organisms that can cause the reduction of ferric iron, which produces reduced by-products and decreases the Eh below the stability of ferric oxides.

Microbes have been found in oil reservoirs and in the geologic column itself to several hundreds and even thousands of feet from the surface. These populations should continue to respond by positive growth if the necessary survival conditions exist for a particular family or species of microbe. If the food is present but other required conditions do not, or cease to, exist, the microbes may not be able to take advantage of the food source. Therefore, the absence or presence of these microbes has to be considered when determining the environmental conditions that cause the removal or generation of iron anomalies derived from hydrocarbons migrating to the surface.

Sampling

Sampling depths vary from the surface down to 10 m; the preferred depth is between 0.2 and 2 m. Surface sampling is not common because of the unreliability of the data and problems caused by near-surface fluctuations in moisture content, weather, and other soil conditions. However, surface sampling must be done when the bedrock is close to, or at, the surface. Deep sampling is avoided because bacteria concentrations have been found to be erratic below the water table. In dry areas, low moisture content restricts population growth. Here the best results have been obtained when sampling depths are between 0.2 and 2 m because the moisture content is usually sufficient and stable over the short term to maintain population growth.

Analysis

There are numerous methods for determining the presence of anomalous hydrocarbon-digesting microbe populations in the soil. Investigators have utilized a wide variety of artificial environments in the laboratory and have also introduced bacteria to the prospective site itself. Almost all the methods require an extended period of time, usually longer than two weeks, before results are achieved. Therefore, microbial surveys do not have rapid turnaround or evaluation times.

One of the earliest microbial methods consisted of placing a sample of soil in a container with hydrocarbons and oxygen. In this environment, the microbes oxidize the hydrocarbons. The possibility of oxygen-consuming bacteria interfering with the petroleum-digesting bacteria is headed off by using control samples that are free of hydrocarbons. This method is not effective in detecting small differences in populations between samples. More complicated variations of this method use a detector for analyzing the change in the hydrocarbon content of the atmosphere in the container.

A method to determine which bacteria are oxidizing the atmosphere and which are not was developed by Davis (1957). He placed a soil sample in a container with oxygen, hydrocarbons, and ^{14}C. During the oxidation process, the common by-products are biological waste and carbon dioxide. Carbon 14 is converted along with nonradioactive carbon to CO_2. Therefore, the greater the population of hydrocarbon-digesting microbes in the sample, the greater the intensity of radioactivity.

Another method was developed that cultured soil microorganisms in a bottle with inorganic nutrients and a gas mixture containing 65% light hydrocarbons, 30% oxygen, and 5% carbon dioxide. If petroleum-digesting microbes are present in the soil, the gas volume will be reduced. Hydrocarbon consumption is assumed to have occurred if the volume of gas is reduced to less than the volume that can be accounted for by complete utilization of the oxygen and carbon dioxide. Gas chromatography can determine which hydrocarbons were used. Figure 9-2 shows how to acquire this type of data. A short traverse utilizing this device extends across the edge of a field. The results can be graphed to indicate the hydrocarbon consumption over the field.

In another method (Fig. 9-3), a sample is collected and placed in a container with an aqueous inorganic salt medium and a hydrocarbon nutrient mixture (Updegraff and Chase, 1958). The time required for bacteria to consume the mixture is compared to established standards. The standards are a fixed group of bacteria consuming x amount of hydrocarbons over a specific period of time. The length of analysis is long enough to eliminate problems with non-petroleum-consuming bacteria that may adapt and metabolize. This period of time is usually greater than 25 days. Maddox (1959) discussed a variation on this method that uses the optimum nutrient mixture to promote growth and that incubates the nutrient and sample for a period of 24 to 72 h under agitated conditions.

Hitzman (1959, 1966) has developed a variety of methods, of which one of the more common is to introduce a specific organic liquid, such as methanol, ethanol, butanol, or propanol, as the sole carbon source in the sample. For propagation to occur, the microorganisms have to use the liquid as a nutrient. It is assumed that hydrocarbon-digesting organisms are the only organisms that can use the nutrient. The greater the number of microbes or the larger the population for a specific alcohol, the more likely it is that an oil and gas deposit is present.

Case Histories

Published microbial survey results are very scarce. Few have entered the literature in the last 20 years.

Methane- and Ethane-Oxidizing Bacteria Studies

Davis (1969) described two examples of surveys comparing methane- and ethane-oxidizing bacteria. Figures 9-4 and 9-5 show a microbial survey in and around an oil-producing area, location unknown. The method used was a mercury manometer culture system with the anomalous values representing 25-mm gas uptake in 10 days for methane and 20 days for ethane. Figure 9-4 indicates the methane-oxidizing bacteria values; numerous anomalies exist both within and outside the limits of the producing field. Figure 9-5 shows that all anomalous values (except one) for the ethane-oxidizing bacteria fall within the oil-producing area.

Figures 9-6 and 9-7 exhibit another example of this type of survey. The oil-producing area is located near the intersec-

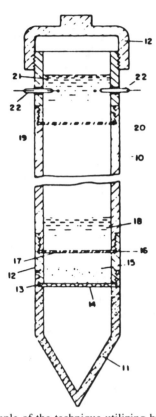

Fig. 9-2 Example of the technique utilizing hydrocarbon digestion. A device for holding a culture of the methane-consuming microorganisms can be inserted into the ground to an appropriate depth and is constructed so that methane and oxygen from the soil can diffuse to the culture to cause growth of microorganisms. More specifically, the device comprises an elongated housing (10) formed from threaded tubing that is connected to a pointed bottom closure member (11) and a cap (12) at the top. The bottom member is made of a porous material such as sintered metal or hard sintered resin so that gases from the earth can diffuse through the wall and enter the housing. The bottom member is threaded to a ring (12), and a screen (14) is held between the two pieces at the shoulder (13). The screen is used for supporting a bed (15) of a suitable absorbent for carbon dioxide, e.g., a hydrous caustic soda supported on asbestos known as Ascarite. The ring is threaded to the tube, and a membrane (17) is held between the abutting shoulders at (16). This membrane serves to support the culture medium (18) that contains the methane-consuming microorganisms. This membrane is made of a material through which both methane and oxygen can readily diffuse. Examples of such material are polyethylene, polypropylene, and other synthetic organic resins. Another membrane (19) is positioned between shoulder (20) of the tube 10 and the top portion of the device. This membrane, which can also be made of polyethylene or other synthetic resins, will allow carbon dioxide to diffuse through it. The membrane is used as a support for an absorbent capable of absorbing carbon dioxide which, in this case, is preferably an aqueous solution of barium hydroxide as shown at (21). A pair of insulated electrodes (22) is positioned in packing glands in the upper wall of the device so as to be immersed in the aqueous solution. These electrodes are employed for determining any changes in the electrical conductivity of the barium hydroxide solution and thus indicate whether any change in the barium hydroxide concentration has occurred (reprinted with permission from Noyes Data Corporation).

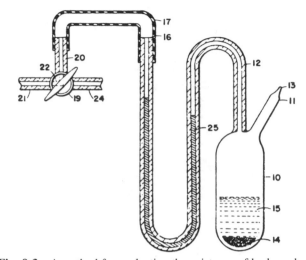

Fig. 9-3 A method for evaluating the existence of hydrocarbon-digesting microbes using inorganic salts. A chamber (10) is provided with a filling tube (11) and manometer tube (12). The filling tube narrows at the end to form a closed tip as shown (13). Prior to introduction of the earth sample (14) and the aqueous inorganic salt medium (15), the tip is open but thereafter is closed by sealing with a flame. The sealed tip and the integration of the manometer tube with the chamber eliminate the need for joints, stoppers, and so forth, which could leak or require lubricants. (The hydrocarbon nutrient could dissolve in the lubricants, which would result in error in measuring the time for the consumption of the nutrient.) To reuse the apparatus, the tip may be resealed. For operation, the apparatus is first cleaned and sterilized. A portion of the earth sample is weighed on sterile paper and is then introduced through the open tip with a sterile powder funnel. Thereafter a measured amount of aqueous inorganic salt medium is introduced through the open tip with a sterile pipette. Following this, the tip is flame-sealed. The chamber (10) is evacuated, which permits the gaseous nutrient mixture to enter. The petroleum hydrocarbon nutrient then comes in contact with this mixture. The chamber can be evacuated by connecting the open end (16) of the manometer tube (12) to (17), which leads to line (20). The end of line 20 connects to a three-way valve or stemcock (19) provided with line (21), which leads to a vacuum pump. The gaseous nutrient mixture is then permitted to enter the chamber by turning the valve to position the channel between line (20) and line (24), which leads to a source of the gaseous nutrient mixture. The pressure of the nutrient mixture should preferably be slightly above atmospheric so that air cannot enter the chamber when the tubing is disconnected from the manometer tube. After disconnection, mercury or other liquid in which the hydrocarbon nutrient is not soluble is poured into the open end (16) of the manometer tube. The apparatus is then maintained at a temperature of about 30°C by placing it in an incubator or water bath. The two levels of the mercury column (25) are read frequently. The time at which the levels begin to change at a more rapid rate than the levels in a control apparatus (containing sterile aqueous medium and gas mixture only) is also recorded, indicating the beginning of consumption of the hydrocarbon nutrient by the bacteria in the sample in the chamber (reprinted with permission from NOYES Data Corporation).

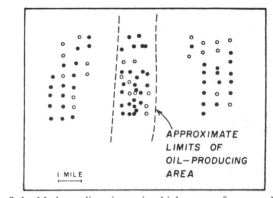

Fig. 9-4 Methane-digesting microbial survey for an unknown location in the United States from (Davis, 1969). The open circles represent areas where methane oxidation is not occurring and the closed circles indicate areas where it is occurring. Methane oxidation is taking place over a large area and is not defining the field (reprinted with permission from Elsevier Scientific Publication).

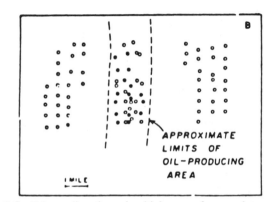

Fig. 9-5 Ethane-digesting microbial survey for an unknown location in the United States (from Davis, 1969). The open circles represent areas where ethane oxidation is lacking, and the closed circles represent areas where oxidation is occurring. Ethane oxidation appears to be taking place mainly within the limits of the field (reprinted with permission from Elsevier Scientific Publication).

tion of Texas, Louisiana, and Arkansas. The method of analysis was again the mercury manometer culture system, and similar criteria were used for anomalous bacterial content. The methane survey indicates anomalous values not only within the producing area but in several other locations across the survey. The ethane results are similar to those in Fig. 9-5 in that the anomalies are almost always associated with the existing producing area. No well data or exact sample locations were provided for either Fig. 9-6 or 9-7 and, thus, there is no clear indication of drilling results either before or after the surveys.

Keota Dome Area

The Keota Dome on the eastern flank of the Forest City Basin is located in Section 19, Township 76 North, Range 9 West, Washington County, Iowa. It was discovered in 1962. The discovery and only producing well pumped 30

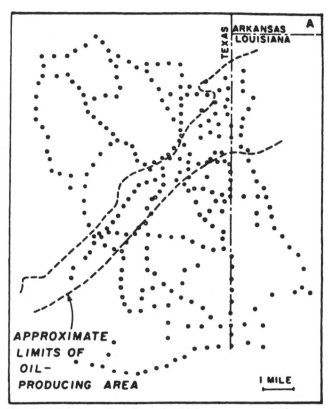

Fig. 9-6 Methane-digesting microbial surveys near the intersection of Texas, Louisiana, and Arkansas (from Davis, 1969). The open circles represent a lack of methane oxidation, and the closed circles represent sites where methane oxidation is occurring. Methane oxidation is taking place in the area of the oil field but is not definitive (reprinted with permission from Elsevier Scientific Publication).

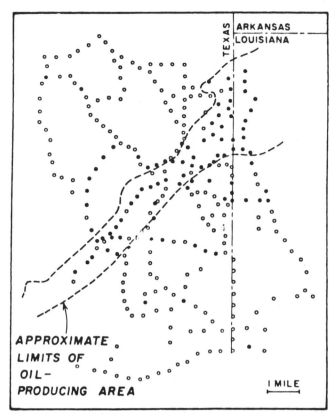

Fig. 9-7 Ethane-digesting microbial surveys near the intersections of Texas, Louisiana, and Arkansas (from Davis, 1969). The open circles represent areas where there is no ethane oxidizing, and the closed circles represent areas where oxidation is taking place. The field is generally the only area where ethane oxidation is occurring (reprinted with permission from Elsevier Scientific Publication).

bbl of oil a day plus water. The well produced 440 bbl of oil before conversion to an observation well for a gas storage facility. The Keota Dome is a Precambrian quartzite erosional remnant that created a topographic high through time in overlying strata. Additional drilling in the area indicated the presence of other features of similar, but not as great, relief.

Prior to the implementation of a microbial survey, iodine and soil-gas surveys were carried out in the same area. Both surveys suggested the presence of anomalous conditions to the southeast, similar to those found at the Keota Dome. A microbial survey was conducted in two phases. A reconnaissance survey indicated anomalous values correlating with the iodine and hydrocarbon surveys (Fig. 9-8). An in-fill survey was then conducted that continued to indicate the presence of an anomaly in the prospect area (Fig. 9-9). At each site, a sample was collected from a depth of 0.2 m and placed on a culture dish with a source of food (butanol). After 10 days, the bacterial population present on each sample dish was counted. A statistical evaluation of the data was performed to calculate the mean and standard deviations necessary for determining anomalous and background values. The anomalous values were highlighted; their relation-

ships (or lack thereof) to each other were compared to determine if extensive or isolated anomalies existed. Subsequent drilling found shows in the Platteville section and Arbuckle Formation. Reinterpretation of all the geochemical data suggested that the anomalous areas, which seemed to be linear, were probably delineating the results of leakage from faults and fracture zones similar to those that have been clearly identified at Keota Dome.

Southern Kansas Study

Beghtel et al. (1987) reported on a microbiological method that was used to predict drilling results in southern Kansas (Fig. 9-10). Samples were taken from select sites, with a predetermined number of samples collected along section boundaries. Hydrocarbon-oxidizing bacteria metabolize the hydrocarbons in steps; the first produces alcohol. If butane is being oxidized, butanol (C_4H_8OH) is produced. The alcohols are toxic to other types of microorganisms and eliminate them from the soil. Soils from the survey area were placed onto a culture dish with a non-nutrient agar

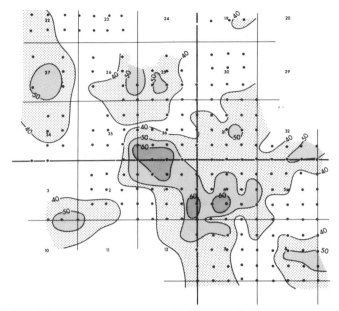

Fig. 9-8 Reconnaissance microbial survey, Keota Dome area, Washington County, Iowa. The value at each sample point represents the number of butanol-eating bacteria counted. Anomalous populations were determined by statistical means; data are smoothed, and the contour values are 70, 80, 90, and 100 counts (reprinted with permission from CST Oil & Gas Corporation).

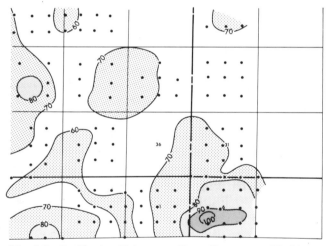

Fig. 9-9 In-fill microbial survey, Keota Dome area, Washington County, Iowa. The value at each sample point represents the number of butanol-eating bacteria counted. Anomalous populations were determined by statistical means, and data were then smoothed. The contour values are 90, 40, 50, and 60 (reprinted with permission from CST Oil & Gas Corporation).

spiked with butanol. If only butanol-tolerant organisms grow, this indicates the presence of hydrocarbon-digesting bacteria.

After a drilling location was staked, 64 samples were collected along the boundaries of a 1-sq. mi. area. The microorganisms were cultured in the agar containing butanol. A confident prediction for the well depends on the number of samples with positive indicators or on the relative

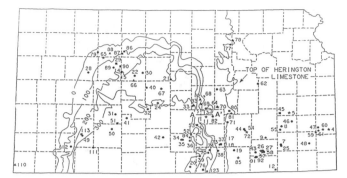

Fig. 9-10 Location map of the surveys carried out by Beghtel et al. (1987) in Kansas. The contours indicate the thickness of the salt deposit (reprinted with permission of the Association of Petroleum Geochemical Explorationists).

size of the colonies growing on the plate. Figure 9-10 shows the 86 drill-site locations in Kansas for which predictions were made. It should be noted that the greatest number of successful predictions were made for wells outside the zero contour line of a Permian evaporite/anhydrite sequence. Klusman (1989) suggested that evaporates may impede the vertical migration of hydrocarbons. Discussions with one of the authors of the paper (Beghtel et al., 1987) suggest that the salt section may partially inhibit and deflect the migrating hydrocarbons as they move toward the surface. The statistical evaluation thus did not recognize that some of the sample points in the evaporite areas may have been anomalous. Subtle anomalies in the data would not be recognized because these values did not achieve statistical significance. The sampling line along the edge of the square-mile area may have inadvertently biased the interpretation of the data. However, this type of approach has merit as long as the user realizes that some undiscovered fields will be missed in the evaluation procedures.

Drilling resulted in the following success rates. Of the 86 wells drilled, 26 were productive. The microbial method predicted that 18 would be productive; of these 18, 13 were actually productive.

Ethane-Oxidizing Microbial Survey

Sleat and Van Dick (1989) presented data for ethane-oxidizing microbial surveys across three oil fields. Figures 9-11 and 9-12 present the microbial surveys across the existing fields. There was a reasonable response, and the results outlined the extent of some of the fields. Figure 9-13 is a microbial survey across another oil field, and it generally confirmed the extent of the field. Figure 9-14 is a repeat of this survey, and it appears that it completely failed to give indications of anomalies that could be clearly associated with the field. All these surveys were single profiles. Without any three-dimensional depth to the survey, it would be erroneous to conclude that the second survey did not detect the

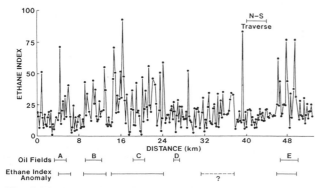

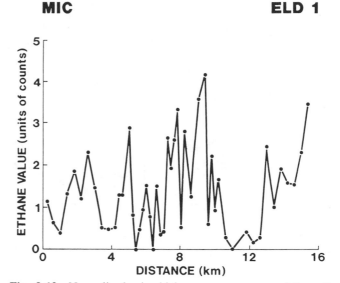

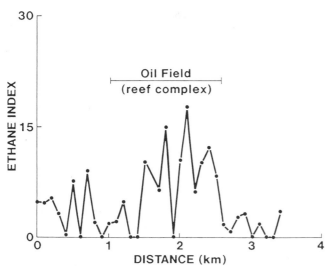

Fig. 9-11 Normalized microbial survey, ethane oxidizers across a series of existing oil fields (Sleat and Van der Dick, 1989). The anomalies are not defined and seem to be a series of isolated points. However, only a single profile was used, which can be misleading or biased or can yield insufficient statistics for analysis of the surface above the oil field.

Fig. 9-13 Normalized microbial survey across an existing oil field using ethane-oxidizing microbes (Sleat and Van der Dick, 1989). The oil field has anomalous values of microbes through two-thirds of its extent. However, only a single profile was utilized.

Fig. 9-12 Normalized microbial survey, ethane oxidizers across a second oil field, which is a reef complex (Sleat and Van der Dick, 1989).

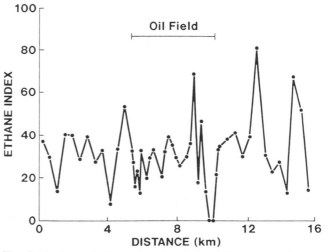

Fig. 9-14 Repeat of the normalized microbial survey in Fig. 9-13 and indications that the oil field was not detected (Sleat and Van der Dick, 1989). However, single profile lines are susceptible to this type of failure, regardless of the type of surface geochemical methods.

field. Because of the variable nature of surface geochemistry, specific sample locations may not always be anomalous from survey to survey, but the field or accumulation is detected because the general overlying surface area is anomalous.

Typical Microbial Survey

A typical current microbial survey is shown in Fig. 9-15. Unfortunately, the author's experience indicates that only a single sample line for microbial data is usually used. An extensive microbial survey is considered one that uses the available road network rather than an evenly spaced sampling grid, which would produce a more reliable interpretation and evaluation.

Summary

The claims for successful microbial surveys abound in the relatively scarce literature available on these surveys. Mogilevski (1953) claimed an 80% success rate, but others have indicated that not all these successes were the result of microbial methods alone. Sealy (1974) and Beghtel et al. (1987) have also claimed a better than 80% rate, but they presented no data. The author's own experience and discussions with various exploration companies indicate that this method can be an effective tool. Problems arise because

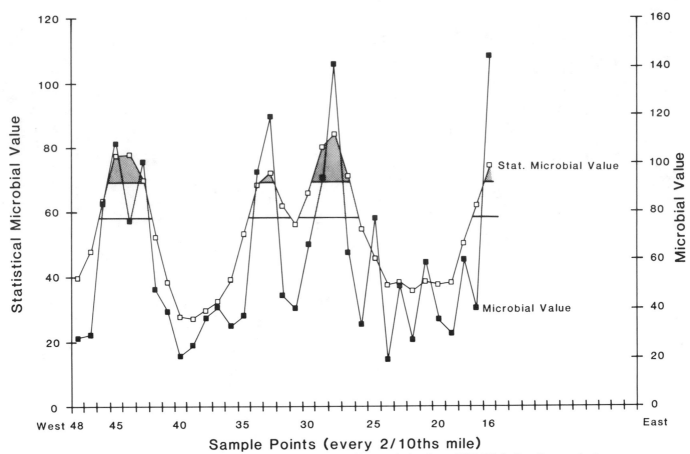

Fig. 9-15 A typical microbial survey as carried out today (printed with permission from CST Oil & Gas Corporation).

there are other factors besides the presence of petroleum that affect microbial population development and growth. Survey design, sample collection, and analytical methods are critical for the success of this type of survey. Statistical methods are used because background and anomalous values vary from area to area and through time. Consequently, the total number of microbes per sample does not have a useful application. Because of this, subtle anomalies hidden within high statistical background values are not recognized as potentially indicative of a petroleum accumulation at depth.

We can conclude that microbial methods are a viable technique for surface exploration of petroleum deposits. The difficulty in using these methods arises from:

1. Determining if critical soil conditions are present that allow petroleum-digesting bacteria to exist. Usually, these conditions are met in many environments, but they may be absent on a site-specific basis or they may have seasonal changes. The iron and sulfide bacteria could be a particular problem in using data collected for magnetic and heavy-metal methods in the shallow soil substrate.

2. Finding and utilizing a method in which the targeted microbes can be grown without interference from other microbes that can adapt to a new food source.

3. The cost of microbial analysis may be prohibitive for some projects.

The microbial methods can add another dimension to exploratory work, but they may not be applicable in all exploration situations or geologic environments.

References

Barton, D. C. (1925). Review of paper concerning "paraffin dirt" by H. B. Milner, *American Association of Petroleum Geologists Bulletin,* Vol. 9, pp. 1118–1121.

Beghtel, F. W., D. O. Hitzman, and K. R. Sundberg (1987). Microbial oil survey technique (MOST) evaluation of new field wildcat wells in Kansas, *Association of Petroleum Geochemical Explorationists Bulletin,* Vol. 3, No. 1, pp. 1–14.

Burke, W. H. and W. G. Meinschein (1955). Carbon 14 dating with a methane proportional counter, *Review of Scientific Instruments,* Vol. 26, pp. 1137–1140.

Coleman, D. D., C. Liu, and K. M. Riley (1988). Microbial methane in the shallow paleozoic sediments and glacial deposits of Illinois, U.S.A. *Chemical Geology,* Vol. 71, pp. 23–40.

Davis, J. B. (1952). Studies on soil samples from a "paraffin dirt" bed, *American Association of Petroleum Geologists Bulletin,* Vol. 36, pp. 2186–2188.

Davis, J. B., (1957). U.S. Patent 2,777, 799, assigned to Socony Mobil Oil Company, Inc.

Davis, J. B. (1967). *Petroleum Microbiology*, Elsevier Publishing Company, New York.

Davis, J. B. (1969). Microbiology in Petroleum Exploration, in *Unconventional Methods in Exploration for Petroleum and Natural Gas I*, Southern Methodist University, Dallas, TX, pp. 139–158.

Davis, J. B. and R. M. Squires (1954). Detection of microbially produced gaseous hydrocarbons other than methane, *Science*, Vol. 119, pp. 381–382.

Fassbinder, J. W. E., H. Stanjek, and H. Vali (1990). Occurrence of magnetic bacteria in soil, *Nature*, Vol. 343, pp. 161–163.

Frobisher, M. (1969). *Fundamentals of Microbiology*, W. B. Saunders, Philadelphia, PA.

Hitzman, D. O. (1959). Prospecting for petroleum deposits (detecting hydrocarbon consuming bacteria colonies by artificial hydrocarbon nutrient culturing), U.S. Patent 2,880,142, assigned to Phillips Petroleum Co.

Hitzman, D. O. (1966). Prospecting for petroleum deposits (using phenolic compounds to test for the existence of hydrocarbon consuming bacteria in soils), U.S. Patent 3,281,333, assigned to Phillips Petroleum Co.

Hunt, J, R. J. Miller, and J. K. Whelan (1980). Formation of C4–C7 hydrocarbons from bacterial degradation of naturally occurring terpenoids, *Nature*, Vol. 288, pp. 577–578.

Juranek, J. (1958). A contribution to the problem of the origin of C_1–C_5 hydrocarbons in samples of soil air in gas survey work, *Czechoslovakian Institute of Petroleum Research Transactions*, Vol. 9, pp. 57–79.

Kartsev, A. A., Z. A. Tabasaranskii, M. I. Subbota, and G. A. Mogilevskii (1959). Geochemical methods of prospecting and exploration for petroleum and natural gases, University of California, Press, Berkeley, CA.

Klusman, R. W. (1989). Surface and Near-Surface Geochemistry in Petroleum Exploration, Short Course, *Rocky Mountain Association of Geologists*, Denver, CO.

Machel, H. G. and E. A. Burton (1991). Chemical and microbial processes causing anomalous magnetization in environments affected by hydrocarbon seepage, *Geophysics*, Vol. 56, No. 5, pp. 598–605.

Maddox, J. (1959). Geomicrobial prospecting (detecting hydrocarbon consuming bacteria by stimulation with inorganic salts and artificial hydrocarbon nutrients); U.S. Patent 2,875,135, assigned to Texaco.

McLaren, A. D. and J. Skujins (1971A). *Soil Biochemistry: Volume I*, Marcel Dekker Inc., New York.

McLaren, A. D. and J. Skujins (1971B). *Soil Biochemistry: Volume II*, Marcel Dekker Inc., New York.

Mogilevski, G. A. (1938). Microbiological investigation with gas surveying, *Razvedka Nedr.*, Vol. 8, pp. 59–68.

Mogilevski, G. A. (1940). The bacterial method of prospecting for oil and natural gases, *Razvedka Nedr.*, Vol. 12, pp. 32–43.

Mogilevski, G. A. (1953). Microbiological Method of Gas and Oil Prospection, *Gostoptekhizdaty*, Moscow.

Parduhn, N. L. (1991). A microbial method of mineral exploration: A case history at the Mesquite Deposit, *Journal of Geochemical Exploration*, Vol. 41, No. 1–2, pp. 137–150.

Price, Leigh C. (1986). A critical overview and proposed working model of surface geochemical exploration, in *Unconventional Method in Exploration, IV*, Southern Methodist University, Dallas, TX, pp. 245–304.

Riese, W. C. and G. B. Michaels (1991). Microbiological indicators of subsurface hydrocarbon accumulations, *American Association of Petroleum Geologists Bulletin Association Round Table* (Abstract), Vol. 75, No. 3, p. 660.

Roeming, S. S. and T. J. Donovan (1985). Correlations among hydrocarbon microseepage, soil chemistry, and uptake of micronutrients by plants, Bell Creek Oil Field, Montana, *Journal of Geochemical Exploration*, Vol. 23, pp. 139–162.

Samtsevich, S. A. 1955. Seasonal aspects and periodicity in the development of soil micro-organisms, *Mikrobiologiya*, Vol. 24, pp. 615–625.

Scientific America (1970). The Biosphere, Scientific American, W. H. Freeman Co., San Francisco.

Sealy, J. R. (1974). A geomicrobial method of prospecting for oil, Part 1, *Oil & Gas Journal*, Vol. 72, pp. 142–146; Part 2, *Oil & Gas Journal*, Vol. 72, pp. 98–102.

Sleat, R. and H. Van der Dick (1989). Soil hydrocarbon-oxidizing activities as a tool for oil and gas prospecting-a new approach. *2nd IGT Symposium on Gas, Oil, Coal and Environmental Biotechnology*, New Orleans, LA, Dec.

Subbota, M. I. (1947A). A complex study of seasonal variation of data in soil gas surveys, *Nefti. Khoz*, Vol. 25, pp. 13–17.

Subbota, M. I. (1947B). Field problems in oil prospecting by the bacterial survey method, *Razvedka Nedr.*, Vol. 13, pp. 20–24.

Thode, H. G., R. K. Wanless, and R. Wallouch (1954). The origin of native sulphur deposits from isotope fractionation studies, *Geochim. Cosmochima, Acta*, Vol. 5, pp. 286–298.

Updegraff, D. M. and H. H. Chase (1958). U.S. Patent 2,861,921; assigned to Socony Mobil Oil Company, Inc.

Veber, V. V. and N. M. Turkel'taub (1958). Gaseous hydrocarbons in recent sediments, *Geol. Nefti*, Vol. 2, pp. 39–44.

10

Helium Methods

Introduction

Anomalous accumulations of helium have been identified in association with petroleum deposits both in the reservoir itself and at the surface. Historically, there have been intermittent periods when helium has been used as a prospecting tool for petroleum. This technique has been useful in finding fault and fracture systems associated with a trap.

Occurrence of Helium and its Isotopes

There are two naturally occurring helium isotopes. The most abundant is ^4He and the less common ^3He. Both isotopes can be of either primordial or radiogenic origin.

The majority of ^4He is radiogenic and is derived from the alpha decay of naturally occurring radionuclides. The main sources of ^4He are from the decay of ^{238}U, ^{232}Th, ^{206}Pb, and ^{208}Pb. As an inert gas, ^4He does not bond with other elements into stable compounds. It will eventually migrate with relatively low velocity from the subsurface into the earth's atmosphere and then out into space with relatively low velocity. The amount of ^4He in the atmosphere is thought to represent a constant equilibrium between production and loss. The normal atmospheric concentration of ^4He is approximately 5.24 ppm and remains constant across the earth's surface despite areas of degassing. It remains in relative equilibrium because the rapid mixing by wind eliminates areas that are anomalous. The crustal abundance of ^4He is approximately 8.2×10^{-13} cm$_3$g (Dyck, 1976), which is calculated from the crustal abundance of U and Th. The total He or ^4He content of the average continental crust should be 9×10^{-3} cm$_3$ g$-$liter, but it is probably much less because of helium's ability to diffuse quickly. Igneous rocks have low volumes of He, whereas shallow sedimentary rocks may have large volumes present. Durrance (1986) noted that very high levels of He are associated with oil and gas accumulations. Sedimentary rocks contain enough uranium and thorium to provide these high He measurements although some probably diffuses from basement rocks or granite.

Klusman and Jaacks (1987) studied the seasonal variations of helium in conjunction with radon and mercury at a Colorado test site for approximately two years (Fig. 10-1). The resulting data indicated that helium production was less during the summer months than in fall and winter. Variations in helium at the surface are probably due to temporary regional or localized degassing rather than to a long-term relationship generated by an underlying petroleum deposit or other source or accumulation of helium.

Generally, ^3He is of primordial origin. It is derived from the decay of lithium and cosmic bombardment while in the atmosphere. This decay also generates He^4.

Helium as an Exploration Tool

The use of He^4 as a possible pathfinder for uranium deposits, geothermal sources, and petroleum accumulations has been documented by Pogorski and Quirt (1981), Butt and Gole (1984), and Roberts (1981). Helium migrates upward

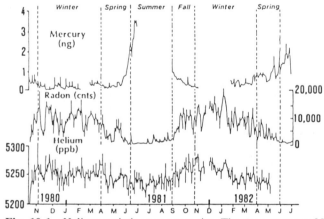

Fig. 10-1 Helium variations at a test site. The area averages 20 in. of rain per year and is located at 2110-m (8,000-ft) elevation in a semiarid region (Klusman and Jaacks, 1987; reprinted with permission of *Journal of Geochemical Exploration*, Elsevier Scientific Publication).

to the surface via diffusion and creates a very dispersed halo. The concentrations tend to be associated with fractured and faulted areas because the migrating He is channeled along these systems. Figure 10-2 is a survey utilizing helium in conjunction with methane and propane across the San Andreas Fault (Jones and Drozd, 1983). The data illustrate that large volumes of migrating helium are being concentrated locally as a result of the presence of the fault zone. However, low concentrations of helium in a specific area may be as critical as high concentrations elsewhere. Gregory and Durrance (1985) found that helium anomalies were affected by the thickness of soil cover. Anomalies occurred in fractured areas overlain by a thin soil cover, but the anomalies disappeared when the fractures were overlain by thick soil cover.

Helium can be channeled laterally away from the source. The effects were described by Holland and Emerson (1979) in a study of the Bush Dome, Potter County, Texas. Bush Dome is used to store 40 billion ft^3 of helium a year for the U.S. Bureau of Mines (Fig. 10-3). The underground storage reservoir is from 970 to 1060 m deep and has an anhydrite cap rock that is 120 m thick. Thirty-one water wells in the immediate area were sampled for helium. The resulting

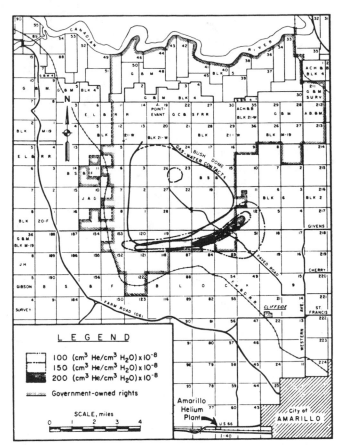

Fig. 10-3 Helium survey from groundwater analysis across the Bush Dome, Cliffside Field, Potter County, Texas. The anomaly lies on the south side of the field with a minor anomaly on the north side of the field (Holland and Emerson, 1979).

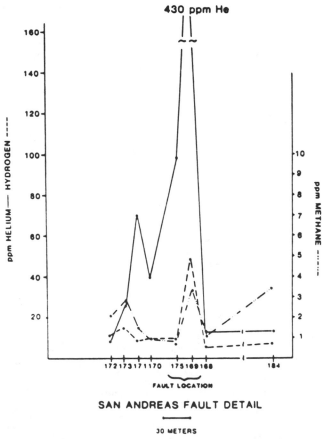

Fig. 10-2 Helium survey across the San Andreas Fault, California. Note the large helium anomaly over the fault. (Jones and Drozd, 1983; reprinted with permission of the American Association of Petroleum Geologists).

contour map, in cubic centimeters of helium per cubic centimeters of water, indicates an anomalous concentration of helium on the south side of the dome (Fig. 10-3). This suggests that migrating helium has been channeled through fracture and fault systems trending toward this area of the dome.

Two soil-gas surveys of 613 and 170 samples were conducted in March and September of 1977. Samples were taken from a depth of 0.6 m adjacent to and across the dome. The samples were extracted using a hollow stainless steel probe and were analyzed at the end of the day. Figures 10-4 and 10-5 are the results of these two surveys. The first survey (March), comprising 613 samples, indicates anomalous areas both on and off the dome (Fig. 10-4). The second survey (September) indicates an anomalous condition in the eastern part of the dome, extending to the northeast (Fig. 10-5). In both surveys, it is difficult to assess what the anomalies are related to. They may be a product of sampling other helium sources or of migration related to the accumulation stored in the dome.

A major concern when helium is used as a prospecting tool is atmospheric contamination during sampling. Gole and Butt (1985) surveyed two gas fields in Western Austra-

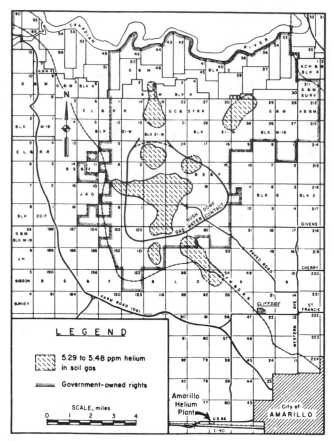

Fig. 10-4 First helium survey in March 1977, obtained by the U.S. Geologic Survey, Bush Dome, Cliffside Field, Potter County, Texas. Note that anomalous areas somewhat correspond to the Bush Dome (Holland and Emerson, 1979).

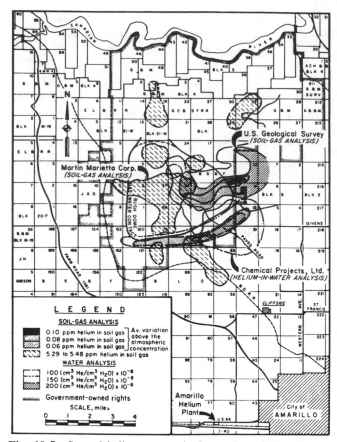

Fig. 10-5 Second helium survey in September 1977, obtained by the U.S. Geologic Survey, Bush Dome, Cliffside Field, Potter County, Texas. Note that the anomalous area lies on the east side of the field and extends over the area that is known to be in the water leg (Holland and Emerson, 1979).

lia, where production is from 3600 to 4425 m. They found that significant dilution of soil-gas samples occurred as sampling approached the surface. This may be due to both atmospheric migration downward and biological processes. Their initial sampling depth was 1 m, but they extended the depth to 7 m because of the contamination. Sampling from greater depths showed that the anomalous areas were consistent in relation to the oil field over a period of four years even though the intensity and shape of the near-surface expression changed.

Holland and Emerson (1979) were able to resolve the problem of atmospheric contamination by measuring neon with helium from a series of samples. Neon, which is also an inert gas, is present only in the atmosphere. Therefore, neon's presence in a soil sample would indicate atmospheric contamination at that particular sampled depth. The ratio of helium to neon became an indicator of contamination. A survey across the Gingin Gas Field in Western Australia applied this concept (Fig. 10-6), and high helium-to-neon ratios were found in association with the field. False helium anomalies were eliminated when the ratio was used.

Roberts (1981) discussed two types of helium anomalies associated with petroleum deposits: halo and apical. The

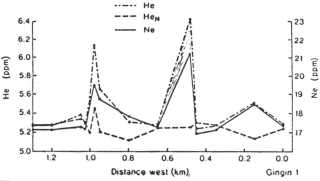

Fig. 10-6 Ratios of normalized helium, and helium to neon along Traverse 5, Gingin Field, Western Australia (Gole and Butt, 1985; reprinted with permission from the American Association of Petroleum Geologists).

Harley Dome, Unita Basin, Utah, depicts an apical helium anomaly associated with a gas reservoir in the Entrada sandstone (Fig. 10-7). The gas in the reservoir is composed of 7% helium and 85% nitrogen. The anomaly varies in intensity from 10% to 100% above background (5.24 ppm). The high helium in the reservoir is the probable cause of the helium anomaly at the surface. Cement Oil Field, Anadarko

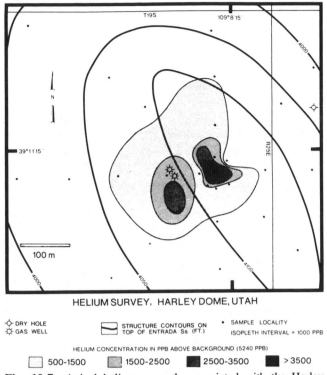

HELIUM SURVEY, HARLEY DOME, UTAH

◇ DRY HOLE   ▭ STRUCTURE CONTOURS ON   • SAMPLE LOCALITY
☼ GAS WELL   TOP OF ENTRADA Ss (FT.)   ISOPLETH INTERVAL = 1000 PPB

HELIUM CONCENTRATION IN PPB ABOVE BACKGROUND (5240 PPB)

☐ 500-1500   ▨ 1500-2500   ■ 2500-3500   ■ >3500

Fig. 10-7 Apical helium anomaly associated with the Harley Dome, Utah. Structure contours are on top of the Entrada sandstone; contour interval is 50 ft. The various patterns represent helium concentrations of 500–1500, 1500–2500, 2500–3500, and greater than 3500 ppb. Sample locations are represented by small dots (Roberts, 1981; reprinted with permission of the Institute for the Study of Earth and Man, Southern Methodist University).

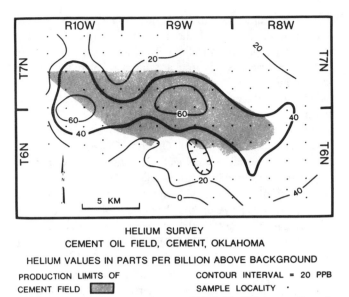

HELIUM SURVEY
CEMENT OIL FIELD, CEMENT, OKLAHOMA
HELIUM VALUES IN PARTS PER BILLION ABOVE BACKGROUND

PRODUCTION LIMITS OF   CONTOUR INTERVAL = 20 PPB
CEMENT FIELD ▨   SAMPLE LOCALITY •

Fig. 10-8 Helium survey, Cement Oil Field, Oklahoma. Sample locations are represented by small dots and the productive limits of the field by the shaded area. Contour interval is 20 ppb (Donovan, 1974; reprinted with permission of the American Association of Petroleum Geologists).

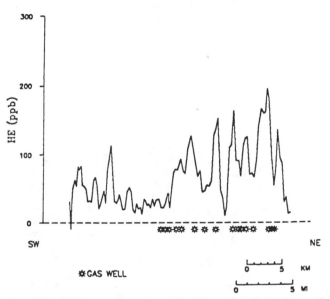

Fig. 10-9 Helium traverse across the Athens Gas Field, Pennsylvania. Note that the most anomalous areas (<100 ppb of He) are related to the field (Roberts and Roen, 1985).

Basin, Oklahoma, has also been cited (Donovan, 1974) as having an apical helium anomaly (Fig. 10-8). Anomalous values, determined in parts per billion above background, in this case were 20, 40, and 60 ppb. The Cement Field anomaly is less than 10% above background.

Roberts and Roen (1985) discussed two helium traverses, one across the Athens Gas Field, Pennsylvania (Fig. 10-9), and another across a suspected lineament zone. No helium anomalies were found in association with the lineament (not presented). Any explanation would be speculative because the survey was a traverse. The gas field indicated anomalous values across most of its extent with almost totally background conditions existing for the rest of the line. Once again, it would be easy to indicate a relationship between the field and the anomaly. However, there is no mention of soil thickness, subsurface configuration, other sources of the helium, or fracturing with respect to the reservoir along the traverse.

The presence of a halo helium anomaly is thought to be caused by vertically migrating hydrocarbons that are oxidized to carbon dioxide and calcite cement in the soil. The calcite precipitates in the pores and theoretically inhibits helium migration to the surface in the center of the field (Roberts, 1981). The edges of the anomaly have less petro-

leum gas migrating to the surface and, subsequently, less oxidation of hydrocarbons occurs. However, cementation in the soil substrate may not be the cause because some fields have an apical helium anomaly whereas others have a halo. Instead, as discussed earlier, the presence of faults and fractures on the edges, at the center, or over the entire deposit may direct migration. The halo may be the result of a concentration of pathways at the edges of the accumulation. An even distribution of pathways will cause a generally dispersed or apical anomaly. Figure 10-10 is an example of a halo anomaly around the Red Wing Creek Oil Field, Williston Basin, North Dakota. The rather large areal anomaly may not necessarily indicate that it is related to oil and gas production as Roberts (1981) contends. There seem to be two linear trends that are more indicative of general faulting.

Helium exploration has been used in a portion of the Michigan Basin that lies in Ontario, Canada. There was reportedly some success in locating fractured reservoirs present in the Ordovician Trenton-Black River group. In the eastern part of the Michigan Basin, linear trending anomalies seem to be indicative of this type of petroleum production. Fractures in the Trenton-Black River are productive dolomitized "chimneys" encased in a lithographic limestone. The migration pathways are basement-derived faults and fractures extending upward into the Ordovician strata. The dolomitized chimneys are extremely narrow but have considerable length.

Roberts (1981) reported the results of 15 helium surveys, 11 across existing fields and 4 across prospects. Six of the surveys indicated a clear definition of the existing fields. Three surveys showed anomalies, but the results indicated that there would have been difficulty in pinpointing the fields if the surveys had been taken prior to drilling. The two remaining fields had no helium anomalies associated with them. Two of the four prospects had anomalies but remained undrilled. The other two did not have associated anomalies; subsequent drilling revealed dry holes. However, the concern here is that two of the eleven fields (approximately 19%) were not detected by helium.

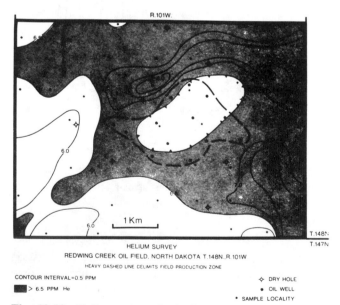

Fig. 10-10 Helium survey, Red Wing Creek Oil Field, North Dakota. The field is represented by the dashed outline. The shaded area represents >6.5 ppm He, and sample locations are represented by small dots (Roberts, 1981; reprinted with permission of the Institute for the Study of Earth and Man, Southern Methodist University).

Sampling

Helium samples are collected primarily by hammering a hollow probe a specified depth, typically 1 to 3 m. A central rod is withdrawn from the hollow center. The hollow area is then purged by withdrawing a gas sample of greater volume with a syringe at the surface. Access to the hollow part of the probe is through a rubber septum. A sample of the gas trapped in the soil is drawn into the probe by the created vacuum. The gas is then extracted and placed in small-diameter copper or lead tubes and sealed. The permeability of the soil determines the depth of sampling. The greater the soil permeability, the greater the sampling depth must be in order to avoid dilution and contamination by the atmosphere.

Analysis must be done quickly because helium is highly mobile and may leak out of the container. There may also be atmospheric contamination by diffusion into the syringe and sample container. Analysis by a mass spectrometer is the most common method. Less sensitive instrumentation can be used in the field and is typically truck-mounted (Friedman and Denton, 1975; Reimer, 1976; Reimer and Denton, 1978; and Roberts, 1981).

There is an alternative form of helium sampling that can be used to avoid problems associated with the free-air probe. An auger is employed to take a whole soil sample, which is placed in a container and sealed. A gas sample is extracted from the container and is then analyzed. This method is similar to the head-space gas methods described in Chapter 5. The resulting helium value must be corrected for known parameters of soil mass, porosity, moisture content, free-gas volume, and concentrations of other gases. The resulting data are evaluated by comparing them to the atmospheric abundance of ^{4}He.

Discussion

There are no case histories in the literature indicating that helium was the main cause or the partial cause of a field discovery. Interpretation is dependent on differences between a known background value, which locally can vary dramatically, and apparent anomalous values. The advantages of helium are: (1) its dispersion in a halo effect around the source, (2) its stability, and (3) its ability to indicate faults and fractures and the petroleum traps associated with them.

The disadvantages of helium are generally considered much greater than the advantages.

1. Helium concentrations may be the result of resources other than petroleum, such as uranium, base metals, and geothermal sources. As an example, Fig. 10-11 indicates the mapping of a uranium roll-front prospect in Weld County, Colorado (Reimer and Otton, 1976; Reimer and Rice, 1977). Weld County also has several highly prolific oil and gas

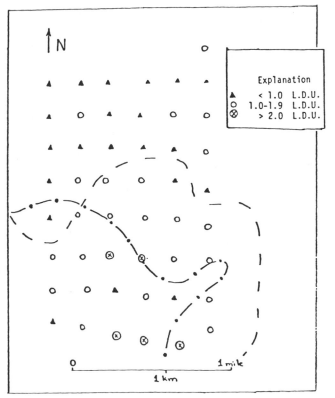

Fig. 10-11 Helium survey across a uranium roll-front deposit, Weld County, Colorado. The dashed and dotted lines represent the mineralized roll-front in the lower and upper sands, respectively. Helium values obtained from leak detection units: ▲<1.0, ○1.0–1.9, ⊗>2.0. Note that the higher values are associated with the roll-front deposit (Reimer and Otton, 1976).

fields covering relatively large areas in the Denver-Julesberg Basin. Therefore, an interpretation of a helium anomaly would be difficult because of the existence and influence of both petroleum and uranium in the same area. Helium has also been found in association with magma degassing, earthquake swarms, and gravity changes (Wakita et al., 1978).

2. Repeatability of a survey is doubtful. Helium values for a given area can be affected by changes in soil, weather, and seasonal conditions. Surveys in the same area at different times of the year have yielded widely varying results. A dry, cracked soil substrate may be purged of helium to the sampled depth or be subject to atmospheric contamination. A sample collected during dry times may actually be a sample of the atmosphere. Conversely, moisture in the soil may act as a seal and inhibit helium migration and contamination. Wind during sampling has been noted to disperse and remove anomalies.

3. Helium sampling and analysis can present problems. Collection via a soil probe must be done with care to avoid atmospheric contamination. Transporting the sample to the laboratory may result in loss or addition of helium by diffusion. Mobile laboratories may be the answer, but the sensitivity and quality of field instruments are generally not sufficient to obtain satisfactory confidence levels.

4. The presence or absence of a surface anomaly is difficult to interpret. Thus, the resulting surface anomalies may exhibit no direct or discernible relationship to the source at depth. Differences between background and postulated anomalous values can be so small that these variations may reflect localized phenomena rather than evidence of petroleum accumulations. The absence of a surface anomaly does not necessarily signify the lack of a source at depth. Rather, there may be no conduits to the surface. Therefore, a clear understanding of bedrock geology is critical.

5. Costs for helium surveys can be similar to, or much greater than, costs for soil-gas surveys.

Summary

Helium methods may be useful in pursuing and delineating faults and fracture systems rather than in specifically determining the presence or absence of petroleum accumulations. Therefore, the reservoir or trap that requires a fault or fracture component is a suitable candidate for the helium method. The problems encountered at the Bush Dome, Texas, indicate the necessity of determining through orientation surveys the sampling depth and the collection method on a site-by-site basis. Other geochemical methods must be used to support and further delineate any anomalies found because helium is not necessarily associated intimately with petroleum.

References

Butt, C. R. M. and M. J. Gole (1984). Helium manometry under Australian conditions—Preliminary results. *Journal of Geochemical Exploration*, Vol. 22, p. 359.

Donovan, T. J. (1974). Petroleum microseepage at Cement, Oklahoma—Evidence and mechanisms, *American Association of Petroleum Geologists Bulletin*, Vol. 58, pp. 429–446.

Durrance, E. M. (1986). *Radioactivity in Geology: Principles and Applications*, John Wiley & Sons, New York, pp. 219–227.

Dyck, W. (1976). The use of helium in mineral exploration, *Journal of Geochemical Exploration*, Vol. 5, pp. 3–20.

Friedman, I. and E. H. Denton (1975). A portable helium sniffer, U.S. Geologic Survey Open-File Report 75-532.

Gole, M.J. and C.R.M. Butt (1985). Biogenic-thermogenic near-surface gas anomaly over Gingin and Bootine gas fields, Western Australia, *American Association of Petroleum Geologists Bulletin*, Vol. 69, No. 12, pp. 2110–2119.

Gregory, R. G. and E. M. Durrance (1985). Helium, carbon dioxide and oxygen soil gas, small scale variations over fractured ground, *Journal of Geochemical Exploration*, Vol. 24, pp. 29–49.

Holland, P. W. and D. E. Emerson (1979). Helium in ground water and soil gas in the vicinity of Bush Dome Reservoir, Cliffside Field, Potter County, TX, U.S. Bureau of Mines Information Circular No. 8807.

Jones, V. T. and R. J. Drozd (1983). Prediction of oil and gas potential by near surface geochemistry, *American Association of Petroleum Geologists Bulletin,* Vol. 67, No., pp. 932–952.

Klusman, R. and J. Jaacks (1987). Environmental influences upon mercury, radon, and helium concentrations in soil gases at a site near Denver, Colorado, *Journal of Geochemical Exploration,* Vol. 27, No. 3, pp. 259–280.

Pogorski, L. A. and G. S. Quirt (1981). Helium manometry in exploring for hydrocarbons: Part I, in *Unconventional Methods in Exploration for Petroleum and Natural Gas, Symposium II,* Southern Methodist University Press, Dallas, TX, pp. 124–135.

Reimer, G. M. (1976). Design and assembly of a portable helium detector for evaluation as a uranium exploration instrument, U.S. Geologic Survey Open-File Report 76-398.

Reimer, G. M. and E. H. Denton (1978). Improved inlet system for U.S. Geological Survey mobile helium detector, U.S. Geologic Survey Open-File Report 78-588.

Reimer, G. M. and J. K. Otton (1976). Helium in soil gas and well water in the vicinity of a uranium deposit, Weld County, Colorado, U.S. Geologic Survey Open-File report No. 76-699.

Reimer, G. M. and R. S. Rice (1977). Line traverse surveys of helium and radon in soil gas as a guide for uranium exploration, Central Weld County, Colorado, U.S. Geologic Survey Open-File report 77-589.

Roberts, A. A. (1981). Helium manometry for exploring for hydrocarbons: Part II, In *Unconventional Methods in Exploration for Petroleum and Natural Gas, Symposium II,* Southern Methodist University Press, Dallas, TX, pp. 136–149.

Roberts, A. A. and J. B. Roen (1985). Near-surface helium anomalies associated with faults and gas accumulations in western Pennsylvania, U.S. Geologic Survey Open File Report 85-546.

Wakita, H., N. Fujii, S. Matsuo, K. Notsu, K. Nago, and, N. Takaoka (1978). "Helium Spots": caused by a diapiric magma from the upper mantle, *Science,* Vol. 200, April 28, pp. 430–432.

pH/Eh Methods

Introduction

Many of the techniques presented in the previous chapters are in some way affected by Eh and pH conditions of the soil and rock strata that are sampled. Only in recent years has research for soil science concentrated intensely on Eh and pH and how they affect chemical and biological processes in the soil substrate. Surface geochemical surveys typically do not measure Eh and pH as standard procedure. Petroleum microseepage into the soil substrate, which causes several chemical reactions and microbial oxidation of hydrocarbons, also causes decreases in Eh and pH. All changes should be measurable depending on such factors as soil mineralogy, clay composition and absorption capacity, amount of organic matter present, humic/fulvic acids present, type of parent material, moisture content, the original Eh and pH, bacterial activity, and climate.

Eh

The Eh of a solution is a measure of the oxidizing or reducing potential (redox) of a chemical system. Oxidation or reduction involves transfer of electrons from one chemical species to another, which can be measured in millivolts or volts. The oxidation-reduction potential is actually an intensity factor trend of a specific half-reaction. The problem is that we cannot actually measure this value from the reduced material but only approximate it. The oxidation-reduction, or redox, potential reflects the equilibrium of ions in solution. Natural systems rarely reach equilibrium. The standard to which all measurements are referenced is the hydrogen electrode, which is arbitrarily set at zero, a pH of zero, and an atmospheric pressure of 1. Oxidation potential is a relative figure whereby:

$$2H^+ + 2e^- = H_2 \quad E = 0.00 \text{ V}$$

The aqueous electron (e^-) is the negative logarithm of the free-electron activity expressing soil oxidizability. Chemical reactions occurring in water with a pH of zero are limited

to redox equations with an Eh of 0 to 1.23 V. The 1.23 V. represents the Eh at which oxygen is liberated from water. Large Eh values tend to indicate the existence of electron-poor or oxidized species. Small Eh values indicate that the species are proton-rich or reduced (Sposito, 1989). Redox, or Eh, values of a system are pH-dependent. Consequently, most theoretical systems are adjusted to a pH of 7. Therefore, it is critical in measuring the Eh of a sample to measure the pH as well.

The scale extends on either side of the zero since the reaction is compared to a hydrogen cell. Reactions that occur on the positive side of the scale are oxidizing, and reduction occurs in the negative direction. Measurement of Eh is done with an Eh meter. The quality of the data depends on the type of probe used. For accurate measurements of Eh, a platinum disk has been found to be more effective than a platinum wire because of the disk's greater area of exposure to the sampling medium (Fausnaugh, personal communication).

pH

pH is a relative expression of the acidity or alkalinity of an aqueous system. pH is based on the concentration of the OH^- and H^+ ions and is the negative logarithm (to the base 10) of the hydrogen ion activity. The scale of pH ranges from zero (acidic) to 14 (alkaline), and the neutral point is 7. The standard pH solution of 7.0 is at 25° C and 1 atm, where the OH^- ions equal the H^+ ions in number.

A simple way to determine the pH of a soil is to create an aqueous mud in the laboratory. Mix 50 g of soil into 50 ml of distilled water, stir constantly for 1 hr, and measure the pH during that time period (Back and Barnes, 1964). More water may be needed if the soil is high in organic matter. Dyes have also been used to measure soil pH directly in the field, but they are less reliable.

The solubility of most elements is sensitive to changes in pH. Only the alkali metals (Na, K, and Rb), alkaline earths (Ca, Mg, and Sr), and acid radicals such as N and

Cl are soluble across the entire pH spectrum found in the soil (6–8.5 pH). Most elements are less mobile in the alkaline environment and generally precipitate out. pH is an important consideration in the adsorption capacity of clays. In the case of montmorillonite, H^+ is preferentially absorbed over metal ions under acidic conditions.

Mineral Stability

Mineral stability in any environment is dependent on, and is a function of, the pH (acidity) and Eh (oxidation potential). We have seen that the introduction of petroleum into the soil environment destabilizes many compounds and increases the solubility of the trace and major elements. Many metal elements are highly soluble in acid solutions but will be precipitated as oxides and hydroxides with increasing pH. We can determine the pH by the equilibrium constant:

$$K^n = \frac{a^{2H^+}}{a^n} = 10^{-7.35}$$

where n equals the pH of the compound. Since $pH = -\log a^{H^+}$, the equation is:

$$\text{Log } a^n = 7.35 - pH$$

Using this equation, we can construct stability/instability diagrams for any minerals based on pH (exclusive of Eh) vs. concentration.

An important control on the mobility of elements is the concentration of electrons in the environment, which is the oxidation-reduction (redox) potential. Many elements occur in one or more valence states, such as Fe^{+2} and Fe^{+3}. This is important in evaluating the overall geochemical environment and determining which minerals will be stable or unstable. The most abundant oxidizing agent at the earth's surface is oxygen. An oxidizing environment is common where the interchange occurs with the atmosphere or as a result of downward groundwater percolation. As free oxygen decreases with depth, the environment becomes progressively less oxidizing and more reducing. The presence of an oxidizing environment is also controlled by the percentage of organic matter in the soil.

Mineral stability can be further evaluated by combining Eh and pH into a diagram that shows the conditions under which various minerals are stable. The Eh-pH diagram is a fundamental part of geochemistry, but its application in surface geochemistry with respect to petroleum is nonexistent. The Eh-pH diagram is typically constructed using 25°C and 1 bar of pressure. Constructing an Eh-pH diagram at different temperatures is straightforward, and a variety of equations are available (Brookings, 1987). Eh-pH diagrams were discussed in general terms in Chapter 7 and 8, but there has been no direct application to petroleum exploration. A general concept of oxidizing and reducing environments, as

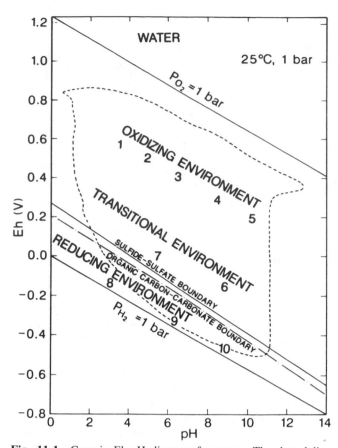

Fig. 11-1 Generic Eh-pH diagram for water. The dotted line represents the range of natural Eh-pH water measurements. Oxidizing, transitional, and reducing environments are indicated. The soil line represents the sulfide-sulfate boundary delineating oxidizing and reducing conditions. The dash-dot line separates organic carbon (below) and carbonate species (above). The large numbers represent different water conditions from specific environments (see text for explanation).

seen with an Eh-pH diagram for water, is demonstrated in Fig. 11-1. This generic Eh-pH diagram assumes that the upper and lower limits of O_2 and H_2 are equal to 1 bar of pressure. The numbers correspond to different conditions in which water is found:

1. Mine waters are normally acidic and oxidizing.

2. Rain, in many parts of the world, is becoming acidic and oxidizing as a result of buildup of CO_2 caused by the burning of fossil fuels.

3. Stream waters typically are near-neutral in pH and are oxidizing. However, these can be modified by rain or mine waters.

4. Ocean water in the near-surface to surface is in equilibrium with CO_2 and $CaCO_3$.

5. Aerated waters in saline environments are typically oxidizing and have a pH that is basic.

6. Groundwaters usually have a pH of 8.4 and are saturated with respect to $CaCO_3$.

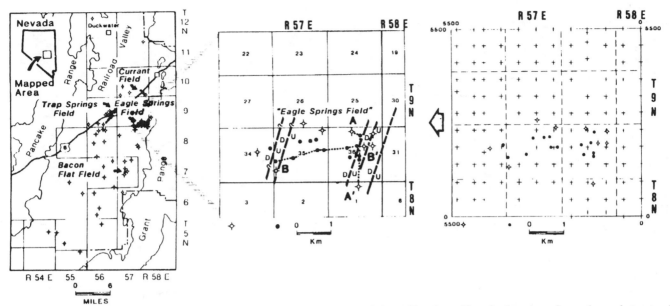

Fig. 11-2 Location of Eagle Springs Oil Field, Railroad Valley, Basin and Range Province, Nevada; location of samples and structural geology (Klusman et al., 1992; reprinted with permission of the American Association of Petroleum Geologists).

7. Bog waters typically are acidic and reducing and thus cross the sulfide-sulfate boundary.

8. Waterlogged soils are acidic and reducing.

9. Euxenic marine waters are reducing and have a near-neutral pH.

10. Waters that are saline and organic-rich are typically reducing and alkaline.

The reader is referred to Garrels and Christ (1965), Brookins (1987), and Krauskopf (1979) for a more expanded discussion of this diagram.

The most important chemical elements affected by soil redox reactions are Fe, Mn, S, O, N, and C. In contaminated soils or in environments where microseepage is occurring, the list would expand to include all the additional petroleum and man-made hydrocarbons as well as any additional metals and nonmetals. If we look at a closed system where microseepage is not occurring, there is a well-defined sequence of reduction half-reactions. The sequence of events can be summarized as follows. Below an Eh of $+11$, O_2 is reduced to H_2O. As the Eh decreases below 8, reduction of NO_{3-} occurs. As the Eh begins to drop to the range of 7 to 5, reduction of Fe and Mn from solid phases occurs. The reduction of Fe and Mn cannot occur until sources of O_2 and NO_{3-} are depleted. As the Eh descends below $+2$, the soils become anoxic, and sulfate reduction by microbial processes occurs. The impact of a reducing or oxidizing environment seems to have a greater impact on methods analyzing for microbes, Fe, Mn, Ni, V, Cu, C, and Hg. The impact of changes in reducing and oxiding conditions on the retention of soil gas and soil-gas reactions with halogens is not well understood.

Eh-pH as an Exploration Tool

The essential element in the use of a geochemical exploration tool is determining under what conditions it does or does not work. From previous discussions, we should be

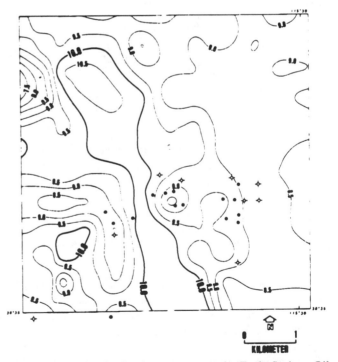

Fig. 11-3 pH results for the survey over the Eagle Springs Oil Field, Railroad Valley, Basin and Range Province, Nevada. The pH study does not indicate any correlation with the areas of production (Klusman et al., 1992; reprinted with permission of the American Association of Petroleum Geologists).

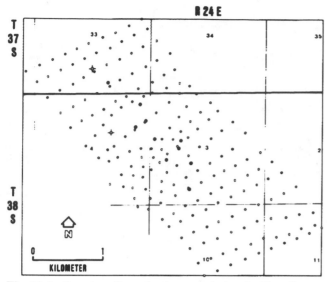

Fig. 11-4 Location of samples (open circles) at the Cave Canyon area, Paradox Basin, Utah. Productive wells are indicated by closed circles (Klusman et al., 1992; reprinted with permission of the American Association of Petroleum Geologists).

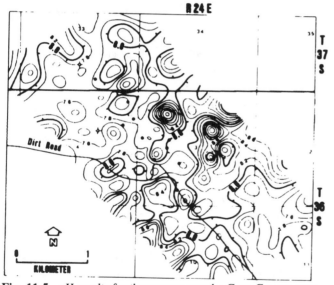

Fig. 11-5 pH results for the survey over the Cave Canyon area, Paradox Basin, Utah. The survey does not indicate any correlation with the area of production (Klusman et al., 1992; reprinted with permission of the American Association of Petroleum Geologists).

able to measure variations in the Eh and pH of the soil environment in the area of petroleum seepage. The problem is, however, that not all petroleum seepage causes a perceptible change in the redox potential of the soil. If the soil is rich in carbonates (alkaline pH) and has only minor amounts of petroleum leaking into it, the difference may not be perceptible or statistically significant. On the other hand, a reducing environment may not be significantly impacted either. Only a small amount of work has been done with Eh-pH compared to soil-gas, iodine, and radiometric surveys, but this work strongly suggests that there are mappable

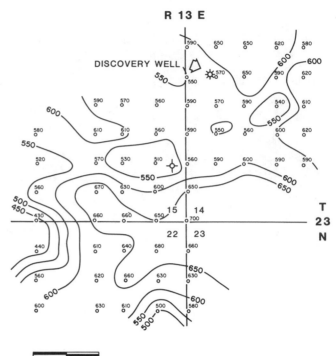

660 FEET

Fig. 11-6 Eh results that led to the discovery of the gas well located in the upper-right area of the map. The well produced gas from a Pennsylvanian Cherokee sandstone. Values are in millivolts. The discovery well lies in a general low. Soil-gas was also used to locate the discovery well (reprinted with permission from Gallagher Research and Development, Inc.).

changes in these chemical characteristics stemming, in some instances, from petroleum-generated seepage.

Klusman et al. (1992) presented a pH map of the soils in relation to the study of trace and major elements in a survey around the Eagle Springs Oil Field, Railroad Valley, Basin and Range Province, Nevada (Figs. 11-2 and 11-3). The pH map was not accompanied by a soil map. The authors concluded that there is no relationship between pH and microseepage in the area. The survey grid is considered widely spaced in sample density based on the small areal extent of the fields. A similar survey was carried out over the Cave Canyon Field, Paradox Basin, Utah (Figs. 11-4 and 11-5). The results indicate a conclusion similar to that of the Eagle Springs area. It is difficult to determine the validity of this method because there are no soil maps showing whether different soils that may have variations in pH occur here. However, along with these surveys are numerous Eh-pH anomalies that are related not to seeping hydrocarbons but to other causes, such as chemical variations in the soil composition, other types of natural resource deposits, water variations, and biological activity.

The use of Eh, pH, and conductivity in conjunction with soil gas has led to the discovery of at least one shallow gas field. The location is in Washington County, Oklahoma, producing from a Cherokee (Pennsylvanian) sandstone reservoir at less then 600 m (2000 ft). Figure 11-6 is the

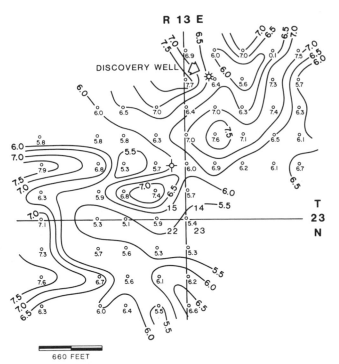

Fig. 11-7 pH results that led to the discovery of the gas well located in the upper-right area of the map. The well produced gas from a Pennsylvanian Cherokee sandstone. Values represent the pH of the soil. The discovery well lies in the center of a halo type of anomaly. Soil-gas was also used to locate the discovery well (reprinted with permission from Gallagher Research and Development, Inc.).

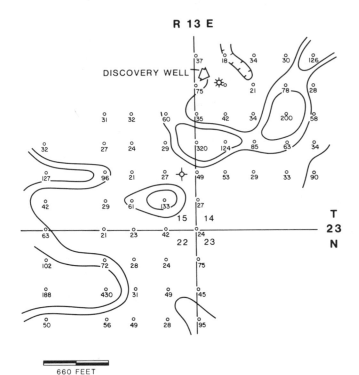

Fig. 11-8 Conductivity results that led to the discovery of the gas well located in the upper-right corner of the map. The well produced gas from a Pennsylvanian Cherokee sandstone. Values are in microsiemens. The discovery well lies in the center of a halo type of anomaly. Soil-gas was also used to locate the discovery well (reprinted with permission from Gallagher Research and Development, Inc.).

Eh data in millivolts. The data indicate a general low that encompasses the discovery and a dry hole. The pH data are presented in Figure 11-7. The pH data are more distinctive in indicating a halo effect around the discovery well. The pH conditions in the halo represent changes from a reducing to an oxidizing environment. This is consistent with the concepts of microseepage. Conductivity data are presented in Fig. 11-8. Conductivity is a measure of the medium's ability to carry an electrical current. Oil is a resistive medium, whereas salt water by its nature is conductive. Therefore, the changes in the Eh/pH, or redox, potential of the soil cause increases or decreases in free ions present. The conductivity data indicate a halo effect around the discovery well. This implies that there is a significant increase in the amounts of salts deposited on the edge of the microseepage area. The conductivity and pH map are in better agreement than the Eh map.

The lack of case histories using Eh-pH as an exploration method makes overall evaluation difficult. Therefore, Eh-

pH cannot be considered a stand-alone technique at this time but serves as an adjunct to soil-gas, iodine, and radiometrics.

References

Back, W. and I. L. Barnes (1964). Equipment for field measurements of electrochemical potentials, U.S. Geologic Survey Professional Paper 424-C.

Brookins, D. G. (1987). *Eh-pH Diagrams for Geochemistry*, Springer-Verlag, New York.

Garrels, R. M. and C. L. Christ (1965). *Minerals, Solution and Equilibra*, Harper and Row, New York.

Klusman, R. W., M. A. Saeed, and M. A. Abu-Ali (1992). The potential use of biogeochemistry in the detection of petroleum microseepage, *American Association of Petroleum Geologists Bulletin*, Vol. 76, No. 6, pp. 851–863.

Krauskopf, K. B. (1979). *Introduction to Geochemistry*, 2nd Ed. McGraw-Hill, New York.

Sposito, G. (1989). *The Chemistry of Soils*, Oxford University Press, New York.

III

Survey Design and Data Analysis

12

Statistical Analysis and Spatial Variation

Discovery commences with the awareness of anomaly, that is, with the recognition that nature has somehow violated the pre-induced expectations that govern normal science

Thomas Kuhn

Introduction

Several statistical techniques are useful in the evaluation of surface and near-surface geochemical data. The numerous statistical methods available to the reader are in a sense limitless in manipulating data for evaluation. However, the present level of understanding of surface geochemical exploration for petroleum precludes using several of the statistical methods and instead focuses on a few simple methods that aid interpretation. Because there may be multiple variables from a surface geochemical survey, analysis typically does not extend beyond (1) simple profiles; (2) use of means, modes, and standard deviations; and (3) contouring of the data. Usually, the raw data from the lab analysis indicate the presence or absence of an "anomaly" without statistical manipulation. Implementing statistical analysis requires obtaining sufficient data to identify that part (if any) of the population that is related to microseepage, and then determining how best to conclude confidently that it is indeed "anomalous" and is the part of the population being explored. Investigators typically use simple statistics. Some have developed proprietary equations, that is, transformations in the form of complex functions, ratios, and filtering, that are supposed to identify data that help isolate the truly anomalous from false indications and background. Actual practice suggests, however, that these manipulations are not universally applicable. Extensive transformations can skew or stretch a normal population data set into appearing anomalous when it is not, which can lead to misinterpretation.

Statistical methods, in any science, are used to determine the presence in a data set of separate, not always distinctive, populations that are not clearly discernible by visual evaluation. The interpretation of data for petroleum or other natural resources seeks populations that are termed *anomalous* and that indicate the presence of the commodity being sought. In general usage, an anomaly is something that departs significantly from the norm. In surface geochemistry, *anomaly* means a deviation from the expected normal range of data for measured amounts of hydrocarbons, iodine, microbes, radioactive minerals, trace and major metals, and helium in the soil. Lightman and Gingerich (1992) discussed the nature of the word *anomaly* and clearly defined the present usage as meaning something that does not fit the existing model. Thus, when the anomaly is incorporated into the model, it is no longer anomalous but part of the normal population. *Anomalous* may therefore be a misnomer in surface geochemistry. Macroseepage and microseepage are clearly part of the hydrodynamic system in any basin, to a degree that depends on the amount of petroleum present. Prolific basins and areas, such as Monterey, the North Sea, Saudi Arabia, Kuwait, Permian, Illinois, and the Gulf Coast, contain tremendous amounts of hydrocarbons today as they did in the past. Numerous geochemical expressions—anomalies—did and do exist at the surface in these areas, and technically they are part of the normal geochemical soil framework. While it may be true that word *anomaly* is incorrectly applied in surface geochemistry, this definition is embedded in the usage of the profession: *An anomaly represents that part of the normal population whose source is the product of petroleum seepage and an integral part of the basin fluid system.*

The statistical methods that are the most appropriate for geochemical interpretation will be presented here and will focus on:

1. Recognition of multiple populations (histograms)

2. Methods used to depict variations in the collected data (probability plots)

3. How to establish background and normal responses and recognize changes in those populations that do not necessarily imply the presence of anomalies (means, modes, and standard deviations)

4. Transformations

5. Use of pattern recognition where classical statistical methods fail

6. General overview and discussion of the different forms of surface expressions

Many of the case histories presented in previous chapters evaluated data by using the simple statistical methods of determining mean, mode, and standard deviation. Histograms were visually analyzed, and then the data were contoured based on the artificial determination of frequency intervals. Statistical analysis of surface geochemical data is used for two reasons: (1) to interpret a data set that exhibits multiple populations and complex patterns that are difficult to correlate with geologic and geophysical data, and (2) to evaluate small data sets. Some have thought that statistical manipulation of data, as applied to a small data set, is really done to enhance and generate an interpretation for erroneous data. Based on the author's experience, this is often true. A third form of evaluation has not been widely used in petroleum exploration: pattern recognition. In large data bases, pattern recognition is more applicable than either simple or complex statistics, which may or may not provide a supporting role. The reader should not misinterpret the last group of statements to mean that statistics do not have a place or are not really useful in surface geochemistry. But there has often been an over reliance on statistics, which can lead to the wrong interpretation. Computer contouring is a similar artificial procedure, even though it can provide an acceptable interpretation with many types of geologic data. Unlike structural and isopach mapping, geochemical values typically demonstrate wider variations over short distances and may not be suited for interpretation with present algorithms. Fractal and fuzzy logic types of analyses may change this but, to the author's knowledge, they have not been applied to surface geochemistry.

More sophisticated methods such as kriging, chi-square distributions, multivariant analysis, linear regression, and autocorrelation have not been used very often in the literature. These methods may be applicable, but they are probably unnecessary if simple statistics will suffice. A brief description of multivariant analysis involving cluster, discriminant, and factor analysis is presented.

The purpose of surface geochemistry in petroleum exploration is to discover abnormal geochemical patterns, or "anomalies," that are related to hydrocarbon seepage and from which to infer the presence of a petroleum accumulation at depth. Determining that a geochemical anomaly is related to petroleum seepage requires more than just unusually high (or low) values of gases, liquids, solids, or compounds in the sampled media. The geochemical anomaly is defined as a departure from values that are considered to be normal background variations or nonpetroleum seepage populations in a geochemical setting. Numerous "geochemical anomalies" may be unrelated to petroleum seepage, and a primary goal of statistical analysis is to identify anomalies resulting from other causes. Success in geochemical interpretation depends on the objective use of statistics.

The goal of statistics is to provide a fundamental starting point from which to evaluate geochemical data and to integrate that data with other types of nonrelated techniques. Statistics will fail if it must provide answers when the data are insufficient or cannot support the geologic/geophysical models.

Application of Statistics to Surface Geochemical Techniques

Statistical methods are a way of recognizing anomalous populations in surface geochemical data. However, not all surface geochemical methods utilize statistics in analysis. Soil-gas and major- and minor-elements data are usually subjected to evaluation by histograms, means and standard deviations, ratios, filtering methods, transformations and, to a lesser extent, chi-square distribution and multivariant analysis. Iodine, helium, microbial, and Eh/pH data evaluations are usually restricted to histograms, mean and standard deviations, and filtering methods. These types of data can be entered as variables in a transformation function involving the previously mentioned gases and elements. Radiometric data analysis varies, but usually point data are evaluated like iodine or helium. Continuous radiometric profile data are typically evaluated based on the slope of the line and visual examination.

Populations and Distributions

The composition of natural materials exhibits a distribution that can be useful in determining whether there are different groups or populations present. The explorationist must be able to determine, in an unbiased manner, whether there are anomalous samples in a survey. All types of geochemical data have a range of values that is termed a *normal distribution*. A range of values can be expected in any sampling program, and it is necessary to decide what fraction, if any, of the samples represents the "anomalous" population sought.

The classic example of a population distribution is in the form of a histogram, which is illustrated in Fig. 12-1 for a hypothetical set of data. The typical histogram is bell-shaped or symmetrical, and a smooth curve can be fitted to the data. Typically, geochemical data result in an asymmetrical histogram (Fig. 12-2). Data containing anomalous values can be skewed to either the right or the left, which gives the impression of a tail. Data skewed to the right are termed a *positive skewness*. This is the more common histogram form in petroleum exploration. Data skewed to the left are called a *negative skewness*. It is not a common form and, when present, can result from data that is more representative of the anomalous population rather than of the background values or of a normal population with relatively low values. Negative anomalies have been noted occasionally in surface geochemical methods exploring for petroleum, but their existence is not well understood. One or more curves fitted to the data indicate that different populations are present.

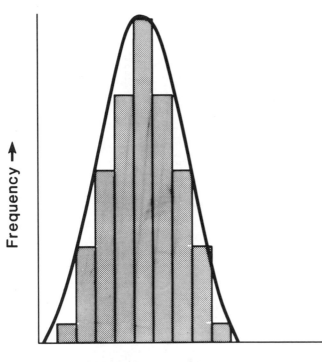

Fig. 12-1 Classic symmetrical population distribution histogram for a hypothetical data set. A curve can be fitted to the data, suggesting a symmetrical form.

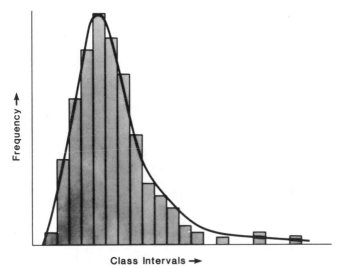

Fig. 12-2 A hypothetical data set that is skewed to the right, termed *positive* and asymmetrical in form.

The histogram *x* axis is typically the concentration scale in either an arithmetic or logarithmic form. Class interval is determined by the overall range of values. A narrow interval allows each class to have a shorter range. A wide variation in data causes the classes to have a wider range. A simple way to determine class interval is to use one-quarter to one-half the standard deviation of the data. The

y axis is the frequency of the number of classes that fall into each interval.

Figure 12-3 is a distribution diagram for iodine concentrations in 273 soil samples collected around the Weaver and Elbridge oil fields, Illinois Basin. The samples are divided into class intervals, and the number of samples in each class are plotted on a bar graph. Frequently, geochemical data with anomalous populations are skewed to the right, have more than one peak, and can be termed *multimodal*. The data, when plotted on a map (Fig. 12-4), indicate that values above 2.0 ppm tend to congregate around the existing field. Values of 1.5 to 1.9 also seem to be associated with existing production. Therefore, we can establish that two bell-shaped populations exist (Fig. 12-3).

Figure 12-5 is a distribution diagram for 221 iodine values in the Denver Basin, Lincoln County, Colorado. In this example, the distribution is clearly skewed to higher concentrations. If the logarithm of the concentration is plotted as in Fig. 12-6, there is a normal distribution. This type of distribution is said to be a *lognormal* population. Most geochemical data exhibit some form of lognormal distribution.

Transformations of data occur when dealing with a histogram that has a skewed tail almost always to the right and only occasionally to the left. Plotting transformations of a skewed histogram as a lognormal population normalizes the data. The lognormal transformation is the only common form present in geology. The angular transformation of bimodal data and the square-root transformation of Poisson data for discrete distributions are rare and generally not helpful in practice (Koch and Link, 1971).

A more elegant way of analyzing distributions is to use a *cumulative frequency* or *probability* plot. The geochemical data from a survey are sorted into increasing orders to determine the range of data. They are then plotted on a special scale called a *cumulative percentage* vs. concentration. The

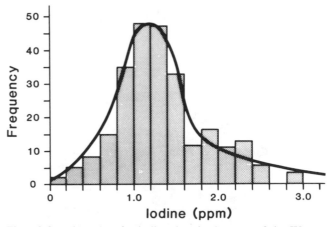

Fig. 12-3 Histogram for iodine data in the area of the Weaver and Elbridge oil fields, Vigo, Clark, and Edgar counties, Illinois Basin, Illinois and Indiana (courtesy of Atoka Exploration Corporation).

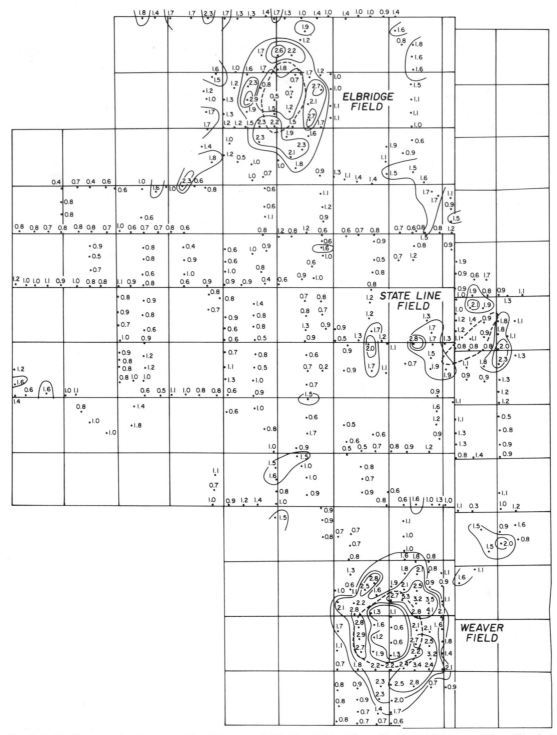

Fig. 12-4 Contoured iodine map for the area of the Weaver and Elbridge fields, Vigo, Clark, and Edgar counties, Illinois Basin, Illinois and Indiana (Allexan et al., 1986; reprinted with permission of the Association of Petroleum Geochemical Explorationists).

cumulative-percentage scale represents the number of samples with values less than a given value converted to a percentage of the total number of samples in the survey.

The lognormal distribution is a property of the data that is characterized by a normal distribution in this form. A lognormal data set is characterized by a clustering of the majority of observations in the data set, with a few large values forming a tail. The formula for a lognormal frequency distribution is:

$$f(\omega) = \frac{1}{\omega\beta\sqrt{2\pi}} \exp\left[-1/2\beta\,(\ln\omega - \alpha)^2\right]$$

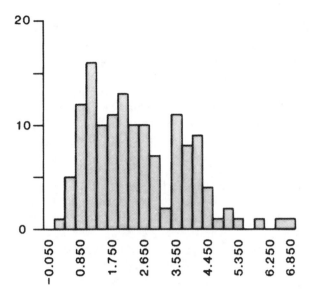

Fig. 12-5 Histogram of organic carbon from surface samples in Township 16 South, Range 57 West, Lincoln County, Denver-Julesberg Basin, Colorado.

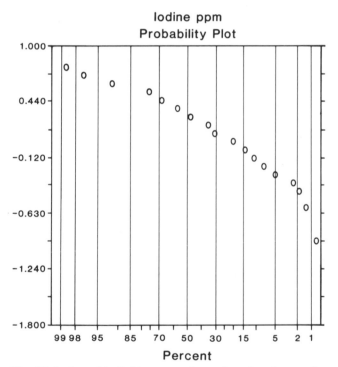

Fig. 12-6 Logarithmic histogram of organic carbon from surface samples in Township 16 South, Range 57 West, Lincoln County, Denver-Julesberg Basin, Colorado.

where α and β^2 are the mean and variance, respectively, of the natural logarithm of φ. The mean of φ is given by the formula:

$$u = \alpha + 1/2\beta^2$$

and the variance of φ by the formula:

$$\alpha^2 = \mu^2 (\varepsilon\beta^2 - 1)$$

These equations are from Koch and Link (1971). If the value of β^2 (variance) is small, so is the skewness, and the distribution will be normal. Lognormal distributions are useful only when the variance is high. Small variances typically do not justify their use.

Figure 12-7 is the cumulative percentage plot for soil-gas propane from a survey in the Sacramento Basin, California. The data are plotted as the cumulative percentage of the propane concentration in the soil gas. The diagram indicates a reasonably good straight line for the bulk of the data. The samples in this concentration range are likely to have a normal range of background values for the area of the survey. The upper group of samples shows an upward deflection, which may be an indication of an anomalous group of samples, an analytical problem, or some other cause. This upper group of samples is the data to which explorationists are likely to direct their attention. The diagram can provide an unbiased way of setting a boundary between anomalous and background samples.

Figure 12-8 is a cumulative-frequency diagram for 663 measurements of soil-gas helium taken at one location over a period of nearly two years (Klusman and Jaacks, 1987). The straight line of the data exhibits a normal distribution. Another group is offset and exhibits a constant low concentration. The authors determined that these sample measurements were invalid because of an analytical problem and that they could not be used in geochemical interpretation. The diagram allows for easy identification of these samples.

Figure 12-9 indicates a histogram of 3500 iodine samples taken in southeastern Colorado, Las Animas Arch. The population distribution indicates a bell-shaped curve with no real skewness, suggesting that there are no anomalous samples present. If 138 samples from a specific area (the Grouse Oil Field, Township 15 South, Range 46 West, Cheyenne County) are segregated out of the histogram, the new histo-

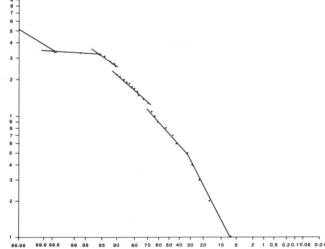

Fig. 12-7 Cumulative-percentage plot for propane from a survey in the Sacramento Basin, California (reprinted with permission from Trinity Oil and Illuminating).

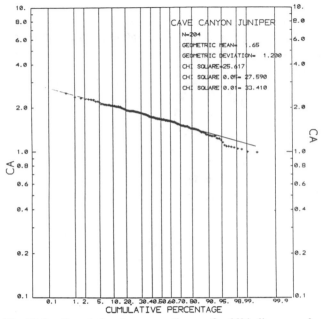

Fig. 12-8 Cumulative-frequency diagram for 663 helium samples taken over a period of two years (from Klusman and Jaacks, 1987).

gram suggests that multiple populations are present (Fig. 12-10). It becomes clearer which data are anomalous and which are not. Therefore, in a large data set, the presence of the anomalies could be difficult to determine using the original histogram.

Probability graphs are a practical tool in the analysis of surface geochemical data because of the normal or lognormal character of such data. One ordinate of the graph is either equal-interval (arithmetic) or logarithmic as required. The other ordinate, the probability scale, is such that a cumulative normal (or lognormal) distribution plot is a straight line (Fig. 12-11). This type of plot is sensitive to departures from normality and therefore to the recognition of combinations of multiple populations.

A cumulative histogram is constructed for purposes of viewing the data from low to high values, or vice versa. Figure 12-12 is an example of two sets of data, one with no indications of an anomalous population (data set A) and the other with more than one anomalous population (data set B). Typically, a line of best fit is applied. Data set A has a straight line easily applied. Data sets with multiple and therefore anomalous populations generally are best fitted

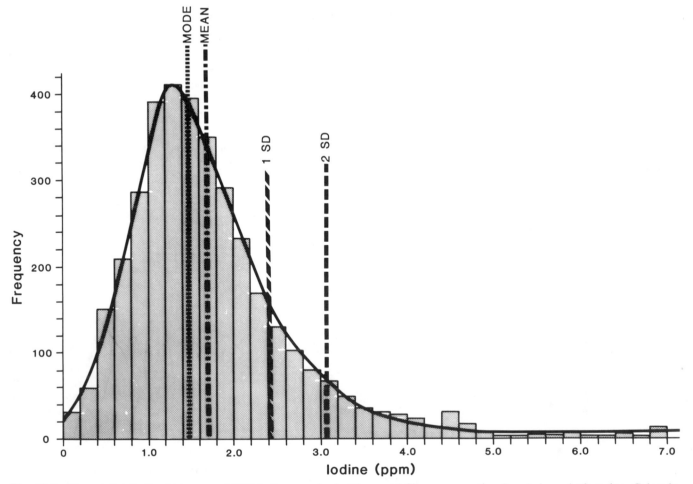

Fig. 12-9 Normal distribution histogram of 3500 iodine samples in Kiowa and Cheyenne counties, Las Animas Arch region, Colorado. Note the location of the mean, mode, and standard deviation (reprinted with permission from Atoka Exploration Corporation).

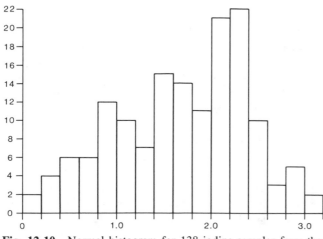

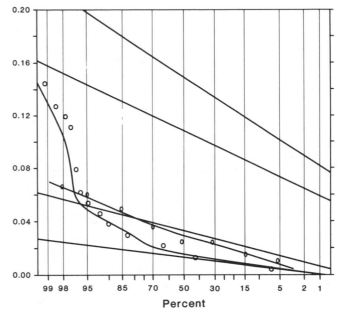

Fig. 12-10 Normal histogram for 138 iodine samples from the Grouse Field area, Cheyenne County, Las Animas Arch region, Colorado. Note that the presence of mean, mode, and two populations (by the two curves) are indicated (reprinted with permission from Atoka Exploration Corporation).

Fig. 12-12 Two data sets plotted on probability paper. Data set A indicates a straight line and no anomalous population. Data set B indicates deflections occurring along the line, which suggests three populations.

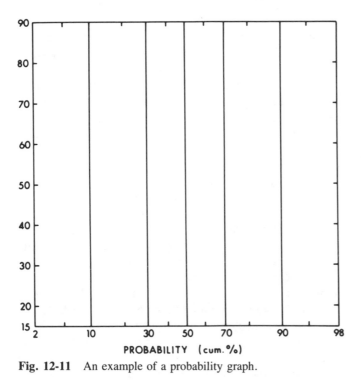

Fig. 12-11 An example of a probability graph.

contouring. This method can supplement or replace the use of simple mean, mode, and standard deviation.

Use of probability plots in mineral exploration is a common method for comparing various elements to determine potential interelement or compound relationships. If the hydrocarbons C_1 through C_6 are petroleum-generated, they typically show the same relationship, with a decreasing amount present toward the heavier end. The exception is when methane is biogenically generated and its relationship with C_2^+ is unclear or nonexistent. Using probability plots for soil gas and trace and major metals could indicate which elements are associated with specific soil types and petroleum-generated hydrocarbons.

Means, Variations, Standard Deviations, and Coefficients of Variance

The beginning of any statistical analysis is determining the central tendency or the mean and the dispersion of a set of samples. Table 12-1 represents the analyses of 52 hydrocarbon samples we have received and for which now need to determine the mean, mode, and standard deviation.

The mean of the data set is defined by the equation:

$$\sum \text{sum} = \frac{N_1 + N_2 \ldots N_i}{T}$$

where Σ = the mean value for the data set, N = the data value of a sample as represented by the subscript, and T = total number of samples taken.

with a curve in some form. The area between curve breaks determines the range and limits of each population. Data set B has three populations as indicated. The population of lower values is assumed to be background. The middle set could be a transition between the lower and higher values, variation in soils, a separate population of background values, or a separate population of values representing microseepage. The third population represents that part of the data that has higher values and is considered the most anomalous. From this graph, the plotted values on the map can be contoured by determining the minimum threshold value for

The mode is determined by placing all the data for a particular hydrocarbon in ascending order from the lowest to highest value. The total number of samples is divided by 2, resulting in the middle or mode number. The samples are placed in ascending order. If the total number of samples in the example data set is 52, then the mode is either the twenty-sixth or twenty-seventh sample value or the average of the sum of the twenty-sixth and twenty-seventh values, depending on personal preference.

Determining the spread of values is important because it leads to the identification of the threshold value that defines the boundary between background and anomaly. *Dispersion* is not usually used in petroleum exploration; another term, the *standard deviation,* is more common. Three other terms that define different parameters about the data are the *range, variance,* and *percentile.*

Variance is the measure of spread in a distribution. The variance is evaluated in the use of the standard deviation, which is its square root. The equation for the standard deviation is:

$$S^2 = \frac{\sum_{i=1} (X_i - \bar{X})^2}{n - 1}$$

where S^2 is the estimate of population variance, $\bar{x}$ is the mean of n items, x_i represents successively the n items, and $n - 1$ represents the degrees of freedom to provide an unbiased estimation of the population. It would seem more logical that variability can be easily measured as simply the average of deviations from the mean. However, simple calculations will indicate that this value will always equal zero. The variance can be further evaluated by the coefficient of variation, which is the ratio of the standard deviation to the mean.

The standard deviation indicates the extent of variation of the population away from the mean. A large variation suggests a wide range of values, indicating possible multiple populations or insufficient sampling. The smaller the deviation, the less the number of anomalous values there are, and the greater the likelihood that the population is normal or background in nature. The determination of what is a small and a large standard deviation depends on the geochemical element or compound involved in analysis. For example, iodine and soil gas typically exhibit wide variations; 2 to 20 times the mean is typical. Helium and trace elements usually vary less than 10% above and below the mean but are considered significant.

The range is determined by the highest and lowest value of the data set. Comparing the range to the mean or central tendency can identify which end of the range the geochemical data set is leaning toward. The range is often misleading, however, because one value can drastically alter it. Typically, large data sets use some other form to measure dispersion, such as the standard deviation.

Comparison of the mode and the mean can reveal the following conditions. If the mean and mode are reasonably close, it indicates that the mean is close to the typical or nonanomalous values sought. Anomalous values will represent a small part of the total population. A large difference between the mode and the mean suggests that the population is skewed by the presence of more than one population. To take this a step further, the mean, the mode, and the standard deviation of propane shown in Table 12-1 suggest close agreement, and a histogram (Fig. 12-13) suggests a relatively normal distribution. However, plotting the data reveals a different view (Fig. 12-14). The data actually would be called the anomalous population, and sampling is insufficient to determine background confidently.

A typical evaluation technique is to use the mean as the exact background value and assume that values above the mean plus one standard deviation are anomalous. This is termed a *threshold value.* The threshold value is an artificial determination of what is anomalous and what is background. But this is a basic and common way to evaluate data and begin an interpretation. Two, three, four, or more standard deviations above the mean can be used if the standard deviation is small and if it seems that background values are overlapping with anomalous ones.

Returning to Table 12-1, we can plot the data on a normal histogram to begin a visual review (Fig. 12-15). Then we use a probability plot (Fig. 12-16). It indicates a general bell-shaped curve for methane and ethane. Propane data indicate the potential for the two populations. We then determine the mean, mode, and standard deviation and highlight values of one, two, three, and four standard deviations in Table 12-1. With these values, we can plot and contour data based on the standard deviation. Using one standard deviation clearly indicates the presence of a large anomalous area. As we progress by using greater and greater standard deviations, the anomalous areas become smaller in areal extent and give us a greater definition of the petroleum accumulation present. The number of standard deviations above the mean that are used depends on the geochemical, geologic, and geophysical models and on the assessment of which anomalous values represent seepage of petroleum from the subsurface.

A second example uses 487 iodine values from the Forest City Basin, Kansas. A histogram indicates multiple populations (Fig. 12-17) with the mean, mode, and standard deviation. Because there are several populations present, the analysis of these types of data becomes more complex and is dealt with in the sections below.

Transformations

A transformation is a function applied to an observation or data set that subsequently defines a new set of data. The application of transformations has been applied in surface

Table 12-1 Fifty-two soil-gas analyses for an unidentified survey area. Anomalous ratios or percentage values are determined based on the mean $\pm \frac{1}{2}$ a standard deviation. Positive anomalous ratios are denoted by an $*$. Negative anomalous ratios are denoted by $+$. Boldface denotes values that are considered anomolies

C_1	C_2	C_3	iC_4	nC_4	
2.67	0.124	0.124	0.078	0.149	
0.86	0.060	0.019	0.023	0.010	
0.64	0.024	0.009	0.014	0.009	
1.62	0.088	0.033	0.011	0.017	
1.61	0.017	0.006	0.012	0.005	
0.90	0.069	0.034	0.007	0.012	
0.79	0.028	0.013	0.011	0.004	
0.80	0.024	0.010	0.009	0.003	
0.74	0.047	0.017	0.013	0.009	
1.60	0.105	0.077	0.015	0.026	
1.43	0.149	0.082	0.021	0.035	
1.47	0.150	0.080	0.026	0.038	
0.86	0.083	0.043	0.028	0.019	
0.82	0.041	0.017	0.017	0.007	
1.34	0.206	0.107	0.009	0.052	
0.63	0.035	0.009	0.033	0.003	
1.18	0.113	0.055	0.019	0.023	
1.54	0.011	0.008	0.019	0.003	
1.29	0.129	0.065	0.011	0.034	
1.31	0.165	0.086	0.019	0.047	
0.46	0.016	0.006	0.045	0.003	
0.87	0.037	0.014	0.008	0.015	
1.24	0.057	0.031	0.008	0.012	
2.19	0.350	**0.193**	0.029	0.090	
2.10	0.284	0.167	0.057	0.066	
1.04	0.141	0.075	0.046	0.029	
0.90	0.115	0.057	0.028	0.014	
0.50	0.029	0.022	0.020	0.007	
0.38	0.019	0.017	0.015	0.008	
1.07	0.140	0.069	0.011	0.045	
3.93	**0.656**	**0.311**	0.019	**0.176**	
4.29	**0.529**	0.029	0.089	**0.143**	
12.73	**4.006**	**0.670**	0.077	**0.971**	
4.20	0.426	**0.299**	**0.722**	0.083	
1.23	0.175	0.098	0.078	0.051	
0.85	0.090	0.060	0.029	0.067	
1.34	0.109	0.080	0.025	0.030	
1.10	0.075	0.054	0.021	0.017	
1.03	0.096	0.131	0.015	0.020	
0.96	0.226	0.106	0.019	0.050	
3.13	0.338	**0.284**	**0.132**	0.111	
1.65	0.107	0.156	0.028	0.029	
2.58	0.273	0.014	0.026	0.099	
0.55	0.300	0.028	0.046	0.018	
0.74	0.039	0.029	0.011	0.010	
0.78	0.038	**0.315**	**0.110**	0.010	
2.23	**0.607**	**0.315**	**0.170**	**0.129**	
0.63	0.048	0.072	0.089	0.024	
1.08	0.259	0.137	0.039	0.058	
1.73	0.460	**0.250**	0.060	0.107	
1.63	0.234	0.160	0.049	0.058	
1.82	**0.561**	0.32	**0.102**	**0.136**	
1.64	0.240	0.099	0.050	0.059	Mean
1.18	0.024	0.020	0.065	0.075	Mode
1.79	0.552	0.119	0.100	0.134	Standard deviation

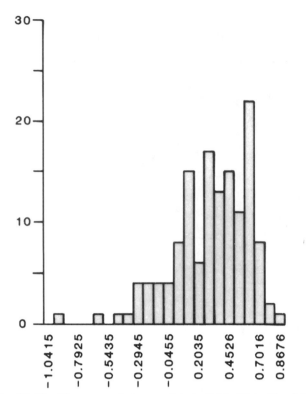

Fig. 12-13 Histogram of propane data in Table 12-1. The data are negatively skewed.

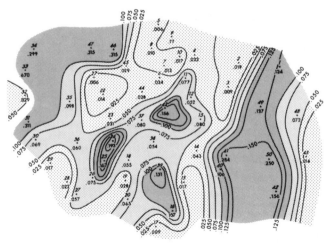

Fig. 12-14 Contouring of the propane data from Table 12-1. Contour interval is 0.25 ppm.

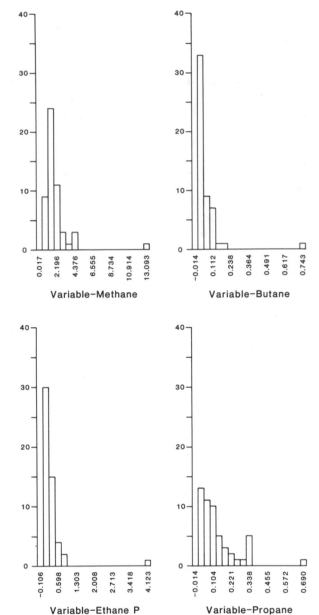

Fig. 12-15 C_1 through C_4 data from Table 12-1 presented in an histogram form. The data were plotted using the Probplot computer program available through the Association of Exploration Geochemists. The intervals were determined by the computer. The data are positively skewed because one large point is clustered to the origin of the graph.

geochemistry for petroleum exploration more as a filter than as a way to yield more definitive results. Transformations are typically applied to nonlinear or lognormal distributed data. A common transformation is to take the log of a data set (lognormal distribution) that is skewed, and the result is a new data set.

Expanding on this, many acquired data sets can be difficult to interpret. For example, soil-gas data across a survey may not show anomalous values for all the various C_1 through C_6 hydrocarbons in any group or specific samples. For a

problem like this, each sample varies in terms of which hydrocarbons are anomalous and may not be comparable between sample sites. A solution has been the transformation. The transformation function is not limited to soil-gas data, nor is it restricted to using only the data acquired with the same method. Hydrocarbon data can be entered into a function that can include iodine, radiometrics, major and minor elements, and microbial values. Therefore, the idea of a transformation is to create a function that brings several

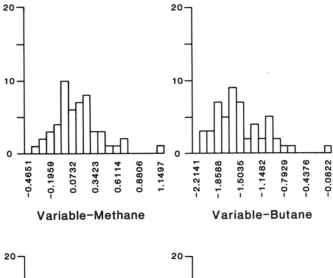

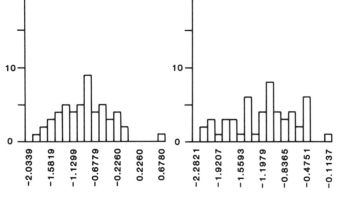

Variable–Ethane P   **Variable–Propane**

Fig. 12-16 Data in Table 12-1 plotted as a lognormal distribution. The data were plotted using the Probplot program available through the Association of Exploration Geochemists. The intervals were determined solely by the computer. Methane and ethane indicate general bell-shaped curves. Propane suggests the presence of more than one population.

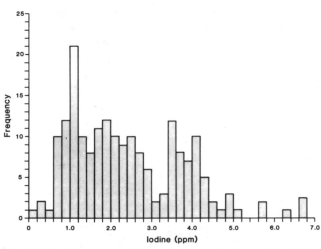

Fig. 12-17 Histogram of iodine data from the Forest City Basin, Kansas. The several peaks indicate numerous populations. Note the location of the mean, standard deviation, and mode.

parameters together to yield a resulting number that can be placed at the sample location and be simply contoured or can be identified as an anomaly or background.

There are three commonly used forms of transformations: (1) ratioing of one hydrocarbon to another hydrocarbon, (2) smoothing the data for any particular variable, and (3) bringing together a variety of variables that are unrelated and converting them into a new variable that is easier to work with.

Ratios

Ratios were discussed in Chapter 5 and can be a powerful tool for soil-gas methods. Ratios or correlation methods are vital in mineral exploration because they can provide a hint of various mineral associations that can be indicative of certain types of deposits. In petroleum, with the exception of trace and major metals, the relationship between different variables is either very definitive, such as between C_1 to C_2, C_2 to C_5, and so forth, or nonexistent, such as C_3 to I or C_4 to Cu.

The use of ratios and their subsequent successful application has been discussed by Jones and Drozd (1983). The ratios of the lighter to the heavy hydrocarbons consistently indicate a decrease in values in the area of petroleum microseepage.

Simple calculation for a ratio is for example:

$$C_3/C_2 \ 424/73 = 6.479$$

or

$$C_3/C_2 \ 73/424 = 0.172$$

where arbitrary values are presented for C_2 and C_3. Either ratio is an acceptable form. Generally, the ratio of C_2 to C_3 or a higher to a lower hydrocarbon will decrease or approach a one-to-one ratio to indicate anomalous conditions. However, this is not always the case as some surveys can have unusually low volumes of a high hydrocarbon; thus, the ratios will yield lower ratios. Establishing a background and threshold values for the ratios is required.

A complex calculation may be, for example:

$$C_4 + C_5 + C_6/C_2 + C_3 = \text{Ratio}$$

Table 12-2 is a set of soil-gas data, acquired by the free-air method from 0.9 m (3 ft) in depth. Both the mean and standard deviations are presented. Table 12-3 is a series of ratios plus percentage data calculated for the data set in Table 12-2. The data that are anomalous in both tables is highlighted. The anomalous values are calculated by using the mean plus one standard deviation to determine the threshold boundary. An alternative approach would be the mean plus or minus half the standard deviation, which will lower the threshold value. This typically is used when a small

Table 12-2 Anomalous ratio or percentage values are determined based on the mean $\pm \frac{1}{2}$ a standard deviation. Positive anomalous ratios are denoted by an $*$. Negative anomalous ratios are denoted by $+$. Boldface indicates values that are anomalous

Sample No.	C_1	C_2	C_3	iC_4	nC_4
1	1,700.00	131.00	15.00	2.00	10.00
2	**6,500.00**	**1,106.00**	50.00	**18.00**	**21.00**
3	1,800.00	211.00	26.00	1.00	4.00
4	2,800.00	227.00	34.00	1.00	6.00
5	4,400.00	925.00	48.00	2.00	6.00
6	5,700.00	**2,115.00**	**168.00**	21.00	**60.00**
7	**6,700.00**	**1,281.00**	82.00	**42.00**	**36.00**
8	**6,000.00**	**1,211.00**	76.00	**43.00**	12.00
9	1,300.00	**3,580.00**	**248.00**	27.00	**60.00**
10	**7,800.00**	**2,329.00**	**164.00**	29.00	**48.00**
11	**6,500.00**	**1,283.00**	82.00	21.00	**36.00**
12	2,500.00	200.00	26.00	1.00	4.00
13	2,200.00	219.00	26.00	2.00	4.00
14	3,700.00	745.00	33.00	1.00	5.00
15	5,400.00	**1,152.00**	62.00	**32.00**	9.00
16	4,300.00	57.00	34.00	5.00	10.00
17	3,200.00	64.00	30.00	1.00	7.00
18	3,900.00	72.00	51.00	3.00	12.00
19	5,200.00	345.00	83.00	**18.00**	**48.00**
20	3,400.00	81.00	36.00	3.00	9.00
21	3,300.00	29.00	28.00	1.00	8.00
22	**6,600.00**	48.00	20.00	1.00	5.00
23	5,200.00	97.00	50.00	3.00	12.00
24	4,400.00	254.00	56.00	9.00	**39.00**
25	3,200.00	67.00	33.00	2.00	9.00
26	4,300.00	260.00	48.00	3.00	15.00
27	3,300.00	61.00	33.00	1.00	9.00
Mean	4,703.70	672.22	60.81	10.85	18.67
Standard deviation	2,282.21	854.61	52.10	13.23	17.66

percentage of data yields very high values and skews the mean and standard deviation toward the anomalous population. This can exclude data from the anomalous group and place it erroneously in the background set. This data set actually represents a data set that is skewed because of a small percentage of high values. The other obvious concern is that there is no clear set of samples that are anomalous overall. Anomalies tend to be limited to certain hydrocarbons and between certain hydrocarbons.

A second table of data is present in Tables 12-4 and 12-5 that were acquired along the same survey as from Table 12-4 using head space soil gas method to 0.9 m (3 ft) in depth. The data set is quite different in that samples of interest tend to be anomalous throughout. However, some of the ratios do not always display anomalous values. Both these surveys resulted in a new field, and seismic was used in exploration as well.

Determining ratios and their subsequent successful application have been discussed by Jones and Drozd (1983). The ratios of the lighter to the heavier hydrocarbons consistently indicate a decrease in values in the area of petroleum microseepage. Figure 12-20 indicates the use of percentages of a

hydrocarbon with respect to the total amount of soil gas detected. In this case, methane shows a marked decrease in the area of the producing field. Propane for the same area shows a marked increase transecting the petroleum accumulation. Ratios have not proved viable in areas where the produced gas from the reservoir is of biogenic origin.

Ratios or correlations between trace and major elements in the seepage typically indicate a relationship similar to that found in the reservoir (Fausnaugh, 1991). But it is difficult to conclude or even to speculate on the relationship between trace element associations found in hydrocarbon seepage and those found in typical soils because little work has been done in this area.

Data Filtering

Various forms of filtering have been employed to manipulate data in order to eliminate "noise" and to determine the anomalous areas from the nonanomalous areas. Noise by definition is surface geochemical data representing isolated or, in certain cases, paired anomalous or nonanomalous

Table 12-3 Anomalous ratio or percentage values are determined based on the mean $\pm \frac{1}{2}$ a standard deviation. Positive anomalous ratios are denoted by $*$; negative anomalous ratios are denoted by $+$. Boldface indicates anomalous values

Sample No.	C_1/C_2	C_2/C_3	C_2/nC_4	$C_1/C_2 \ldots nC_4$	C_3/nC_4	$\%C_2$	$\%C_3$
1	12.98	8.73	**13.10+**	10.76	**1.50+**	0.07	0.008
2	5.88	**22.12***	52.67	5.44	**2.38+**	**0.14***	0.006
3	8.53	8.12	52.75	7.44	**6.50***	0.10	**0.013***
4	12.33	6.68	37.83	10.45	**5.67***	**0.07***	0.011
5	4.76	**19.27***	**154.17***	4.49	**8.00***	**0.17***	0.009
6	2.70	**12.59***	35.25	2.41	2.80	**0.26**	**0.021**
7	5.23	**15.62***	35.58	4.65	2.28	**0.16**	0.010
8	4.95	**15.93***	**100.92***	4.47	**6.33***	**0.16**	0.010
9	3.63	**14.44***	59.67	3.32	4.13	**0.21**	**0.015**
10	3.35	**14.20***	48.52	3.04	3.42	**0.22**	**0.016**
11	5.07	**15.65**	35.64	4.57	2.28	**0.16**	0.010
12	12.50	7.69	50.00	10.82	**6.50***	0.07	0.010
13	10.05	8.42	54.75	8.76	**6.50***	0.09	0.011
14	4.97	**22.58***	**149.00***	4.72	**6.60***	**0.17**	0.017
15	4.69	**18.58***	**128.00***	4.30	**6.89***	**0.17**	0.009
16	**75.44***	**1.68+**	**5.70+**	**40.57***	3.40	0.01	0.008
17	**50.00***	2.13	**9.14+**	**31.37***	4.29	0.02	0.009
18	**54.17***	**1.41+**	**6.00+**	**28.26***	4.25	0.02	0.013
19	15.07	4.16	**7.19+**	10.53	**1.73+**	0.06	**0.015**
20	41.98	2.25	**9.00+**	26.36	4.00	0.02	0.010
21	**113.79***	**1.04+**	**3.63+**	**50.00***	3.50	0.01	0.008
22	**137.50***	2.40	**9.60+**	**89.19***	4.00	0.01	0.003
23	**53.61***	**1.94+**	**8.08+**	**32.10**	4.17	0.02	0.009
24	17.32	4.54	**6.51+**	12.29	**1.44+**	0.05	0.012
25	**47.76***	**2.03+**	**7.44+**	**28.83***	3.67	0.02	0.010
26	16.54	5.42	**17.33+**	13.19	3.20	0.06	0.010
27	**54.10***	**1.85+**	**6.78+**	**31.73***	3.67	0.02	0.010
Mean	28.85	8.94	40.90	17.93	4.19	0.09	0.010
Standard deviation	34.44	6.90	43.30	19.11	1.80	0.08	0.003

values that cause a disrupting presence. Noise data can obscure and confuse analysis by including additional area with the true anomalous area, or it can create a false anomaly. Noise that is anomalous or that suggests that it may be anomalous can be inadvertently and subsequently included in the interpretation. Noise usually has nothing to do with the anomaly being defined. The source of the noise can be sample contamination, laboratory contamination, changes in soil chemistry, and minor accumulations of petroleum seeping or present on the surface. Filtering of noise can take a variety of forms from arbitrarily determining background and anomalous values to actual mathematical manipulation.

The three-point moving average is a typical method employed for traverse data. Figure 12-18 shows the plotted raw microbial data. To filter the data, every three points are added together and then divided by 3:

$$\frac{N_1 + N_2 + N_3}{3} = N_{2R}$$

N_{2R} becomes the new value for N_2. The resulting data are recontoured or new profiles are presented (Fig. 12-19). The end two stations are eliminated in this process.

The nine-point filter is a variation on the same theme but is applicable only to evenly spaced grids. Figure 12-20 is the raw data (point radiometrics in this case). The equation for the filter is:

$$N_{1R} = \frac{\begin{array}{c} N_2 + N_3 + N_4 \\ + \\ N_5 + N_1 + N_6 \\ + \\ N_7 + N_8 + N_9 \end{array}}{9}$$

The resulting value is placed at the center or N_{1R} sample point. The next point is calculated:

$$N_{1R} = \frac{\begin{array}{c} N_3 + N_4 + N_{10} \\ + \\ N_1 + N_6 + N_{11} \\ + \\ N_8 + N_9 + N_{12} \end{array}}{9}$$

The data are then replotted on the map and contoured (Fig. 12-21). The outer sample points can be either eliminated or

Table 12-4

Sample No.	C_1	C_2	C_3	nC_4	iC_4	C_5
1	350.00	54.00	34.00	5.00	19.00	12.00
2	1,132.00	157.00	101.00	56.00	42.00	44.00
3	246.00	37.00	12.00	7.00	9.00	3.00
4	278.00	42.00	15.00	8.00	15.00	7.00
5	122.00	46.00	12.00	42.00	18.00	9.00
6	1,985.00	235.00	118.00	76.00	78.00	42.00
7	2,103.00	444.00	156.00	54.00	92.00	51.00
8	1,877.00	331.00	144.00	83.00	101.00	43.00
9	1,345.00	336.00	171.00	112.00	73.00	27.00
10	1,734.00	379.00	112.00	92.00	42.00	93.00
11	1,458.00	287.00	181.00	68.00	88.00	87.00
12	122.00	43.00	12.00	9.00	9.00	2.00
13	143.00	33.00	16.00	7.00	19.00	0.00
14	156.00	27.00	19.00	11.00	6.00	1.00
15	145.00	51.00	12.00	17.00	7.00	9.00
16	183.00	59.00	13.00	10.00	18.00	2.00
17	202.00	87.00	11.00	9.00	14.00	7.00
18	117.00	88.00	12.00	6.00	12.00	3.00
19	345.00	157.00	67.00	56.00	89.00	89.00
20	109.00	54.00	12.00	2.00	3.00	0.00
21	131.00	67.00	21.00	11.00	7.00	0.00
22	142.00	83.00	22.00	9.00	11.00	0.00
23	132.00	81.00	12.00	6.00	4.00	0.00
24	245.00	199.00	34.00	63.00	32.00	51.00
25	188.00	172.00	45.00	51.00	61.00	53.00
26	201.00	162.00	32.00	43.00	31.00	61.00
27	110.00	52.00	22.00	7.00	4.00	0.00
Mean	566.70	139.37	52.52	34.07	33.48	25.78

used as an artificial set of data points that are one sample spacing out. The value commonly placed at these artificial sample points is the mean or mode number for the data set. A sample set of 8 by 8 sample grid (64 points) will result in new values at only 36 points. Therefore, small grids will not be usable with this technique. Another caution: If the anomalous areas lie at the edges of the grid, they may be eliminated by filtering. The subsequent statistics will then determine threshold values that are too low. The reverse is true with data that are essentially anomalous.

Both the three- and nine-point filters attempt to eliminate areas that are termed noisy and to pinpoint only those areas that are "truly anomalous." All types of geochemical data seem to work well with these two methods when sufficient data are present and when both anomalous and background data are well represented.

Another form of filtering was presented by Saunders et al. (1991), who established an arbitrary determination of the mean or mode. This determination is based on experience and is used in conjunction with small data sets. All data below this arbitrary value are declared background and are discarded. Typically, the mean, mode, and standard deviation are determined with the aid of a histogram, a cumulative or probability plot, or sophisticated statistics designed for this purpose. Based on these data, there is an arbitrary determination of what the anomalous threshold is. This type of analysis ignores the mean and possibly the mode. The mode may represent background data and be useful. In certain cases, to determine the intensity of the arbitrarily defined anomalies, the mode is used as a barometer of the value of the anomaly. This form of filter is based more on experience with a specific type of geochemical data than on any form of mathematical solution. This type of filter is not recommended because it is subjective and open to criticism.

Functions

The first step in utilizing a transformation function is to develop one from data obtained across one or more existing fields. These fields or models can be restricted to a small area or a basin or may be used worldwide. The function must be consistently applied across all the models established. Otherwise, it is limited to those areas where it applies.

A simple transformation might be:

$$\frac{(C_1 + C_2 + C_3 + C_4 + C_5 + C_6)/2}{(C_1 + C_2)} = \mathbf{T}$$

Table 12-5 Anomalous ratio or percentage values are determined based on the mean $\pm \frac{1}{2}$ a standard deviation. Positive anomalous ratios are denoted by an $*$; negative anomalous ratios are denoted by $+$. Boldface indicates anomalous values

Sample No.	C_1/C_2	C_2/C_3	C_2/nC_4	$C_1/C_2 \dots C_6$	C_3/nC_4	$+C_2$
1	**6.48***	**1.59+**	**10.80***	**2.82***	**6.80***	0.11
2	**7.21***	**1.55+**	2.80	**2.75***	1.80	0.10
3	**6.65***	3.08	5.29	**3.62***	1.71	0.12
4	**6.62***	2.80	5.25	**3.20***	1.88	0.12
5	2.65	3.83	**1.10+**	**0.95+**	**0.29+**	0.18
6	**8.45***	**1.99+**	**3.09+**	**3.42***	1.55	0.09
7	4.74	2.85	8.22	**2.50***	**2.89***	0.15
8	**5.67***	**2.30+**	3.99	**2.57***	1.73	0.13
9	4.00	**1.96+**	**3.00+**	1.82	1.53	0.16
10	4.58	3.38	4.12	**2.24***	**1.22+**	0.15
11	5.08	**1.59+**	4.22	1.94	2.66	0.13
12	2.84	3.58	4.78	1.58	1.33	0.22
13	4.33	**2.06+**	4.71	1.91	2.29	0.15
14	**5.78***	**1.42+**	**2.45+**	**2.44***	1.73	0.12
15	2.84	4.25	**3.00+**	1.48	**0.71+**	0.21
16	3.10	4.54	5.90	1.79	1.30	0.21
17	2.32	**7.91***	**9.67***	1.58	**1.22+**	**0.26***
18	**1.33+**	**7.33***	**14.67***	**0.97+**	2.00	0.37
19	2.20	**2.34+**	2.80	**0.69+**	**1.20+**	0.19
20	**2.02+**	4.50	**27.00***	1.54	**6.00***	**0.30***
21	**1.96+**	3.19	6.09	1.20	1.91	**0.28***
22	**1.71+**	3.77	9.22	1.14	2.44	**0.31***
23	**1.63+**	**6.75**	13.50	1.27	2.00	**0.34***
24	**1.23+**	**5.85**	3.16	**0.63**	**0.54+**	**0.31***
25	**1.09+**	3.82	3.37	**0.45**	**0.88+**	**0.28***
26	**1.24+**	5.06	3.77	**0.59**	**0.74+**	**0.30***
27	**2.12+**	**2.36+**	7.43	1.29	**3.14***	**0.27***
Mean	3.70	3.54	6.42	1.79	1.98	0.21
Standard deviation	2.11	1.75	5.23	0.87	1.43	0.08

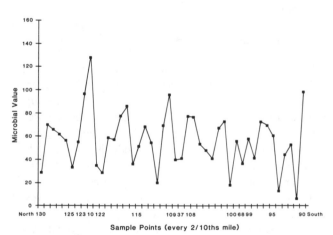

Fig. 12-18 Raw microbial data, based on counts of butanol-digesting bacteria.

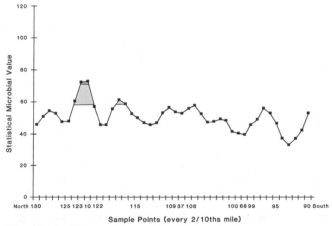

Fig. 12-19 Three-point moving average for the data from Fig. 12-21, with the statistically anomalous values shaded. The equation is $(N_1 + N_2 + N_3)/3 = N_{2R}$; the subscript R designates the new value of N_2 in the new data set. The general equation is $(N_x + N_x+1 + N_x+2)/3 = N_{XR}$

173

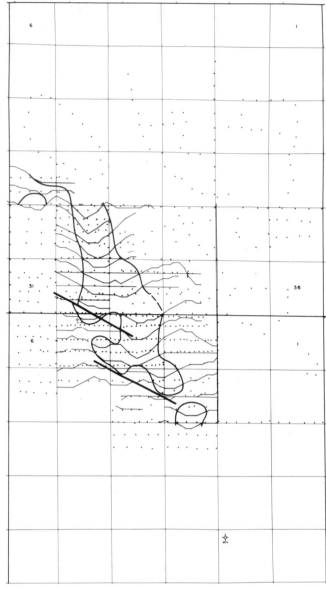

Fig. 12-20 Raw radiometric point data from southeastern Colorado. The data are presented in profile form. The shaded area indicates the anomaly.

Essentially, all values for a particular sample are added, divided by 2, and divided by the sum of $C_1 + C_2$, which results in the value **T**. The value **T** can be placed on a map and contoured, or it may simply be compared to a set of data. The model data set will have certain values of **T** that represent anomalous values or background.

Bernand et al. (1977) presented a transformation to discriminate between biogenic and thermogenic gases. The equation

$$\frac{C_1}{(C_2 + C_3)}$$

yields a ratio. If the ratio is 10, then the gases are thermogenic. If the ratio is 10^3 to 10^5, then the gases are biogenic. Carbon 13 was also used in the evaluation as a confirmation tool.

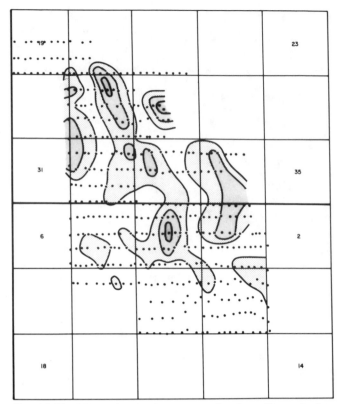

Fig. 12-21 Nine-point moving average for the radiometric data from Fig. 12-20. Negative values represent the anomalous area, especially below -2. The zero or positive values represent non-anomalous areas. New mean, modes, and standard deviations are determined, resulting in new threshold values. The light-gray shaded area represents values that have exceeded the threshold from background to anomaly. The medium-gray shaded area represents the values that have exceeded another threshold point.

Transformation can include variables that are not related, for example:

$$\% \text{ Carbonate} \times \frac{(C_2 + C_3 + C_4 + C_5)\, I}{(C_2)^2} = R$$

Table 12-6 is set of data utilizing this transformation.

Transformations have to be used with caution. They will not eliminate problems with data that were poorly acquired. They should not and do not change or enhance the absence or presence of leakage. This form of statistical manipulation has been used to predict dry vs. producing wells and, in some cases, to locate the depth to the potential reservoir. There are numerous transformation functions available and, in use, many of them are closely guarded secrets. However, they can often act as another piece of unnecessary information, and simply overlaying the raw surface geochemical data maps will produce the same result.

Evaluation of Profiles

The simplest and most effective way to evaluate geochemical data is with profiles. Continuous radiometric profiles

Table 12-6 Data utilizing a transformation of nonrelated variables. Positive anomalies are indicated by ∗. Boldface indicate anomalous values.

Sample No.	C_2	C_3	iC_4	C_5	% Carbonate	Iodine	R
1	15.53	9.07	**2.55***	**11.31***	7.20	1.80	0.089
2	16.59	7.53	**2.53***	**9.74***	6.10	2.10	0.063
3	9.90	6.36	0.45	1.86	3.10	1.10	0.172
4	15.23	7.39	**2.12***	6.65	9.20	**3.10***	0.044
5	**27.01***	8.78	0.64	1.84	7.10	2.40	0.022
6	12.68	4.64	0.54	1.05	**12.40***	2.70	0.044
7	16.90	6.65	0.16	1.44	6.80	**2.90***	0.030
8	6.46	2.57	0.69	0.84	1.20	1.10	0.230
9	19.54	6.77	0.50	1.31	5.60	0.90	0.082
10	7.95	4.04	1.62	0.58	4.40	0.80	**0.281***
11	22.89	**10.03***	**2.04***	6.71	**11.10***	**3.70***	0.021
12	8.84	6.04	**2.92***	**15.31***	9.90	1.00	**0.422***
13	11.01	7.98	**2.27***	**12.23***	7.30	1.20	**0.230***
14	8.71	7.26	0.57	**14.46***	8.10	0.90	**0.454***
15	18.77	6.11	0.53	1.01	**11.10***	**3.60***	0.021
16	19.04	7.98	0.45	1.88	**12.90***	2.50	0.032
17	**30.83***	**11.18***	0.29	2.52	4.70	2.70	0.017
18	23.24	**9.64***	0.49	2.37	9.80	1.20	0.055
19	**29.65***	**9.71***	0.24	3.05	**13.10***	1.80	0.027
20	15.15	6.20	1.25	6.71	7.70	1.70	0.075
Mean	16.8	7.3	1.14	5.14	7.9	1.96	0.121
Standard deviation	7.00	2.07	.9	4.79	3.19	.92	0.131

lend themselves readily to this method and can determine the presence of an anomaly through the slope-of-line technique. Ground level is assumed to be zero degrees. If there were no radioactivity or if it were uniform, the results would be a horizontal line. Because radioactivity fluctuates, the departure from the horizontal will vary by ±90 deg. Typical radiometric anomalies that reflect microseepage are a depression in the radiometric data. Therefore, the percentage of departure from the horizontal determines the significance of the radiometric anomaly. The extent or length of the radiometric depression is also a factor. Figure 12-22 is a typical radiometric profile, with soil types and topography plotted beneath. The anomaly over the Clifford Field is indicated, and there was no obvious relationship between it and the soil and topography. The abrupt change in slope of the radiometric count on the west side of the anomalous depression, followed by an overall gradual increase to the east, can be construed as a typical anomaly associated with petroleum seepage. The east side of the line actually has a steeper slope. The slope of the line from the point where the radiometric count changes to where it flattens out is determined. In this case, it is 50%. The slope can be as little as 10%, but usually the value is greater than 20%. Higher numbers are preferable. The Clifford Field has a slope of greater than 40% on the west and east sides.

Another example indicates an anomaly associated with an existing petroleum accumulation (Fig. 12-23). The anomaly has depressions and peaks, some of which may be related to the topography and soil changes. It is doubtful that this anomaly would have been drilled because of the coincidental changes in soils and topography.

Soil-gas profiles are often used for presentation purposes. Usually an example of the best result is presented, representing a slice of the total data set. The mean and standard deviations are determined for the entire data set and are plotted on the profile appropriately (Fig. 12-24). The y axis shows the amount of hydrocarbons present, usually in parts per million or billion. The x axis indicates the station locations for the survey line chosen. The profile can be viewed without the establishment of any threshold values, or threshold values and color can be added. In addition, several gases can be plotted on the same profile to evaluate the relationship between the hydrocarbons. Another form of profile evaluation is to present several gases along the same grid line in a three-dimensional form (Fig. 12-25).

Other types of surface geochemical data, such as helium, iodine, and other elements, can be presented in similar fashion. This type of evaluation is limited and should not be used solely to determine the viability of a project.

Percentages

A simple form of soil-gas analysis is to determine the percentage of each hydrocarbon in the sample with respect to the total gas present. This is based on the concept that

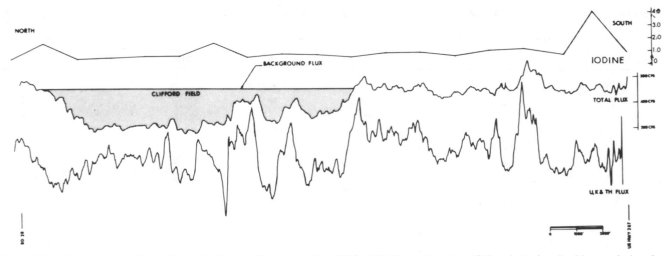

Fig. 12-22 Continuous radiometric and iodine profiles across the Clifford Field, southeastern Colorado (printed with permission from CST Oil & Gas Corporation). The slope of the line method is used and indicates that both east and west sides have a slope exceeding 50%.

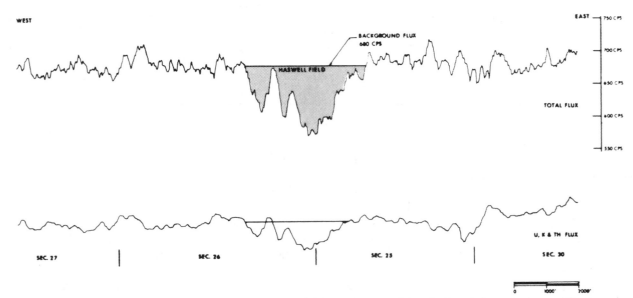

Fig. 12-23 Continuous radiometric profile across the Haswell Field, southeastern Colorado, Las Animas Arch area. The low associated with the field is colored red (printed with permission from CST Oil & Gas Corporation).

soil-gas volumes over areas that are leaking will exhibit a decrease in methane and an increase in ethane and heavier hydrocarbons. Because of their biogenic origin, ethylene and propylene are not used in the equation. Typically, only petroleum-generated hydrocarbons are used; these are methane, ethane, propane, isobutane, n-butane, hexane, and pentane. This form of data analysis usually indicates that methane decreases with respect to the other hydrocarbons when microseepage is detected (Fig. 12-26). Even though ethane increases in volume, its percentage increase (with respect to total gas) with each successively heavier hydrocarbon. Specifically, the heavier the hydrocarbon, the greater the overall increase in volume. The equation for determining the percentage of a gas in a sample is:

For total hydrocarbons:

$$C_n\% = \frac{C_n}{Sum\,(C_1 + \cdots C_x)}$$

For all hydrocarbons excluding methane:

$$C_n\% = \frac{C_n}{Sum\,(C_2 + \cdots C_x)}$$

C_n represents the hydrocarbon whose percentage of the total gas volume is being determined. C_x represents the heaviest hydrocarbon detected.

Figure 12-27 is an example of the use of percentages at the Wehking Field, Atchison County, Kansas. The field produces gas that is 96% to 98% methane from the McClouth sandstone. The remaining gas content is predominantly CO_2 or N_2 with minor amounts of heavier hydrocarbons. The

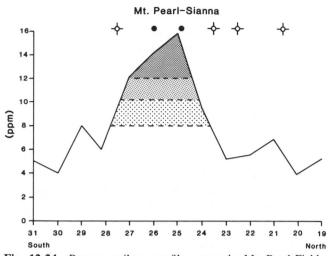

Fig. 12-24 Propane soil-gas profile across the Mt. Pearl Field, Cheyenne County, Las Animas Arch area, Colorado. The *x* axis shows the station locations. The *y* axis indicates propane in parts per million.

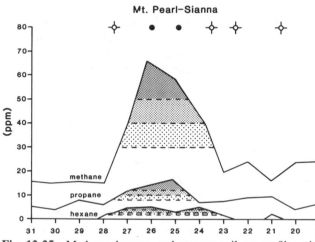

Fig. 12-25 Methane, hexane, and propane soil-gas profiles with threshold values across the Mt. Pearl Field, Cheyenne County, Las Animas Arch area, Colorado (printed with permission from Atoka Exploration). The *x* axis shows the station numbers and locations. The *y* axis indicates hydrocarbons in parts per million. Note the decrease in methane but the increase in propane (C3) and hexane (C6) at station 21.

reservoir contains oil and a tar that acts as the impermeable zone between the gas and water zones. The percentage of methane decreases in and around the field, going from over 90% to about 86% to 87%.

The Shubert Field, Richardson County, Nebraska (Fig. 12-28), produces oil from the Hunton Limestone. The percentage of methane decreases from over 90% to less than 70% in the productive area. The percentage of hexane rises dramatically in the area of the producing field (Fig. 12-29).

Another form of data analysis is to eliminate the methane results from the equation, sum C_2 through C_4 or C_6, and then evaluate the percentages (Fig. 12-30). This determination is

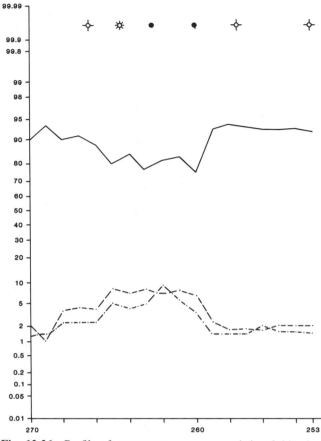

Fig. 12-26 Profile of percentages across an existing field using methane (solid line), hexane (.———.), and pentane (.—.). Note the decrease in methane and the increase in hexane and pentane. The location of the wells in the field is indicated at the top of the figure.

valuable in areas where methane values have a wide range and seem to be related to biological activity. In this case, ethane will typically decrease with respect to the other hydrocarbons.

Normalization

Data are normalized when two or more surveys need to be integrated and evaluated for the same study area. Two types of surveys are normalized: (1) surveys that overlap and (2) surveys that do not. A typical problem arises when geochemical data from two surveys are collected from the same area at different times and all values need to be treated as one data set. Because of variations in daily or hourly gas migration, soil-gas surveys usually have the problem of noncomparability. To overcome this, the mean of each survey is determined. One of the means is selected as the standard. The difference between the two means is the value that will be subtracted or added to each of the values of the other survey. These new values are then statistically evaluated and contoured as if the two surveys were collected

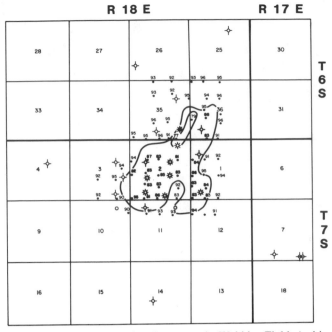

Fig. 12-27   Percentage of methane over the Wehking Field, Atchison County, Forest City Basin, Kansas.

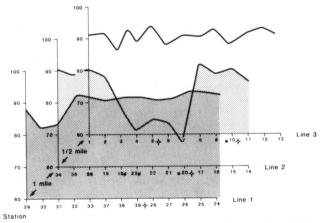

Fig. 12-28   Percentage of methane over the Shubert Field, Richardson County, Forest City Basin, Nebraska. Three survey lines are presented in profile form.

the mean as discussed earlier. The second survey values are adjusted in order to be evaluated with the first survey values.

Normalization is also used when soil or bedrock boundaries are crossed, which causes a change in the intensity of data for background and anomalous values. If the boundary of change is known, the data can be normalized with respect to one soil or bedrock area so that contouring and evaluation are easier.

Normalization is an artificial correction that intentionally or unknowingly ignores several variables that can occur between surveys. Changes that may occur are collection and analysis methods, soil and environment conditions, seasonal and diurnal periods, and leaking soil-gas volumes.

Multivariant Analysis

Multivariant analysis is usually applied when a large data set is available with a large number of parameters existing for each sample. This form of analysis is often used when data sets are acquired under different sampling conditions, when soil conditions have been ignored, or when different surveys must be compared. These methods are best described in Davis (1973) and Rose et al. (1983). The inherent problem with the use of these methods of data reduction is reflected in the old saying, "garbage in, garbage out." Multivariant analysis will not save a poorly collected or analyzed survey and will not necessarily make interpretation easier. What follows is a brief summary.

Multivariant analysis uses matrix algebra. A matrix is a rectangular array of numbers that looks exactly like a table of data. Here we consider the matrix to be a single unit rather than individual entries. Individual numbers in a matrix are identified by subscripts. The first subscript identifies the row and the second the column. The numbers may represent sums of observations, terms in a series of simultaneous equations, variances and covariances (the joint variation of two variables around a common mean), or any set of numbers. The matrix needs to be of equal numbers, or there will be left-over elements, and the operation cannot be completed.

The multivariant analysis uses groups or clusters of data. Many naturally occurring spatial distribution show a pronounced tendency toward clustering. Most clustered distributions are typically regarded as combinations of two or more simpler distributions. One of the distributions describes the pattern of individual points around the center of the cluster and the other the location of the center of the cluster. Multivariant analysis attempts to identify the location of clusters and the members that belong in a heterogeneous data set.

Cluster Analysis

Cluster analysis is used when a model or control survey is not available and thus a comparison cannot be made.

at the same time and under the same conditions. A variation for overlapping survey areas is to average the two values at each station that had duplicate sampling; this average value is then recorded for that station. Values from any collection points not duplicated between surveys are discarded. The main problem is that it is often difficult to decide if the data should be normalized.

Another form of normalization is used when two survey areas do not overlap. Several sample points from the first survey must also be collected during the second survey. The resulting data from these duplicate samples can be directly compared to the samples from the first survey. The difference in the data will again be represented as a difference in

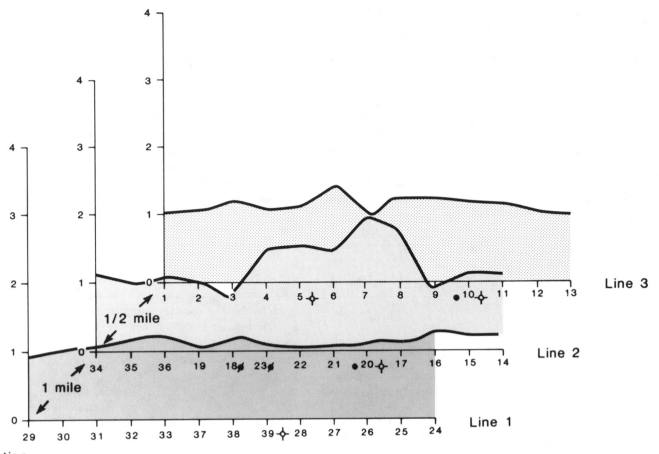

Fig. 12-29 Percentage of hexane with methane in the equation over the Shubert Field, Richardson County, Forest City Basin, Nebraska.

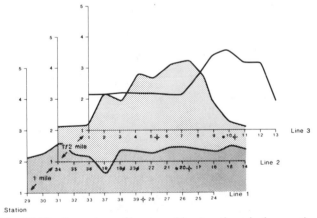

Fig. 12-30 Percentage of hexane without methane in the equation over the Shubert Field, Richardson County, Forest City Basin, Nebraska.

This form of analysis performs classification by assigning observations to groups so that each group is homogeneous and distinct from every other group. The various methods for this type of analysis can be classified into four types: (1) mutual similarity, (2) partitioning methods, (3) hierarchical clustering, and (4) arbitrary origin methods (Davis, 1973). With mutual similarity methods, observations having a com- mon similarity to other observations are grouped together. The data are computed on a matrix of $n \times n$, and those observations having a close intercorrelation will be close to $+1$. Those observations having no correlation will have a much lower value. Partitioning methods attempt to find areas in the population that are poorly represented and to separate the more densely populated areas. Hierarchical clustering is the most common form of cluster analysis. The method calculates the similarities between all the pairs and recalcu- lates the matrix by averaging the similarities that the com- bined observations have with other observations. The com- putations continue until a matrix of 2×2 is achieved and a dendogram can be constructed. The arbitrary origin method establishes a set of arbitrary starting points in which each observation is compared. In the three previous methods, the matrix was $n \times n$, where n represents real observations. This matrix uses $k \times n$, where k represents arbitrary points. A more detailed description of cluster analysis is presented by Davis (1973) and Rose et al. (1983).

Discriminant Analysis

Discriminant analysis attempts to classify a data set into a predetermined number of groups or clusters. Typically, a simple linear discriminant function transforms the original data set into a single discriminant score. The score represents the transformed observation position along a line defined by the linear discriminant function. The discriminant function therefore reduces the multivariant problem into a problem that represents only one variable. The problem is solved by having two clusters of data so that we find the transform function that gives a minimum ratio of variance between a pair of multivariant means. The resulting score can undergo the test of significance for the separation between the two groups. Applying this test requires that five assumptions be valid: (1) the probability of an unknown observation belonging to either group is equal, (2) the observations chosen in either group are made at random, (3) variables are normally distributed within each group, (4) the variances of the groups are equal in size, and (5) none of the observations were misclassified. The use of the discriminant function is closely related to multiple regression and trend surface analysis.

Factor Analysis

Factor analysis represents a series of computational procedures that attempt to reveal a simple underlying structure presumed to be present within a set of multivariate observations. This form of analysis is expressed in the form of a pattern that is present in the variance and covariance and in the related similarities between observations. Factor analysis is best applied when it is assumed that parameters are varying across a survey area; the analysis can determine which parameters are varying. This method is best used with a data set that has been derived from an exploration and model survey.

Pattern Recognition

As previously discussed, the term *anomaly* is applied to that part of the population that represents the set of data being sought. In the acquisition of small data sets based on statistics, there will usually be an anomalous population and probably more than one. However, the word *anomaly* loses its meaning when a large data set is used because the anomalous population is an integral part of the total population. Determination of threshold values becomes arbitrary, and boundaries between the populations become blurred. Therefore, statistical methods fail to recognize these anomalous populations clearly because these methods are attempting to evaluate a normal population. Consequently, this form of evaluation, called *pattern recognition,* plays a significant role in surface geochemical interpretation.

Pattern recognition is used in identifying many geologic variables, and it seeks to determine specific patterns of economic interest. This technique is applicable to all forms of

geologic, geochemical, and geophysical information. The data that produce the patterns are, in reality, not "anomalous" but are part of the general population. Geologists have used pattern recognition to identify porosity, channels, reefs, fractures, as well as other types of patterns or trends that are part of the normal geologic section. For example, a Pennsylvanian cyclothem is a recognized geologic pattern that is of economic importance, and its specific geologic parts, or subsets, are fundamentally part of the total normal population. Statistical methods applied to one of these geologic parts (such as coal) in the cyclothem will not pinpoint it as an "anomalous" population within the total cyclothem but rather as part of a "pattern" that can be recognized and exploited.

A second example is a simple isopach map (Fig. 12-31) of the thickness of Mississippian carbonates in a petroleum-producing area. There are 96 wells, of which 53 are productive and 43 are dry holes. The contours represent an isopach of the producing formation. Therefore, all wells presented reach the base of the Mississippian. It is clear that there is a major thinning associated with the producing area. This is a repetitive pattern of thinning related to potential petroleum production. But, statistically, these data are not anomalous with respect to the data, indicating a bias toward the producing thin areas (Fig. 12-32). The data show positive skewness, and there seem to be two populations present. The means and mode lie in the center of the data, and one standard deviation in either direction would not define all the population. We would have to plot the data visually and contour it to determine that producing areas are related to thinning because the simple use of statistics would not be sufficient to identify the populations.

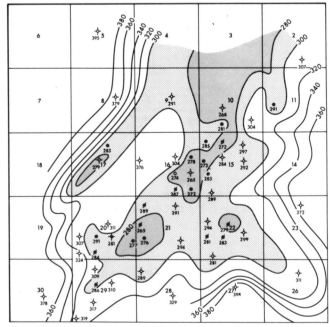

Fig. 12-31 Isopach map of Mississippian carbonates indicating a thinning across an oil field.

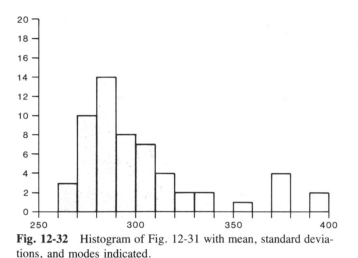

Fig. 12-32 Histogram of Fig. 12-31 with mean, standard deviations, and modes indicated.

Pattern recognition methods are specifically utilized in surface geochemistry for evaluation of the K-V fingerprint and fluorescence methods. Because the type of data obtained with these methods are qualitative and not quantitative, this form of analysis is the most viable alternative at this time. Some attempt has been made to determine relative quantities of the hydrocarbons from the data. However, these resulting numbers are relative and cannot be considered real. The problem with this will be clear in a moment. Both these methods, as discussed in Chapter 5, yield a series of peaks that are either individually visible or form a continuous profile representing the types of hydrocarbons present. Samples for comparison are typically dry holes and producers that yield hydrocarbons being sought. The pattern of the test or prospect survey samples is compared to these models. Typically, there is rarely a perfect fit. But, this is not important. What is important is that the pattern of the test samples fits one of the models with reasonable confidence. This method requires several models of dry holes under different conditions, dry holes with shows, and producers with different types of petroleum and combinations of gas and oil ratios. However, even when we have determined a group of samples with a similar anomalous pattern to the model, we have no way of determining volumes. To reiterate what has been stated earlier by some authors, in most cases volumes are important; therefore, we cannot verify with the fluorescence and K-V fingerprint methods whether an anomaly is minor or major leakage.

The application of pattern recognition in surface geochemistry requires placing the data on the map and determining the appropriate contour intervals, based on previous experience and knowledge of a particular area. In the past, when geochemical data and experience were limited, statistical analyses were the only way to evaluate the data "safely." Gallagher (1984), Duchscherer (1984), and Moriarty (1990), to name a few, have presented geochemical data sets that are so small that they cannot be interpreted in any other way. The recent GERT study by Calhoun (1991) is a

case in point. Several of the methods in this study had less than 20 samples per prospect using a variety of methods. However, in the same study, Landsat methods were used that had an infinite number of data points. Comparing the Landsat to the surface geochemical techniques creates an unfair bias in favor of the satellite technology. Although the effort was well intentioned, the results are of little value and, in fact, could lead people to believe that geochemistry can be effectively applied on this very limited scale.

Establishing a large data base creates the most important component of pattern recognition interpretation, enhancing the evaluation of surface geochemical data and diminishing reliance on statistical methods. Statistically determined "anomalies" may not actually be anomalous, whereas anomalous points may be incorrectly defined as background. Pattern recognition can be one of the few ways to recognize and avoid these problems. Establishing background levels, as documented by Levineson (1980), Koch and Link (1971), and Rose et al. (1983), is far too important to be left to the hope that an isolated sample area is representative.

An example of the failure of statistical methods to determine the presence of an anomaly is the Jace Oil Field, Cheyenne County, Colorado, Las Animas Arch area (presented in Chapter 5). A probability plot of the iodine, pentane, and hexane data implies the presence of a single population (Figs. 12-33–12-35). A summary of basic statistics reveals no large standard deviations. There is little difference between the mode and the mean, and the number of anomalous values present is small. A clear relationship between the anomalous area and production is indicated in Figs. 5-31 and 5-32. As a matter of fact, a comparison of the hydrocarbon data using the minimum contour interval of

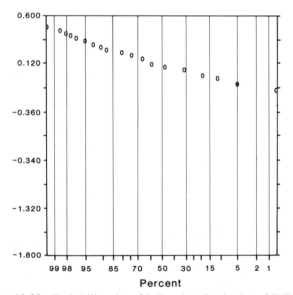

Fig. 12-33 Probability plot of the iodine data for the Jace Oil Field, Kiowa County, Las Animas Arch area, Colorado (printed with permission from Atoka Exploration). Note the relatively uniform distribution that masks the presence of an anomaly.

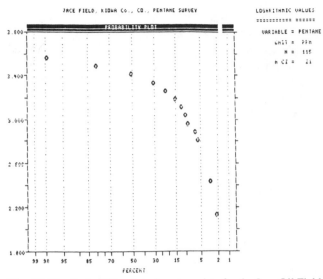

Fig. 12-34 Probability plot of pentane data for the Jace Oil Field, Kiowa County, Las Animas Arch area, Colorado.

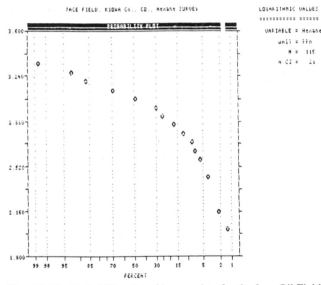

Fig. 12-35 Probability plot of hexane data for the Jace Oil Field, Kiowa County, Las Animas Arch area, Colorado.

parts per billion shows a high correlation, but dry holes indicate a 100% correlation with hydrocarbon values below parts per billion of hexane.

The initial conclusion of the original contractor who surveyed this area was that there was no anomaly. Two other contractors using radiometrics and head-space soil-gas techniques also failed to identify any anomalies prior to discovery. However, drilling results nullified that conclusion (production was established) and suggested that even though statistics are a good first screen, the data still need to be plotted and compared to existing models before dismissal.

Another example from southeastern Colorado is the project with the more than 3500 iodine data points previously presented in Fig. 12-9. The data suggest an asymmetric bell-

shaped curve where the mean is 1.8, the standard deviation is .4, and the mode is 2.0 ppm. Figure 12-36 is contoured on the basis of statistical analysis using three standard deviations above the mean. However, a visual review of the contoured data suggests that the actual background values are less than would be indicated by the mean. The actual threshold value is probably 2.0 ppm, based on the use of models and the observation of the geochemical signature of existing fields in the survey. Reinterpretation of the data (Fig. 12-37) suggests the presence of a series of sinuous anomalies. The Morrow sandstone (Pennsylvanian), which is a productive channel deposit in this area, typically displays geochemical anomalies such as these. If the iodine data are integrated with a structure map on top of the Mississippian (Fig. 12-38), the geochemical data indicate anomalies lying in areas of scour on this surface. Morrow channels are typically found in valleys eroded down into the Mississippian carbonates.

A third example brings together the useful features of pattern recognition and models. An iodine survey was carried out in an unexplored region of the Forest City Basin (Fig. 12-39). The Viola (Ordovician) and Hunton (Dev-

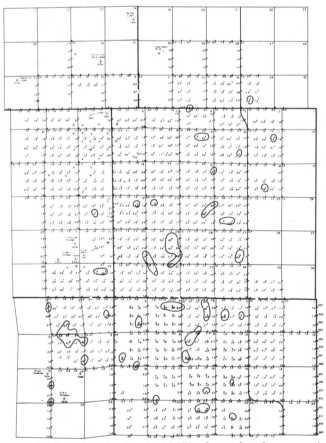

Fig. 12-36 Thirty-five hundred iodine data points for an area in Cheyenne and Kiowa counties, Las Animas Arch area, Colorado, contoured on the basis of statistical analysis. Statistical analysis indicates that all values greater than 3.0 ppm were anomalous (reprinted with permission from Atoka Exploration Corporation).

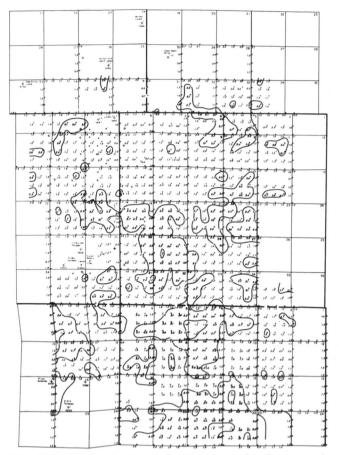

Fig. 12-37 Thirty-five hundred iodine data points for an area in Cheyenne and Kiowa counties, Las Animas Arch area, Colorado, contoured on the basis of pattern recognition analysis. This analysis suggests that the threshold value is 2.0 ppm, which is used to identify anomalous areas. The target is the Pennsylvanian Morrow channel sandstone. Several anomalous areas depict the sinuous apical anomalies associated with the Morrow (reprinted with permission from CST Oil & Gas Corporation).

onian-Silurian) formations were the potential targets. The mean is 1.6 ppm, the standard deviation is .2 ppm, and the mode is 2.1 ppm. The data were contoured on the basis of three standard deviations above background. Only a few anomalies existed that indicated potential for further work. The data were compared to several models from existing fields in the Forest City Basin, such as Figs. 7-24 and 12-40. Only the model from the Wehking Field suggested a similarity in anomaly intensity and size. However, the Wehking Field produces from the McClouth (Pennsylvanian) sandstone, which is not present in the survey area. The anomalies also had a linear nature, suggesting that they may be fault-related. Soil-gas, radiometric, and microbial surveys confirmed the iodine anomalies. The anomalies that did exist coincided with projected structural highs and were subsequently drilled. Shows of oil were encountered in the Viola section, but no production was established. For the Forest City Basin, the anomaly's intensity and areal extent constitute a recognizable pattern that is useful in determining

the viability of a geochemical anomaly for further exploration.

The intensity of an anomaly can be very useful. Figure 12-41 is a survey located in Clinton County, Illinois Basin, Illinois, in a locality where Silurian reefs are the main exploration target. If the survey area is divided into halves, a north half and a south half, and if each half is statistically analyzed separately, Figs. 12-42 and 12-43 result. The two halves of the survey suggest an anomaly that exceeds three standard deviations from the mean (Table 12-7). In the southern half of the survey, the anomaly is related to the Boulder East Field, discovered in 1941. The northern anomaly was found to coincide with a projected subsurface closure. The northern anomaly was subsequently drilled, and several wells had shows, but no commercial production was established. If we treat the two halves as a whole, the intensity of the prospect is less than that of the existing and somewhat depleted field. We conclude that intensity is a recognizable pattern and, if used with prospects and models, would have prevented drilling of the anomaly in the north half of the survey.

Intensity is not always a useful pattern. The Friday Field (Fig. 7-14) had an iodine anomaly that led to a successful discovery. This type of anomaly has been used as a basis for exploration for a long time. When it was compared to the anomaly associated with the Dolley Field (Fig. 7-16) prior to discovery, there was a hesitancy to recommend drilling. The intensity of the anomaly was relatively minor compared to background. However, subsequent drilling has proved the Dolley Field a much more productive field and with a greater areal extent than the Friday Field. Both fields produce from the D sandstone, but the pay thickness at Friday is 25 ft and at Dolley less than 6 ft. One can conclude that the intensity of an anomaly at the surface in this area may have a relationship to the thickness of the productive reservoirs or pay. However, it is more likely that a number of factors are involved. Explorationists in the Denver Basin have been influenced by these two models to use pattern recognition rather than relying exclusively on statistics as the absolute determinate of prospect viability.

False Anomalies

The objective of any surface geochemical survey is to recognize anomalous microseepage related to petroleum accumulations at depth. Therefore a major concern is recognizing "false anomalies" that stem from other sources. It is unfortunate that the majority of the literature discusses anomalies that represent success rather than identifying anomalies that represent failures or that mislead interpretation. Such data would be very valuable.

False anomalies can be categorized into those caused by:

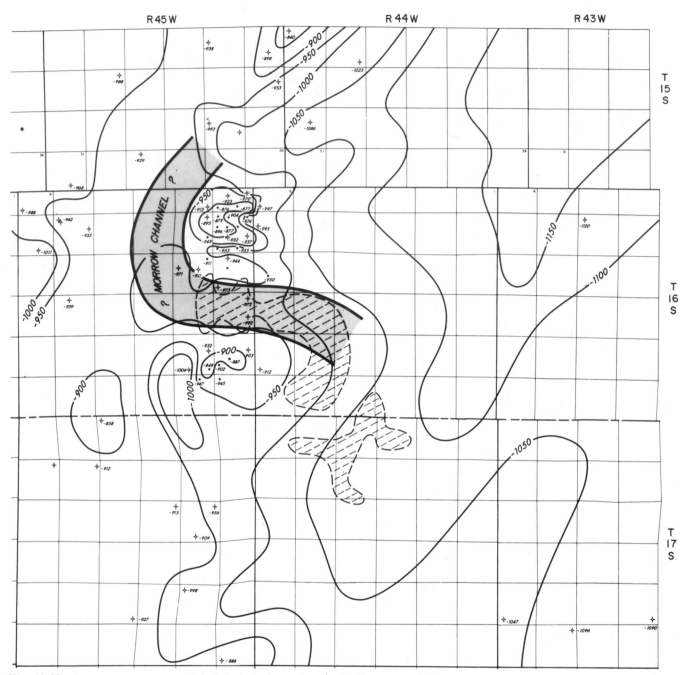

Fig. 12-38 Structure contour map of the Mississippian carbonates in Cheyenne and Kiowa counties, Las Animas Arch area, Colorado. Contour interval is 50 ft.

1. Collection
2. Analysis
3. Contaminants
4. Statistics
5. Other sources

The false anomalies caused during collection generally can be identified during analysis and interpretation or by repeating select sample points. The use of multiple methods or methods that acquire multiple parameters (soil-gas) usually can identify spurious data. The same is true for false anomalies resulting from poor analytical quality assurance.

False anomalies caused by contaminants, such as a petroleum spill at the surface or road material mixed in with a soil sample, may not always be clearly identified. These types of anomalies are usually one-point aberrations and are repeated or discarded. Certain methods, such as soil-gas, can identify a contaminant problem such as the presence of refined petroleum products. Usually, the anomalous values

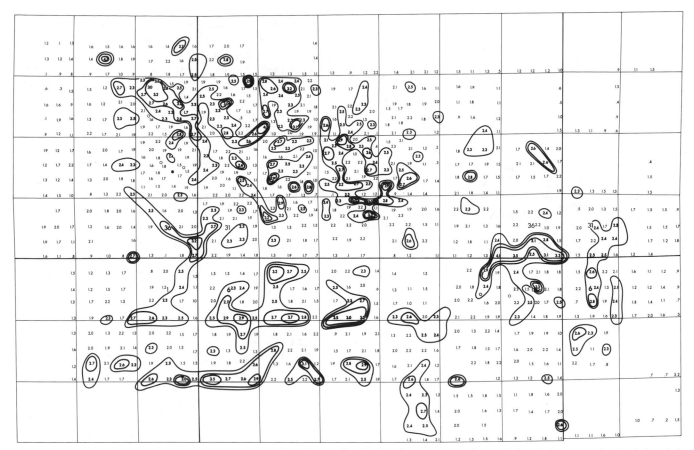

Fig. 12-39 An iodine survey is an unexplored area of the Forest City Basin. Contour intervals are 2.3 to 2.5 ppm (yellow), 2.6 to 2.8 ppm (orange), and >2.9 (red).

are for only one or two hydrocarbons and not for the entire group being analyzed. The amounts of the anomalous hydrocarbon related to a contaminant will probably be excessively high.

Statistical false anomalies are typically caused by noise in the data. Several statistical methods are available that provide robust means by which to identify these types of anomalies. The problem is one of multivariate statistics. To use this form of statistics, we must establish the criteria for deciding what type of anomaly we are attempting to identify. A variety of methods can be used as defined by Andrews et al. (1972), Conover et al. (1977), Bement et al. (1977), and Koch and Link (1971). All the methods described by these authors attempt to eliminate anomalous data that, in all probability, are due to chance and to identify the data that are interrelated.

The "other sources" group is probably the most difficult to recognize. The first four categories can usually be identified by general analysis of the data or by interpretation. However, false anomalies caused by other sources are a major problem in implementing a surface geochemical survey. Oehler and Sternberg (1984) discussed a false anomaly that was found by IP (electrical) methods in the study of the Ashland Field, Oklahoma, and a test case called the Salt

Draw prospect, Texas. A clear model was established with the Ashland Field in which the authors concluded that the electrical anomaly associated with the field was caused by the formation of pyrite, marcasite, and carbonate cements in the shallow surface by vertically migrating hydrocarbons. When a similar investigation was conducted at the Salt Draw prospect, an anomaly was indeed found that was similar in character. However, subsequent investigation and the use of soil-gas methods in conjunction with the IP indicated that the electrical anomaly was related to bedrock exposure of a particular strata. The increases in the pyrite were found to be related to mudstones and limestones, and ^{13}C analysis of the head-space soil gas indicated a different composition at the surface than in the target formation.

Two conclusions can be arrived at from the Ashland case study:

1. No matter what method is used, it must be confirmed by other methods before a decision is made to lease or drill. This is a good approach to eliminating false anomalies.

2. The use of models can be critical in avoiding misinterpretation of data.

The problem with other sources of false anomalies is that their causes—atmosphere, soil, and bedrock—can be

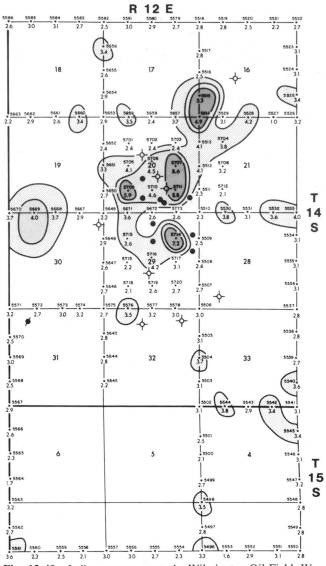

Fig. 12-40 Iodine survey across the Wilmington Oil Field, Wabaunsee County, Forest City Basin, Kansas.

identified only in the field. Soil profiling and the use of bedrock maps can eliminate some of the false anomalies by clearly indicating the potential source. In some cases, false anomalies have no apparent source or cause and may be due to microseepage generated by minor accumulations at depth. These types of anomalies may be impossible to identify and eliminate.

Types of Surface Patterns

The surface expression of a geochemical anomaly and the interpretation of its "value" are what the explorationist is attempting to determine. Integration of the geochemical data with other exploration methods and the final interpretation results in the decision to drill or not to drill. Many authors have cited the existence of halo anomalies. Horvitz

(1950), Price (1986), Duchscherer (1984), Kartsev et al., (1959), and Sokolov (1936) discussed two other types: apical and linear. A halo anomaly has the higher values at the edges of the petroleum deposit, and lower values similar to background or low anomalous values are found in the center of the anomaly (Figs. 5-22, 7-14, 8-18, and 10-4).

Soli (1957) believed that halo anomalies were caused by intense bacterial activity in the uppermost soil layers that destroyed most of the vertically migrating hydrocarbons over the center of the underlying petroleum accumulation. The amount of migrating hydrocarbons would decrease away from the petroleum accumulation and therefore could not support bacterial activity. Price (1986) reiterated this line of thinking with no new evidence or examples. It has been found, however, that halo and apical anomalies detected with the same methods can appear in the same geologic basin and, in some cases, are adjacent to each other. Restricting interpretation to halo-type anomalies has led to many dry holes. This limited procedure fails to address that the data are a statistical sampling of the surface seepage that is controlled by fractures, faults, and so forth. The sampling pattern may not necessarily give an accurate or actual reflection of the surface expression of the geochemical anomaly. Therefore, the data may indicate an anomaly, but its form may be an artifact of the sampling density or procedure.

Moderately to totally depleted fields seem generally to have halolike anomalies. This suggests that the halo may be the product of a lack of hydrocarbons over the productive part of the reservoir as a result of depletion. Therefore, it is possible that the original halo anomalies may have been the result of surveys over existing fields.

An apical anomaly (Fig. 5-29, 5-67, 7-17, 10-7, and 12-44) is one that lies directly over the accumulation and exhibits higher values in the center, with the values decreasing outward. However, some explorationists interpret halos as being apical. This comes from the presence of higher values forming a halo and lower anomalous values in the center. This distinction may be one of semantics, but it continues to cause confusion.

Linear anomalies (Fig. 12-45) are generally associated with faulting or fracturing. Recent interest in fracture plays for horizontal drilling suggests that, in some cases, these traps represent very narrow but highly productive reservoirs. Linear anomalies have essentially been ignored and are addressed only when they affect the halo or apical interpretation for a petroleum accumulation.

Whether the resulting anomaly is a halo, apical, linear, or some combination thereof may not be important because all these types of anomalies are likely to be present in any geologic basin. Price (1986) described the halo anomaly as the result of bacterial activity at its greatest intensity over the center of the petroleum deposit and then decreasing outward. Thus, in the halo area, bacterial activity is at a minimum. If bacterial activity is consistent with the amount of food supply, then there should be no soil-gas, iodine,

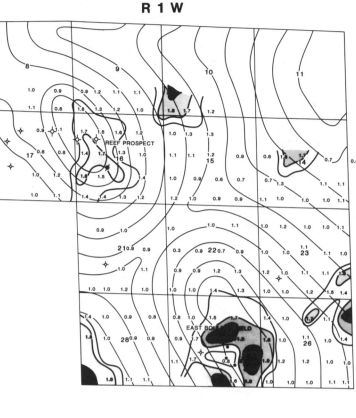

Fig. 12-41 Iodine survey and structure map across a suspected reef and the Boulder East Field, Clinton County, Illinois Basin, Illinois. Contour intervals are 1.5 to 1.9 (light gray), 2.0 to 2.4 (medium gray), and >2.5 (dark gray) (reprinted with permission from CST Oil & Gas Corporation).

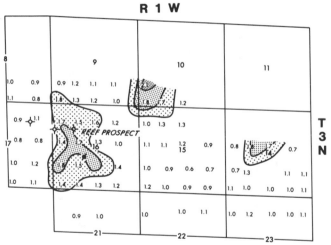

Fig. 12-42 North half of the iodine survey across a suspected reef and the Boulder East Field, Clinton County, Illinois Basin, Illinois. Contour intervals are 1.4 to 1.6 ppm (light gray), 1.7 to 1.9 ppm (medium gray), and >2.0 ppm (dark gray) (reprinted with permission from CST Oil & Gas Corporation).

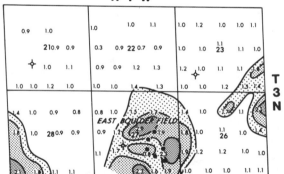

Fig. 12-43 South half of the iodine survey across a suspected reef and the Boulder East Field, Clinton County, Illinois Basin, Illinois. Contour intervals are 1.4 to 1.6 (light gray), 1.7 to 1.9 ppm (medium gray), and >2.0 ppm (dark gray) (reprinted with permission from CST Oil & Gas Corporation).

Table 12-7 Means, standard deviations, and modes for the Boulder East area, Clinton County, Illinois Basin. All values in ppm. 22.

	North Half	South Half	Total Survey
Mean	1.1	1.2	1.2
Standard deviation	0.3	0.3	0.3
Median	1.1	1.1	1.0
No. of samples	78	82	160

radiometric, or metals anomalies at all. Many reported microbial anomalies are apical. Further, surface geochemical anomalies do vary over time, suggesting that the supply of hydrocarbons is fluctuating for the bacteria. Changes in hydrocarbon flux should be reflected by changes in bacteria population but with some delay. However, there has been no indication in the literature of bacterial flux.

Further interpretative studies of anomalies suggest other

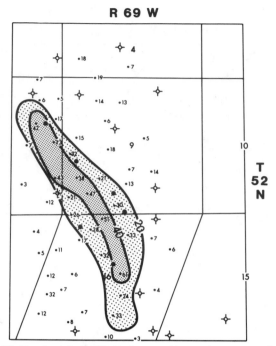

Fig. 12-44 An apical soil-gas (propane) anomaly over an existing field located in Township 52 North, Range 69 West, Powder River Basin, Wyoming. Values are in parts per billion. Contour intervals are 20 and 30 ppb (reprinted with permission from Trinity Oil and Illuminating).

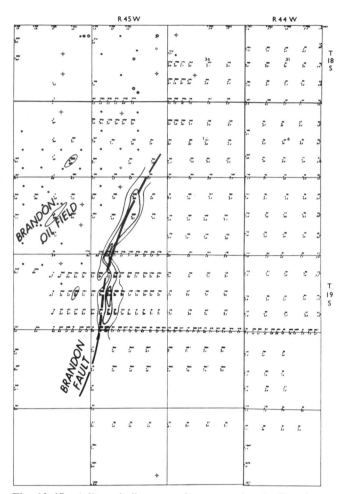

Fig. 12-45 A linear iodine anomaly representing the Brandon or Freeze Out Fault that bounds the east side of the Brandon Field, Kiowa County, Las Animas Arch, Colorado. The iodine is in parts per million, and part of the data is not contoured in order to enhance presentation.

problems to consider. Values in the center of a halo should not be anomalous. There are apical anomalies with low anomalous values interspersed with higher ones. Factors such as migration pathways (previously discussed) and sampling density and grid patterns (discussed in Chapter 13) artificially create the presence of a halo or apical anomaly.

The various anomalies can be shown to be a function of sampling density. An anomaly may seem to be linear or apical or a halo based on widely spaced samples. But, as the density of the grid increases, it becomes questionable which type of anomaly is present. In another case, a halo anomaly may be observed, but more detailed work results in the appearance of two apical anomalies. Therefore, these specific terms almost have to be abandoned in favor of the general term *anomaly*. There is no doubt that a classic example of each type of anomaly exists, but it is more likely that innumerable permutations and combinations of all three exist as well. The literature has generally pigeonholed the interpretations, forgetting that each geochemical anomaly is a product of the interactions of similar geologic and geochemical processes. But local modification by site-specific and unique characteristics will always be tantamount to rigid man-made interpretation and sampling procedures.

Instead of categorizing an anomaly as one form or another, it is probably better to define the expression as an anomalous area that is a surface geochemical manifestation of petroleum at depth. To type the anomaly exactly would require the delineation of the petroleum accumulation by methods other

than surface geochemical techniques. This is called *integrated exploration*.

Data Presentation

Surface geochemical data presentation has several forms. The most common form for the petroleum geologist is the contour map. This type of presentation map is derived from the petroleum geologist's reliance on subsurface geologic maps that are almost always contoured to depict the thinning/thickening or topography of a particular formation, bed, group, or lithologic member. Consequently, surface geochemical data are also contoured in the petroleum industry. However, this may be an inaccurate reflection of the type of data we are dealing with. Unlike subsurface data, which usually do not vary if resampled, geochemical data can change within short distances and through time at the same sample site. Even though the exact geochemical amounts are not repeatable, the assumption is that the areas of background and anomalies remain relatively the same over an indefinite period of time. This contour map presentation makes the intensity of the data one of the critical factors in interpretation. Contour maps can be presented in a two-dimensional form (flat) or a three-dimensional form depending on the data.

Exploration geochemistry utilizes a second form of data presentation that is a product of the collection technique. It has been recognized that: (1) in the case of mineral deposits, various elements (which can also be termed *ore*) vary dramatically across short distances, and (2) in many cases, geochemical units are not mappable as with subsurface geologic data. Thus, the method was developed of increasingly enlarging a symbol with increasing volume of a specific geochemical element or compound. Figure 12-46 is an example from an ore deposit of this type of data presentation. Figure 12-47 illustrates a similar technique for radon results from the Moreland Hills of Ohio (Banks and Ghahremani, 1987). The values are displayed with a symbol based on less than one standard deviation, between one and three standard deviations, and more than three standard deviations above the background mean.

Another variation for presenting this data is outlining only

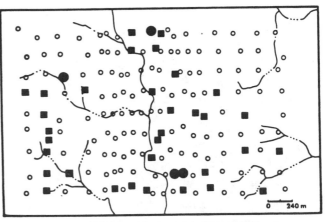

○ Radon activity within one std. dev. of background mean
■ Radon activity between one and three std. dev. above background mean
● Radon activity more than three std. dev. above background mean

Fig. 12-47 Radon survey across the Moreland Hills of Ohio (Banks and Ghahremani, 1987; reprinted with permission of the Association of Petroleum Geochemical Explorationists).

the anomalous areas (Fig. 12-36). This assumes that the intensity of the data is not a critical factor in affecting interpretation.

Profiles are another favorite form of data presentation, especially when dealing with only a single line or widely spaced lines of data. In some cases, several lines are presented in a three-dimensional picture, giving the visual impact of a contour map (Fig. 12-48).

Computers can create various forms of presentation that would be difficult or technically impossible to do by hand. One form is to eliminate all the data on a three-dimensional contour map that are not anomalous (Fig. 12-49). Sometimes the data are manipulated statistically to enhance anomalies that are not distinctive with respect to background by taking the log of each value.

Data presentation is critical not only to the interpreter but also to managers and other exploration personnel who review or critique the results. Consequently, the data should be presented in a form that is clear and acceptable to all. Regardless of the presentation form, what it implies cannot vary.

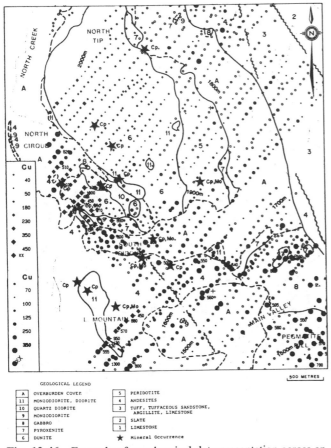

GEOLOGICAL LEGEND

A	OVERBURDEN COVER	5	PERIDOTITE
11	MONZODIORITE, DIORITE	4	ANDESITES
10	QUARTZ DIORITE	3	TUFF, TUFFACEOUS SANDSTONE, ARGILLITE, LIMESTONE
9	MONZODIORITE	2	SLATE
8	GABBRO	1	LIMESTONE
7	PYROXENITE	★	Mineral Occurrence
6	DUNITE		

Fig. 12-46 Example of geochemical data presentation across an ore deposit (Fletcher et al., 1986).

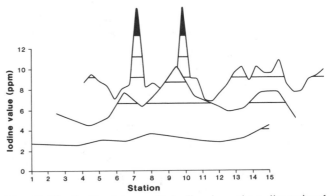

Fig. 12-48 Profile data, multiple line in a three-dimensional display.

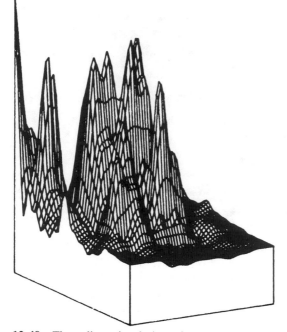

Fig. 12-49 Three-dimensional view of computer-enhanced data using the anomalous areas only.

Thus, if there are no anomalies, manipulating the presentation will not change that condition. Correspondingly, the presence of anomalies must be recognized, and the form of presentation must be acceptable.

Summary

Statistical and pattern recognition methods are an essential part of the evaluation of any surface geochemical data in order to determine the presence or absence of anomalous areas related to petroleum microseepage. The use of all these methods is dependent on the quality of data collection and analysis, the density of sampling, and the number of samples in a data set. Other variables that are addressed more subjectively are the effects of soil, weather, and topography, which usually can be identified and those samples normalized or discarded. The overuse of statistics to support the presence of an anomaly when biased data lead to this type of conclusion seems to be a continuing problem. Experience and the use of pattern recognition methods help to minimize this problem.

References

Adams, C. W. (1990). Jace and Moore-Johnson Fields, in *Morrow Sandstones of Southeast Colorado and Adjacent Areas*, Rocky Mountain Association of Geology, Denver, CO, pp. 157–164.

Allexan, S., J. Fausnaugh, C. Goudge, and S. Tedesco (1986). The use of iodine in geochemical exploration for hydrocarbons, *Association of Petroleum Geochemical Explorationists Bulletin*, Vol. 2, No. 1, pp. 71–93.

Andrews, D. F., P. J. Bickel, F. R. Hampel, P. J. Huber, W. H. Rogers, and J. W. Tukey (1972). *Robust Estimates of Location*, Princeton University Press, Princeton, NJ.

Banks, P. O. and D. T. Ghahremani (1987). Fracture control of radon leakage, Moreland Hills, Ohio: Implications for shale gas exploration, *Association of Petroleum Geochemical Explorationists Bulletin*, Vol. 3, pp. 40–55.

Bement, T. R., D. V. Susco, D. E. Whitman, and R. K. Zeigler (1977). Geostatistics project of the National Uranium Resource Evaluation Program, Los Alamos Scientific Laboratory Progress Report LA-6804-PR.

Bernand, B. B., J. M. Brooks, and W. M. Sachett (1977). A geochemical model for characterization of hydrocarbon gas sources in marine sediments, Offshore Technology Conference (May), OTC 293, pp. 435–438.

Calhoun, G. G. (1991). How 12 geochemical methods fared in GERT Project in Permian Basin, *Oil and Gas Journal*, May 13, pp. 62–68.

Conover, W. H., T. R. Bement, and R. L. Iman (1977). On a method for detecting clusters of possible uranium deposits, *Proceedings of the Department of Energy Statistics Symposium*, Oct. 26–28, Richland, WA, pp. 33–37.

Davis, J. C. (1973). *Statistics and Data Analysis in Geology*, John Wiley & Sons, New York.

Duchscherer, W. Jr. (1984). *Geochemical Hydrocarbon Prospecting*, Pennwell Books, Tulsa, OK.

Fausnaugh, J. (1991). Personal communication, Denver, CO.

Fletcher, W. K., S. J. Hoffman, M. B. Mehrtens, A. J. Sinclair, and I. Thomson (1986). *Exploration Geochemistry: Design and Interpretation of Soil Surveys*, Society of Economic Geologists, Vol. 3, p. 180.

Gallagher, A. V. (1984). Iodine: A pathfinder for petroleum deposits, in *Unconventional Methods in Exploration III*, Southern Methodist University Press, Dallas, TX, pp. 48–159.

Horvitz, L. (1950). Chemical methods, in *Exploration Geophysics*, 2nd Ed., Trija Publishing Co., Newport Beach, CA, pp. 938–965.

Horvitz, L. (1954). Near-surface hydrocarbons and petroleum accumulation at depth, *Mining Engineering*, Vol. 6, pp. 1205–1209.

Jones, V. and R. J. Drozd (1983). Predictions of oil and gas potential by near-surface geochemistry, *American Association of Petroleum Geologists Bulletin*, Vol. 67, pp. 932–952.

Kartsev, A. A., Z. A. Tabasaranskii, M. I. Subota, and G. A. Mogilevskii (1959). *Geochemical Methods of Prospecting and Exploration for Petroleum and Natural Gas*, University of California Press, Berkeley, CA.

Klusman, R. W. and J. A. Jaacks (1987). Environmental influences upon mercury, radon and helium concentrations in soil gases at a site near Denver, Colorado, *Journal of Geochemical Exploration*, Vol. 27, pp. 259–280.

Koch, G. S. and R. F. Link (1971). *Statistical Analysis of Geological Data*, John Wiley & Sons, New York.

Kuhn, T. S. (1970). *The Structure of Scientific Revolutions*, 2nd Ed., University of Chicago Press, Chicago.

Levineson, A. A. (1980). *Introduction to Exploration Geochemistry,* Applied Publishing, Wilmette, IL.

Lightman, R. A. and O. Gingerich (1992). When do anomalies begin? *Science,* Vol. 255, pp. 690–695.

Moriarty, B. J. (1990). Stockholm Northwest Extension: Effective integration of geochemical of geochemical, geological, and seismic data. In Morrow sandstones of southeast Colorado and adjacent areas, Rocky Mountain Association of Geology, Denver, CO, pp. 143–152.

Oehler, D. Z. and B. K. Sternberg (1984). Seepage-induced anomalies, "false" anomalies, and implications for electrical prospecting, *American Association of Petroleum Geologists Bulletin,* Vol. 68, No. 9, p. 1121.

Price, L. (1986). A critical overview and proposed working model of surface geochemical exploration, in *Unconventional Methods in Exploration, IV* Southern Methodist University Press, pp. 245–304.

Rose, A. W., H. E. Hawkes, and J. S. Webb (1983). *Geochemistry in Mineral Exploration,* Academic Press, New York.

Saunders, D. F., K. R. Burson, and C. K. Thompson (1991). Observed relation of soil magnetic susceptibility and soil gas hydrocarbon analysis to subsurface hydrocarbon accumulations, *American Association of Petroleum Geologists Bulletin,* Vol. 75, No. 3 (March), pp. 389–408.

Sokolov, V. A. (1936). *Gas Surveying,* Gostoptekhizdat, Moscow.

Soli, G. G. (1957). Microorganisms and geochemical methods of oil prospecting, *American Association of Petroleum Geologists Bulletin,* Vol. 41, pp. 135–145.

Weart, R. C. and G. Heimberg (1981). Exploration radiometrics: Postsurvey drilling results, in *Unconventional Methods in Exploration for Petroleum and Natural Gas II,* Southern Methodist University Press, Dallas, TX, pp. 116–123.

Wheeler, D. M., A. J. Scott, V. J. Coringrato, and P. E. Devine (1990). Stratigraphy and depositional history of the Morrow Formation, southeast Colorado and southwestern Kansas, in *Morrow Sandstones of Southeast Colorado and Adjacent Areas,* Rocky Mountain Association of Geology, Denver, CO, pp. 9–36.

13

Grids, Surveys, Models, and Economics

Introduction

Survey design is one of the most important aspects of an effective surface geochemical program. The basic objective is to optimize target identification by interpreting changes in diagnostic elements and compounds. There are several factors that enter into survey planning that will influence exploration decisions and affect the outcome. A series of questions outlining objectives that the survey needs to address should be prepared prior to data acquisition. These questions will serve as the framework for designing and conducting the survey effectively. When objectives and requirements are decided before data collection is begun, the survey has an increased chance of success. Success is defined as determining whether or not a geochemical anomaly exists.

Surveys can be divided into three types: orientation, reconnaissance, and detail. Surveys sample soil, rock, water, or the atmosphere. The majority of onshore surveys deal with soils and, to a lesser degree, with airborne collection methods. Offshore methods sample either the water or the bottom sediment. Therefore, surface geochemical survey design is applicable to, and similar for, all sampling media.

Experimental Design

An experimental design determines the procedure for systematically gathering data. Experimental designs are typically thought to apply to laboratory experiments, but developing a geochemical survey framework is also an experimental design. The purpose of the design is to evaluate different kinds of variability. In order to use this method, there must be a clear understanding of what is to be achieved. The variables, or factors and ranges, must be identified. The size of the project to be sampled determines the number of factors and their ranges. Then an experimental design for the number and pattern of observations may be determined sequentially. The pattern is defined by the distance between samples and is termed *sample spacing*. The pattern may be random or one of many systematic types. Typical variables

exhibit a spatial pattern, but time must also be added with surface geochemistry. Geochemical conditions change with time, sometimes in seconds and sometimes in decades, but usually in days or months. The statistical analysis of these patterns is restricted to the limits and parameters of the survey. A usable survey is achieved only through carefully specified objectives and predetermination of the methods by which those aims will be achieved. The design cannot be based on arbitrary judgments. The survey results in observations that can be statistically evaluated and that typically have a portion that is random in nature.

The survey is defined by the required pattern of observations. In mathematics and laboratory experiments, determining the best way to collect data is based on the functional form of the mathematical model and its relationship to the coordinate system. Likewise, in surface geochemical surveys, the collection of the data depends on the minimum areal extent and areal size of the likely target. When the surface area of the target is known, such as in an existing field or model, it is easy to fit the survey to the target site. If the location and size of the target are unknown, the survey design becomes more uncertain. After data have been collected in the defined area, it usually becomes clear whether the pattern of collection was sufficient to define the target adequately or to satisfy the objectives. Instead of random sampling, a sequential pattern may have to be developed. There are many forms of sequential sampling; a few are presented in Fig. 13-1.

Sampling is usually conducted in a systematic pattern because it provides the most thorough coverage across an area of interest. The problem with systematic patterns is that they ignore the periodicity of some types of data. For example, if electricity consumption in Denver, Colorado, is measured only at every twelfth hour, the conclusion about the rate of consumption will be unrealistically low. The sampling or pattern of observations will have missed the times when larger volumes of electricity are used, and this would be an experimental error. Changes in weather during a single day can cause fluctuations in gas being vented from the soil as well as changes in the gamma radiation emitted.

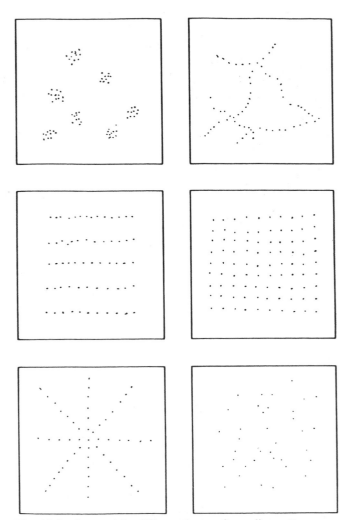

Fig. 13-1 Some of the different forms of sampling patterns.

By resampling several points during the day, we can adjust differences in the data for periodicity. Methods may be influenced by periodicity on a daily, monthly, or yearly basis. If a systematic pattern is adopted (defined partly on existing geologic and geophysical leads, information or models), a degree of sampling randomness and repetition must be introduced during collection, sample preparation, and chemical analysis. This allows the survey to determine the range of statistical error. Usually, this is accomplished by resampling points at different times, by selecting sample points that are inconsistent with the pattern, or by obtaining a specific number of randomly selected samples for repeat laboratory analysis. All geochemical and microbial methods must have randomly selected sample sites that are resampled to establish checks and balances for the survey design. Another check is the use of duplicate and spike samples in the analysis phase of any program.

Cluster sampling, unlike random or systematic sampling, selects a specific number of sites and then collects several samples at each site. Three or more samples are usually collected close to one another but are analyzed separately

in order to check the overall variability of a technique such as soil-gas or iodine. The need to take multiple samples is reduced if there is close agreement among results. If there is wide variation in the results, adding more sample sites is appropriate.

The following particulars should be considered when choosing a pattern of observation that will allow the survey to succeed: (1) the precision and accuracy of the design; (2) the types and distributional form of statistics to be used and the number of observations needed to achieve significance; (3) the variability associated with the surface geochemical method to be used, which will impact on the number and pattern of observations (samples) to be collected; (4) the number of duplicate samples to be collected, prepared, and analyzed that are necessary to establish confidence in the results of the experimental design survey; (5) in the absence of statistics, the number of samples needed to utilize pattern recognition.

Search Area, Targets, and Grid Interrelationships

The search area of the grid is interrelated with the target size and shape. The development of search theory resulted from the need to find effective means of systematically searching an unspecified area of the ocean in the hunt for U-boats during World War II. Fortunately for the explorationist, the petroleum reservoir is not moving at 10 knots in an unknown direction. We can mathematically determine the effectiveness of the pattern and the density of the grid points to achieve success in finding the target. This discussion will be limited to search on a square grid. Other grids are available, but the mathematical models to determine their viability, potential effectiveness, and interpretation become so complex that their usefulness decreases.

We start with a set of necessary assumptions: (1) There are one or more targets in the search area; (2) the targets have size and shape; (3) the relationship between the targets is random; and (4) the targets cover a certain percentage of the total search area.

The first case deals with a single target. If the target is circular, it will always be found if the grid spacing is smaller than $r \times \sqrt{2}$, where r is the radius of the circle. Figure 13-2 diagrams the probability of detecting a circular target of a specific size based on a specific spacing density. The smaller the target diameter, the smaller the grid spacing must be in order to have a high probability of detecting the target.

If the target is elliptical, Fig. 13-3 indicates the probability of detection based on grid spacing where the major axis is two to four times the minor axis. This is summarized in the equation by Drew (1967) and in tables developed by Savinskii (1965) and Singer and Wickman (1969). Other targets of different shapes and sizes can be mathematically constructed

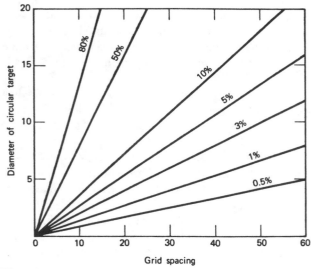

Fig. 13-2 The probability of detection of a specific target size vs. a specific sample density for a circular target (Koch and Link, 1971; reprinted with permission of Dover Publications).

and graphed. Detecting a target becomes difficult when it is sinuous or linear; the width and length are possibly unpredictable. Instead of using mathematical constructs, it is more efficient to use surveys of existing fields or models. Figure 13-4 (see color plate) illustrates an example of a grid design that will miss the target. The target is 160 acres or 400 m × 400 m (1320 × 1320 ft) in areal extent. A grid spacing of 1.6 km (5280 ft) will probably not detect the target the majority of the time, but it may detect it by chance a small percentage of the time. If the spacing is reduced to 800 m (2640 ft) between sample points, the success of detection increases (Fig. 13-5). When the spacing is reduced to 400 m (1320 ft intervals), the risk of missing the target becomes small (Fig. 13-6).

When multiple targets are present, the total target area must be considered, as well as the overlap area (if any). For example, if four targets overlap, one sample point that is anomalous could find them all but would not define the shape, extent, size, or any specific characteristics. Therefore, the construction of a grid to define an area of multiple targets has to be carefully planned. Figure 13-7 displays a grid spacing of 1600 m (5280 ft) with three targets in the search area. Although the location of the targets must be shown for discussion, we will assume their location is unknown. The surface geochemical anomaly encompasses only part of the total area of the three targets. Based on the present data, there is no indication that two or more separate anomalies exist. Figure 13-8 has a grid density of 800 m (2640 ft). All three anomalies have technically been detected but not defined. Figure 13-9 is a permutation of this grid, but one of the targets is still not detected. In Figure 13-10 the grid spacing has been reduced to 400 m (1320 ft), and all five anomalies are present. If only one of the targets was being sought, it would be difficult to separate that anomaly

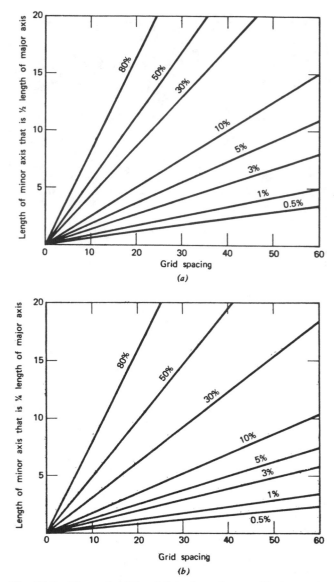

Fig. 13-3 The probability of detection of a specific target size vs. specific sample density for an elliptical target (Koch and Link, 1971; reprinted with permission of Dover Publications).

from the other four. In some cases, possibly to reduce costs, a permutation of the square grid is used (Fig. 13-11; see color plate). This survey can lead to the wrong conclusions about the relationship between the targets and the geochemical anomalies.

The grid spacing or density that is established can lead to misinterpretation and drilling failures. An example is presented in Figs. 13-12 and 13-13. The survey was designed to seek targets based on an assumed average-size field, three sections in size (1920 acres), the long axis two to five times greater than the minor axis, and probably sinuous. The survey indicated that there were at least two potential targets of that size, based on the grid spacing. Therefore, it was assumed that drilling in the center of either of these anomalies would be successful. However, subsequent in-fill sam-

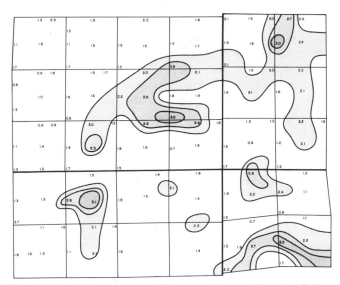

Fig. 13-12 Iodine survey on 800-m spacing, southeastern Colorado. The contour intervals are 2.0 to 2.4 ppm (yellow), 2.5 to 2.9 ppm (orange), and >3.0 ppm (red). The original target size was assumed to be 1 km wide by 7 km long. Possible anomalous areas exist.

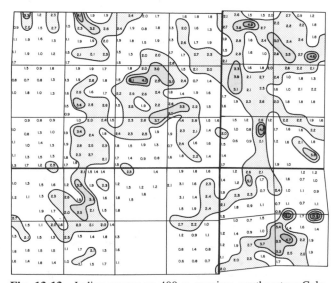

Fig. 13-13 Iodine survey on 400-m spacing, southeastern Colorado. The contour intervals are 2.0 to 2.4 ppm (yellow), 2.5 to 2.9 ppm (orange), and >3.0 ppm (red). The original target size of 1 km by 7 km clearly does not exist. Because there are indications of several small anomalies, either the areas will be abandoned or the geologic concept reevaluated.

pling indicated there were five smaller potential targets of 640 acres or less (Fig. 13-14). Are there two larger anomalies that are merely broken up, or are there five or more smaller targets? This figure emphasizes the importance of clearly defining the field size and designing the grid prior to sample collection. The next section discusses the importance of models that help define the sample density for an effective survey.

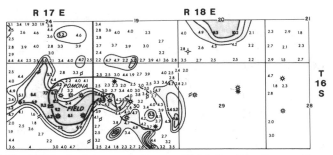

Fig. 13-14 Iodine survey, Pomona Field, Franklin County, Kansas, in which the survey geochemical anomaly is much larger than the field. The contour interval is 4.0 to 4.9 ppm (yellow), 5.0 to 5.9 ppm (orange), and >6.0 (red). The anomaly located in Section 20 defines the southern edge of the Pomona Central Field (reprinted with permission from CST Oil & Gas Corporation).

Model Surveys

A model assists in the design of a survey area or experimental design by using data acquired from the area of a known petroleum accumulation. Preferably, the field is one that is economically productive and recently discovered. Its data will be used as a guide in evaluating other data obtained from similarly prospective areas. The establishment of models is important, especially when a surface geochemical method is used for the first time or is utilized in a new and unfamiliar area. There are numerous examples in the literature of surface geochemical anomalies associated with existing fields, and occasionally there are published anomalies that outlined a field prior to discovery. In many cases, these published surveys present no actual data and thus are not very useful. Data that are presented consist generally of less than 100 sample points or a small number of profiles across a field. These usually do not give an accurate reflection of the surface geochemical anomaly's characteristics and relationship to the accumulation at depth. Therefore, a model must be established (more than one is preferred) prior to implementing a survey across areas that are undrilled. It must have a sufficient, if not excess, number of samples to use for comparison purposes.

There has been a reluctance in the petroleum industry to survey existing fields to clearly characterize the responses and performance of a specific surface geochemical method or a specific series of them. The assumption is that it is a waste of resources because the techniques will see the existing field. However, much survey data can be gathered from an existing field. How many producing wells fall inside vs. outside the geochemical anomaly? How many dry holes fall inside or outside the geochemical anomaly? Are there soil conditions that inhibit or optimize the use of a method or methods being used? A survey method that cannot detect an existing field should be replaced by another technique. The soil conditions may inhibit its use, or the model may be too near the end of its productive life to be useful. A significant surface geochemical response may no longer exist because of pressure depletion.

Models create a frame of reference. Many contractors will present an idealized model that shows their technique under the best situations and conditions. Explorationists can use such models as a starting point but will need to develop their own models that represent more complex geologic, sampling, soil, and weather conditions. For example, radiometrics are notorious for not detecting an existing field when traverses occur parallel to strike, when the ground is wet, or when a weather front moves through. Old or depleted fields that indicate excellent anomalies should be resurveyed. The explorationist needs to develop models with a high confidence level. This means that acquisition, collection, and interpretation conditions and parameters must be clearly defined.

Several criteria should be followed so that single and multiple models are useful for comparative purposes:

1. The model should be created from the most recent discovery possible. The older the field, the more likely it is that pressure depletion has occurred and that the surface geochemical anomaly is not as significant in intensity (especially for a single model) as it was before discovery.

2. The models should be between the minimum and the average size of the targeted accumulation.

3. Sufficient background data should be collected to indicate clearly the range of thresholds between anomalous and background values. These criteria must be expanded for multiple models.

4. Soil conditions that restrict the use of a particular method should be determined.

5. Models should range from simple to geologically complex in order to determine the impact on surface geochemical anomalies (difficult to use with one model).

6. Models should be surveyed at different times of year under different weather conditions.

7. Fields discovered by surface geochemical methods should be resurveyed after a period of time to gain additional data for postdiscovery models.

The single model generally limits the ability to compare different surface geochemical anomalies in the same prospective areas. Multiple surface geochemical models allow for more effective field evaluation by offering broader comparisons and providing a wider range of geologic and soil scenarios. Some large reservoirs may have subdued surface geochemical anomalies that look similar to those of smaller fields. A large and intense surface geochemical anomaly may not necessarily mean a large or even a productive reservoir. Some fields in a particular area may be faulted or have different geometries than other fields. Figure 13-14 is an example of a large geochemical anomaly associated

with a small productive field, and this is typical for petroleum traps in this basin. In Fig. 7-20, the anomaly defined only the center and most productive part of the reservoir and not the wells with less than 2 ft of effective pay. This illustrates that a potential target may be more clearly identified by using several models rather than only one. The types, intensity, and shapes of surface expressions are discussed in Chapter 12.

Orientation Surveys

Orientation surveys are generally not used in the petroleum industry because they are not designed for target identification or delineation of a prospect. These surveys characterize the prospect area by determining general soil characteristics, defining collection problems that might be encountered, observing man-made or natural phenomena that might affect a technique, and investigating any relevant surface geologic features. These surveys can identify soil conditions that may mislead the exploration program or that determine that a specific technique is unusable for a particular area. The approach used in the past has been to move right into the reconnaissance or detailed stage. Consequently, it may never be known if a misinterpretation occurred because a particular soil horizon or soil type either suppressed or enhanced a geochemical parameter.

Reconnaissance Surveys

Reconnaissance surveys vary in size and extent. The purpose of a reconnaissance survey is not to pinpoint a target but to determine the presence of possible targets through minimal sampling. The target size is important. It will dictate whether the reconnaissance spacing design can detect by chance possible targets for further delineation. A reconnaissance survey is designed to be quick and cost-effective. In many cases, this means using the existing road grid that is on one-mile section lines in many parts of the United States. (A section in the United States is usually 1.6 km (5280 ft) by 1.6 km (5280 ft) on a square grid.) With this road grid, the approximate location of the samples can be easily and quickly marked on maps using natural or man-made landmarks for reference. This procedure is usually sufficient to indicate the presence of fields greater than 100 acres (Fig. 13-15). The probability of detecting smaller fields will be reduced.

In areas with minimal or no road grid, the reconnaissance survey may lose its cost and time advantages for locating sample points. Therefore, a detailed survey may be necessary and may be more economical from the beginning. If the samples can be collected and analyzed on site, then one advantage of the reconnaissance survey will remain.

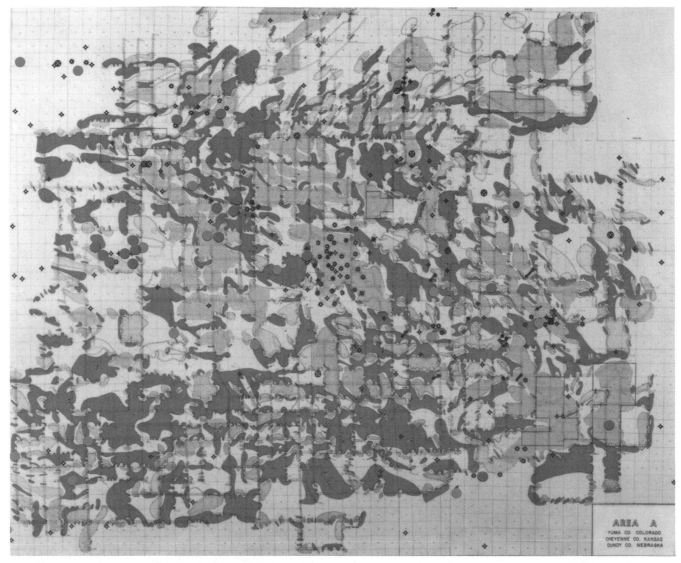

Fig. 13-15 An example of a reconnaissance radiometrics survey along roads in the Beecher Island Area, Las Animas Arch, eastern Colorado. The productive sands lie at 450 m (1500 ft) to 840 m (2700 ft). Drilling results indicated a 71.1% accurate prediction prior to the survey and 67.9% accurate postsurvey results (Weart and Heimberg, 1981; reprinted with permission of the Institute for the Study of Earth and Man, Southern Methodist University).

Figure 13-16 is an example of a reconnaissance propane survey in Railroad Valley, Nevada. Sample spacing is 400 m along survey lines that mainly transect existing oil fields or known structures. The Trap Springs and Eagle Springs oil fields indicate anomalous conditions. The structural anomalies tend to have only one survey line across or adjacent to them. Anomalies are indicated in association with these structures.

Detailed Surveys

There are two different types of detailed surveys: exploration and development. The exploration survey is the most common because surface geochemical methods are more often used to find new fields than to delineate additional sites in old ones. This survey typically pursues targets that may cover a large area and can be defined without interference from existing production. In certain cases, an existing field may be pursued as an exploration rather than a development target because there is the potential for an untapped reservoir to lie above or below the existing productive zone.

The development survey tries to delineate specific well sites or a possible field extension of limited rather than infinite size. This survey is usually advantageous only when a field is near depletion and there may be parts of the reservoir that have not been in communication with the producing portion of the field. General development surveys are not

RAILROAD VALLEY, NEVADA

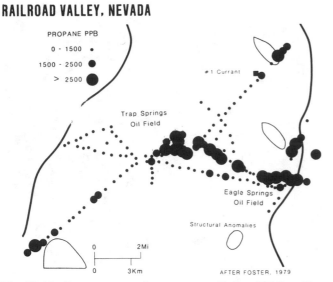

Fig. 13-16 Propane reconnaissance survey, Railroad Valley, Nevada. The fields and structural anomalies are outlined within open circles (Foster, 1979; reprinted with permission of the Rocky Mountain Association of Geologists).

effective because leakage from the reservoir to the soil substrate is dependent on the fracture system that connects them. In certain cases, there seems to be a near-vertical connection between the reservoir and the surface, and well site locations can be predicted with confidence. In the majority of situations, this does not occur. Faults and fractures extend away from the reservoir, or minor accumulations at the edges of the field have intense anomalies. These situations lead to misinterpretation and to dry holes (Fig. 13-17). However, some anomalies clearly define the center of the accumulation, and the edges of the productive area are represented by background data.

Economic Comparison and Discussion of Certain Surface Geochemical Methods

The cost of surface geochemistry surveys is highly variable among the different methods and even among different contractors using the same methods. The expense of a sample has two variables: cost of collection and cost of analysis. Both can be highly variable and not necessarily equivalent. For example, a method may have a low collection cost but a high analysis cost. The lowest cost per sample is one objective in implementing a survey.

The collection of the sample seems to be the easiest part. However, quality collection for certain methods is a major factor in the success or failure of that technique. Collection starts with an analysis of the suitability of the technique to the terrain, soil depth, soil types, weather, season, access, and quality of the collection crew. Typically, surveys that

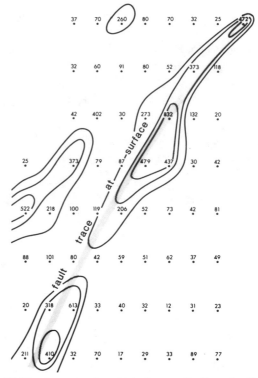

Fig. 13-17 A survey in which the results (in ppb ethane) are affected by faults and fractures.

are conducted along roads will be inexpensive compared to surveys in mountainous terrain. Surveys that can use a vehicle off-road will cost less than those that require walking the survey area. Table 13-1 (covering onshore methods only) illustrates a general comparison of collection costs for the various techniques with respect to the number of samples that can be collected per day.

Table 13-1 Collection Costs

	$0	$25	$50
Direct (soil-gas) methods:			
Free-air		————————	
Head-space		————————	
Acid extraction		————————	
Deabsorption		————————	
Fluorescence		———————	
K-V fingerprint			————————
Time delay			————————
Indirect methods:			
Radiometrics—continuous	—————		
Radiometrics—point	—————————		
Iodine	—————		
Microbial		———————————	
Helium		————————————	
Trace and major Elements		—————————	
Eh/pH	—————		

Table 13-1 indicates that collection costs range rather widely. Costs are increased with difficulty of terrain, access problems, poor weather, wet soil conditions, and inexperienced crew members. The indirect methods have the lowest cost on a per-sample basis because the samples need less special care during collection. An exception is helium, which is in the vapor phase and, therefore, is under the same constraints as other soil-gas methods.

Cost of analysis is also highly variable as Table 15-2 illustrates. The table figures represent the range of fixed costs for the laboratory and do not represent what a laboratory would actually charge. Other variables affect cost. A laboratory would not have enough samples to perform analysis on a continuous basis; downtime is costly. The cost of the free-air method decreases significantly if samples are analyzed on site. Expenses are also affected by the type of equipment used. The majority of soil-gas methods use a gas chromatograph and, for one method, a mass spectrometer. Analysis of soil gas by a plasma spectrometer will be more expensive than by a chromatograph. Radiometric analysis costs can be considered part of the collection expense. Analysis of trace elements by a plasma spectrometer will be significantly more expensive than by atomic absorption methods.

The wide variation in costs for soil-gas analysis does not reflect major differences in procedures but rather consumer ignorance in shopping around. As more and more surveys are carried out, the average cost of setting up and running the analytical equipment will continue to fall to, and below, the low end of the present ranges. Indirect methods show a wide variability. Iodine and major and trace elements are lower in cost because they can be analyzed by labs in the mining industry where competition is healthy. Radiometric costs are variable because of a situation similar to soil-gas. Radiometric equipment is relatively inexpensive to buy but, without experienced personnel, it can be very ineffective.

Tables 13-1 and 13-2 are only guides and do not show absolute values. Collection costs will probably not change through time. The development of new types of equipment could lower the cost for any particular method in the future. On the other hand, the cost could rise with the development of more accurate and sensitive equipment, but the increase could be justified by the higher-quality data obtained.

Airborne, offshore, and some of the more exotic forms of sampling are not discussed because their costs are typically higher and are difficult to compare to the onshore methods.

Several human factors have to be evaluated in order to maintain consistency in terms of collection and analytical costs. The following criteria need to be addressed prior to survey implementation:

1. The experience of the contractor and laboratory personnel, the number of years in business, the type of method, and the quality of the instrumentation should be determined. The contractor should be experienced in using the equipment and should verify that any contaminants or problems related to the survey are identified and mitigated at all times.

2. Sampling procedures should be carefully scrutinized to determine the extent of care taken in collecting each sample.

3. During survey collection, soil changes and soil types should be noted by the field personnel. Some techniques, like iodine, are not impacted by soil changes as are soil-gas and radiometrics. The impact on other methods is not as clear, or research is currently insufficient to reach a conclusion. But even if there is no impact by changes in soil composition, it is easier to collect these data while sampling than to try to verify them later.

4. There should be repeatability of the results. Any particular technique should be repeatable within the parameters of that specific method. In other words, a free-air soil-gas anomaly, even though not repeatable in terms of similar anomalous amounts, should have the same areal extent if sampled at a later date.

Experience in interpretation is a two-edged sword. The experience of a contractor or end user in one area may prove ineffective and disastrous in another. Interpretation costs are the smallest part of the survey budgets, but the interpretation is the most critical element in the survey, assuming that collection and analysis were done correctly. Variations in interpretation have many causes, usually related to the application of the technique or to the interpretation criteria used. Most contractors, collection crews, and laboratory personnel

Table 13-2 Analysis costs

	0	100
Soil-gas methods:		
Free-air		————
Head-space		————
Acid extraction		————
Deabsorption		————
Fluorescence		————
K-V fingerprint		——
Time delay		——
Indirect methods:		
Radiometrics—continuous	—	
Radiometrics—point	—	
Iodine	——	
Microbial	———	
Helium	———	
Trace and major Elements	—	
EH/pH	—	

can collect and analyze data with a high level of confidence. Few of these personnel, and even the end users, can consistently and successfully interpret the data and confidently advise during the decision-making process. Any interpretation of a survey should be supported by models; soil, climatic, and collection problems should be understood and addressed; and the basis of the interpretation should be supported with a reasonable use of statistics. The end user needs to know as much as, if not more than, the contractor. The difficulty has been the lack of numerous and usable case histories under a variety of conditions. This situation is changing for the better.

Any technique or technology is expensive when it fails to find petroleum. Any technique, not only surface geochemical methods, is inexpensive when it consistently reduces risk by avoiding areas that will yield dry holes and targeting areas that have a strong potential for trapped hydrocarbons.

Exploration Philosophy

Surface geochemical methods are most effective as part of an integrated program that includes geophysical methods. Collins et al., (1992) presented an integrated approach that utilized surface geochemistry first as a screening tool and then as a follow-up when an anomaly was found. This targeted an area for seismic shooting. We can classify this in two categories: (1) petroleum seepage identification (surface geochemistry), and (2) trap identification (seismic). Despite some claims for very specific cases, seismic is not at this time a direct hydrocarbon indicator. Therefore, surface geochemistry is a good complement to seismic. If there is no leakage but a trap has been found, historical data have shown that the risk is exceptionally high. The reverse is not as overwhelmingly true because, in many cases, the seismic may be unable to define a potential trap. However, a geochemical anomaly without a specific geologic or geophysical target should be considered high-risk also.

An ideal exploration project leading to a prospect determination begins with a geochemical reconnaissance survey on a regional basis. This defines areas that are similar to the models and that need to undergo denser sampling. After more detailed sampling, areas of limited geochemical extent are eliminated and those that continue to show potential are further defined. The final survey clearly delineates the remaining anomalous areas that will be further refined by other geologic and geophysical methods. As mentioned in Chapter 4, this is a fractal concept.

A hierarchy of geochemical methods must be established because of many techniques that can be used. The indirect methods provide an inexpensive screen that can be followed up by direct methods. This allows the more costly soil-gas to be site-specific and minimizes the various influences that must be considered. However, the indirect methods can often allow bypassing the use of soil-gas methods and can lead directly to seismic.

Summary

The purpose of the geochemical survey must be determined prior to actually acquiring the data. Defining objectives will resolve whether a reconnaissance or detailed survey is appropriate.

The search method is the product of an experimental design (the survey) that effectively collects a sufficient number of observations (samples) to detect the target with an adequate level of confidence. The experimental design that results in a grid is built on target shape, size, and extent and on the number of models that are developed. The models are complicated if: (1) more than one target is present; (2) the targets overlap; and (3) the targets differ in intensity, shape, and size compared to the geochemical expression. A successful surface geochemical program attempts to design the search so that it minimizes these problems before the data are collected. This eliminates the need to redesign the survey to collect new data.

References

Collins, B. I., S. A. Tedesco, and W. F. Martin (1992). Integrated petroleum project evaluation—Three examples from the Denver Basin, Colorado, *Journal of Geochemical Exploration*, Vol. 43, pp. 67–89.

Drew, L. J. (1967). Grid-drilling exploration and its application to the search for petroleum, *Economic Geology*, Vol. 62, pp. 698–710.

Foster, N. H. (1979). Geomorphic exploration used in the discovery of Trap Spring Oil Field, Nye County, Nevada, in G. W. Newman and H. D. Goode, eds., *Basin and Range Symposium and Great Basin Field Conference*, Rocky Mountain Association of Geologists, Denver, CO, pp. 477–486.

Koch, G. S. and R. F. Link (1971). *Statistical Analysis of Geological Data: Volume II*, Dover Publications, New York.

Levinson, A. A. (1980). *Introduction to Exploration Geochemistry*, Applied Publishing Ltd., Wilmette, IL.

Savinskii, I. D. (1965). Probability tables for locating elliptical underground masses with a rectangular grid, Consultants Bureau, New York.

Singer, D. A. and F. E. Wickman (1969). Probability tables for locating elliptical targets with square, rectangular, and hexagonal point-nets: Mineral Science Experimental Station, Pennsylvania State University Special Publication, Pennsylvania State University, University Park, PA, 1–69.

Weart, R. C. and G. Heimberg (1981). Exploration radiometrics: Postsurvey drilling results, ed. B. M. Gottlieb, *Unconventional Methods in Exploration for Petroleum and Natural Gas III*, Southern Methodist University Press, Dallas, TX, pp. 116–123.

14

Summary

Surface geochemistry is still an evolving exploration tool. Gains in its acceptance have resulted from an expanding data base, supplemented by increasing success, rather than from any significant technological breakthrough. This book was written to bring together several related techniques and present them in an objective format, supported with case histories and methods to improve analytic interpretation. The conclusion is that these techniques can increase and enhance the success of an exploration program. The methods have often been criticized on the grounds that most published American studies have been written by contractors or promoters, whose published work may lack objectivity and whose motivation may be self-interest. University researchers have published reviews or investigated new geochemical concepts but, for the most part, their work has contradicted or ignored field results. End users or explorationists have either not published or the surveys they presented are unconvincing for lack of hard data. In the geochemical literature, a considerable amount of criticism has been directed against their conclusions, but there are no surveys to prove them wrong. If we look further at the factors that have slowed the rate of acceptance of surface geochemistry, they can be summarized as a lack of understanding of the near-surface and surface environments, poor interpretation skills, proven cohesive migration theory as an industry requirement, unproven or unsubstantiated claims and, until the 1970s, inadequate detection limits of analytic technology.

Understanding the surface and near-surface environment is critical to implementing an effective survey and determining the conditions under which geochemical methods will or will not work. The mining industry uses geochemical exploration technology on a regular basis, focusing on the near-surface environment of the survey area by determining and defining the ongoing chemical nature and processes of the soil. Changes in soil composition, clay types, Eh, pH, and moisture content are all evaluated. Their analytical methods are standardized, the results are repeatable, and laboratory problems can be identified and corrected quickly. The geochemical segment of the petroleum industry has not only been slow to address these conditions but, when it has,
it has done so inconsistently. False anomalies do exist, and they are generated either by the components of the soil itself or by other sources. Investigating the soils prior to, during, and after the survey can identify these problems so that they may be avoided during the survey or removed in analysis.

Problems related to the near-surface environment continue to require study. The density and location of pathways for migrating hydrocarbons must be determined. Classifying anomalies as always halo or always apical is slowly fading from practice. It is now generally understood that there are infinite combinations of different types of anomalies. Another outdated assertion is that geochemical methods can outline the entire field and all potentially productive locations. Recent data have proved this wrong. The best use of surface geochemistry is not as a panacea for exploration but as a guide or screening method for more definitive methods such as seismic. In an area where a specific geologic stratum (such as a sandstone) contains hydrocarbons only one out of five times, surface geochemistry has the potential to eliminate the four water-filled prospects. This does not mean that the fifth will be economic or productive; it means that the fifth does contain petroleum. In an area where success is one well discovering a new field for every 20 wells drilled, surface geochemistry has generally been very successful in predicting the truly dry holes. If surface geochemistry can eliminate some percentage of every 20 wells drilled, this is a significant success.

Evaluation and interpretation of surface geochemical data have been a center of controversy. Conclusions from numerous surveys, both for petroleum and other mineral deposits, suggest that success is based on using statistical methods. However, many of the published examples have represented best-case scenarios. Geochemical interpretation is often subjective rather than objective. The geochemical data are forced to fit the location of the petroleum accumulation. With the greater acceptance of surface geochemical methods comes the realization that small data sets are not satisfactory and can be very misleading and a waste of economic resources. Even though the methods are relatively inexpensive, the results are costly if they continually lead to dry holes. Increasing the number of samples in a data set and adding numerous models result in a significantly higher confidence level and possibly increased success rates.

The need for a unified microseep theory seems to be passing. Increasing evidence suggests that all petroleum reservoirs leak to some extent. The source and end effect can be considered a significant and repeating coincidence, or the relationship is such that it is no longer important to determine the mechanism of migration. There is no doubt that the phenomenon requires further investigation, but this should not impede the acceptance of the techniques.

The claims of many contractors and some oil companies about the success of surface geochemical methods have seemed outrageous at times. But there has been a general consistency in these claims and their continual ability to achieve success over other exploration tools. One of the problems is that the petroleum industry looks at achievement in terms of success rate. A contractor usually does not make the decision to drill or not to drill. This judgment is made by the end user. Therefore, a contractor does not have a success rate, but merely repeat business.

A serious consideration in using any exploration method and justifying its use is the amount of published material substantiating its achievements through field discoveries. In conjunction with this are the type and amount of data acquired in a survey and presented in the literature. Historically, the published geochemical data bases have been generally insufficient to make an objective decision. Published seismic successes are usually supported by displayed lines of data, waveform anomalies, and isochron and structure maps. Surface geochemical data, on the other hand, are typically shown as outlines of anomalies, profiles, or vague maps from results of statistical manipulations. This trend is changing because end users have the right to know and understand what values are being called anomalous and background and how the contoured product was achieved.

Surface geochemistry has often been relegated to the role of the last tool used prior to drilling. At this late date in the project, geochemical results are generally ignored, and exploration or drilling continues. It is difficult for a company or an individual who has already spent considerable money on the project to cancel it if the surface geochemistry results conclude that no anomalous conditions exist. Reluctance to believe in the effectiveness of the tool is a product of a variety of factors from a management perspective that the contractor cannot control. This is not to say that a contractor cannot misinterpret data. Exploration for mineral resources is a series of decisions in a hierarchy that result in the discovery (or failure) of a geologic anomaly of economic importance. The study of this type of decision making is outside the realm of this book. However, any tool used results in data that affect these decision. The data are objective, assuming they were collected and analyzed properly. The interpretation of the data can be considered subjective even when every attempt is made to be as objective as possible. Therefore, the data do not lie, but the decisions based on the data may result in the wrong conclusions. Surface geochemistry needs to be used at the beginning of an exploratory program with follow-up at the middle and possibly at the end of the ongoing project.

Responses by proponents and opponents of surface geochemical methods have been as emotional as many political issues. That we are dealing with a scientific investigation suggests that there should be a reasonable, orderly investigative approach, but this has not been the case. Several reviews of surface geochemical methods have been written in anger, either for or against, and were opinionated without extensive factual support data. There are explorationists who feel that seismic is not a useful tool, even today, despite numerous successes. In certain basins or provinces, seismic has yet to be proved effective on a consistent basis. The development of any exploration tool in its early years is a constant battle. However, acceptance of surface geochemical methods seems to have taken an exceptionally long time.

We must also consider technological advances. Until the 1970s, analytical methods were slowly evolving. The capability of laboratory technology greatly increased with the advent of serious concern for the environment, the development of environmental geology, and the establishment of environmental regulations that required the detection of ever more minute amounts of many chemicals. The laboratory advances spilled over into the petroleum industry and began a continuing trend of acquiring higher-quality surface geochemical data. The growth of environmental geology also has brought a greater understanding of the soil. This is probably a more important advance than laboratory methods because it has forced explorationists to focus on the geochemical environment and its ongoing chemical and biological processes, its effects on soil-gas and other methods, and the need for more efficient survey techniques.

Surface geochemical methods would not have been effective early in the development of the petroleum industry because exploration enjoyed high success rates simply by drilling areas with multiple stacked and overlapping productive reservoirs. The numerous geochemical anomalies would have been indecipherable not only with the technology of the times but with today's methods as well.

The best role for surface geochemical methods may be in regional evaluations of unexplored basins, in areas where there are single or widely scattered reservoirs, or in relatively mature productive areas where new concepts can be investigated. Internationally, the fields of the size being pursued today would lend themselves to surface geochemical methods. The number of samples needed would be fewer in relative terms than the number needed for exploring for small targets in the United States. As discussed previously, surface geochemical methods are more of a screening tool in that they attempt to define areas of leakage as targets for subsequent seismic methods. Drilling geochemical anomalies that are not confirmed by other methods is viable, but the success rate seems to be significantly less than with an integrated approach. Consequently, geochemical methods are especially suited to eliminating areas that do not indicate any seepage and that present a high risk of drilling a dry hole.

The use of geochemical methods to define an exact drilling location has had limited effectiveness. The increasing amount of data in the literature and subsequent drilling clearly indicate that surface geochemical anomalies can either approximate an outline of the field or suggest the presence of a petroleum accumulation, but they cannot accurately define the edges of the field, and the anomaly may be offset. The application of these methods to development drilling has not proved definitive enough to warrant their replacing sophisticated seismic or traditional subsurface mapping techniques. Strong geochemical anomalies encountered in old fields might be indications of additional zones in which petroleum has accumulated or evidence of compartments in existing reservoirs that were isolated and not drained.

Choosing a technique is dependent upon several variables, which can be summarized as soil conditions, topography, and budget considerations. Major and minor elements, delta C, helium, and Eh/pH methods have some merit but may reflect processes other than seeping hydrocarbons, and there are soil scenarios in which they would prove ineffective. This leaves soil-gas, radiometrics, iodine, and microbial methods as the best tools available. Radiometrics and microbial techniques have limited use in some areas, but they can be either first-pass or follow-up tools. Soil-gas and iodine remain the two methods that consistently get good results. Soil-gas has a myriad of forms, but free-air, head-space, and acid extraction have been researched and implemented enough so that they can provide very effective means of exploring for petroleum. The remaining onshore methods have not been adequately verified. The limited data in the literature suggest that there are some problems; whether they can be overcome is something to be determined by future research.

Costs are a real concern in the use of several of these methods. Some of the soil-gas techniques are extremely expensive, samples are hard to collect, and do not provide enough additional data to justify their cost. Some of the non-soil-gas techniques are very inexpensive, and a small percentage of these are extremely effective. A combination of indirect and direct methods is recommended in order to have checks and balances in prospect evaluation. The real cost of any survey is not in the analysis but in sample collection, the costs of which can vary widely. Sample costs

are inherently related to the difficulty of collecting the sample, the time spent at each sample site, the terrain, and the weather. The experience and competence of the sampling crew are critical to the survey's success. Soil changes, manmade objects, recent soil disruption, vegetation changes, and wet or dry conditions should be noted during the survey. Inexperienced or poorly trained crews are usually not very observant.

Mechanized sample collection equipment is controversial. Proponents of mechanization indicate that their advantage is collecting samples with large volumes of gas from greater depths. However, mechanized sample collection can cause anomalies and be extremely costly. During the initial evaluation, the explorationist may consider an orientation survey using several different methods to determine each technique's suitability and cost. The objective of any survey is to evaluate an area effectively at a reasonable cost. Therefore, mechanized methods in numerous situations will have no advantage and, in some cases, can be uneconomic compared to nonmechanized equipment.

The effectiveness of surface geochemical methods depends on choosing a technique that will be able to prove or disprove the presence of microseepage in an area. The design and sample density of the survey are critical in proving this. An insufficient number of samples or unreasonable areal distribution downgrades the ability of statistics to provide a reasonable basis on which to interpret the data. Large data sets will eliminate the need to rely on statistics as the sole means of interpretation and will utilize pattern recognition methods instead. Insufficient sample density can cause an interpretation to be meaningless and can mislead in either direction by the drilling of a dry hole or by not recognizing, and therefore missing, a potential accumulation. Survey density is not only determined by the size of the perceived target but also by the use of models that help guide exploration.

Interpretation skills are critical to making a decision from the results of a survey. It is of paramount importance to be able to determine when a survey is inadequate to answer the questions posed by the experimental design about the existence or nonexistence of an anomaly. What statistical or nonstatistical methods to use and how to recognize an anomaly are skills learned through experience; they are difficult to teach. This book has shown, through discussion and examples, the directions to pursue and how to achieve some level of confidence in the interpretation of the data. By no means all geochemical surface expressions and soil problems have been presented here, but the reader has been given the basics on which to build.

Index